CISM COURSES AND LECTURES

The series presents lecture notes, monographs, edited works and proceedings in the field of Mechanics, Engineering, Computer Science and Applied Mathematics.
Purpose of the series is to make known in the international scientific and technical community results obtained in some of the activities organized by CISM, the International Centre for Mechanical Sciences.

INTERNATIONAL CENTRE FOR MECHANICAL SCIENCES

COURSES AND LECTURES - No. 381

ROMANSY 11

THEORY AND PRACTICE OF ROBOTS AND MANIPULATORS

PROCEEDINGS OF THE ELEVENTH CISM-IFToMM SYMPOSIUM

EDITED BY

A. MORECKI
WARSAW UNIVERSITY OF TECHNOLOGY

G. BIANCHI
POLYTECHNIC OF MILAN

C. RZYMKOWSKI
WARSAW UNIVERSITY OF TECHNOLOGY

Springer-Verlag Wien GmbH

Le spese di stampa di questo volume sono in parte coperte da
contributi del Consiglio Nazionale delle Ricerche.

This volume contains 264 illustrations

In order to make this volume available as economically and as
rapidly as possible the authors' typescripts have been
reproduced in their original forms. This method unfortunately
has its typographical limitations but it is hoped that they in no
way district the reader.

ISBN 978-3-211-82903-5   ISBN 978-3-7091-2666-0 (eBook)
DOI 10.1007/978-3-7091-2666-0

PREFACE

The CISM-IFToMM RoManSy Symposia have played a dynamic role in the development of the theory and practice of robotics. The proceedings of the eleven symposia to date present a world view of the state of the art, including a unique record of the results achieved in central and eastern Europe.

RoManSy'96, held July 1-4,1996 in Udine, Italy was attended by 59 participants from 13 countries.

The proceedings of this eleventh edition of RoManSy focus mainly on problems of mechanical engineering and control.

In his opening lecture, O. Khatib presented an overview of the ongoing efforts, at Stanford University, for the development of mobile manipulation systems and summarised the basic models and methodologies for their analysis and control.

The 46 papers illustrate significant contributions in mechanics (10 papers), synthesis and design (4), walking machines and mobile robots (13), biomechanical aspects of robots and manipulators (2), control of motion (11), sensing and machine intelligence (3) and applications and performance evaluation (3). They appear here in the order and form in which they were presented in the various working sessions.

The next RoManSy'98 will be held in Paris, France in July 1998.

A. Morecki

G. Bianchi

C. Rzymkowski

CONTENTS

PREFACE

OPENING LECTURE:

Mobile Manipulator Systems
O. Khatib 3

CHAPTER I: **Mechanics - I**

Towards Reducing Thruster-Flexibility Interactions in Space Robots
E. Martin, E. Papadopoulos, J. Angeles 13

Modeling and Optimization of the Tube-Crawling Robot
F.L. Chernousko, N.N. Bolotnik 21

An Investigation of a Quality Index for the Stability of In-Parallel Planar
Platform Devices
J. Lee, J. Duffy, M. Keler 27

Workspaces of Planar Parallel Manipulators
J-P. Merlet, C.M. Gosselin, N. Mouly 37

Nonlinear Control of a Parallel Robot Including Motor Dynamics
L. Beji, A. Abichou, P. Joli, M. Pascal 45

An Inverse Force Analysis of the Spherical 3-DOF Parallel Manipulator with
Three Linear Actuators Considered as Spring System
J. Knapczyk, G. Tora 53

CHAPTER II: **Mechanics - II**

Experimental Research of Dynamics of an Elephant Trunk Type Elastic Manipulator
R. Cieślak, A. Morecki 63

Vibration Suppressing for Hydraulically Driven Large Redundant Manipulators
M. Schneider, M. Hiller 73

Nonlinear Dynamics of Computer Controlled Machines
G. Stépán, E. Enikov, T. Müller 81

Identification and Compensation of Gear Friction for Modeling of Robots
M. Daemi, B. Heimann 89

CHAPTER III: **Synthesis and Design**

Designing Manipulators for Both Kinematic and Dynamic Isotropic Properties
R. Matone, B.Roth 99

Group Theoretical Synthesis of Binary Manipulators
G.S. Chirikjian 107

Distributed Kinematic Design from Task Specification
S. Regnier, F. B. Ouezdou, P. Bidaud 115

Optimal Design of a Shape Memory Alloy Actuator for Microgrippers
N. Troisfontaine, Ph. Bidaud 123

CHAPTER IV: **Walking Machines and Mobile Robots - I**

On Direct-Search Optimization of Biped Walking
P. Kiriazov, W. Schiehlen 133

Development of a Biped Walking Robot Adapting to the Real World
-- Realization of Dynamic Biped Walking Adapting to the Humans' Living Floor
J. Yamaguchi, A. Takanishi 141

Simulation of an Anthropomorphic Running Robot
(Using the Multibody Code MECHANICA Motion)
H. De Man, D. Lefeber, J. Vermeulen, B. Verrelst 149

Gait Control Method of Robot and Biological Systems by Use of HOJO-Brain
Y. Sankai 157

Gait Rhythm Generators of a Two Legged Walking Machine
T. Zielińska 165

CHAPTER V: **Walking Machines and Mobile Robots - II**

Development of a Mechanical Simulation of Human Walking
K. Kędzior, A. Morecki, M. Wojtyra, T. Zagrajek, T. Zielińska,
A. Goswami, M. Waldron, K. Waldron 175

Mechanical Design of an Anthropomorphic Electropneumatic Walking Robot
G. Figliolini, M. Ceccarelli 189

Energetic Analysis of Three Classical Structures of Walking Robots Leg
O. Bruneau, F. Ben Ouezdou 197

Dynamic Control of Wheeled Mobile Robot Using Sliding Mode
P. Ruaux, G. Bourdon, S. Delaplace 205

Simulation Study of a Group-Behavior of Micro-Robots on the Model
of Insect's Behavior
K. Matsushima, Y. Hayashibara, K. Nakamoto 213

CHAPTER VI: **Walking Machines and Mobile Robots - III**

Active Coordination of Robotic Terrain-Adaptive Wheeled Vehicles
for Power Minimization
S.C. Venkataraman, S.V. Sreenivasan, K.J. Waldron 223

Pulse Width Modulated Control for the Dynamic Tracking of Wheeled
Mobile Robots
X. Feng, S.A. Velinsky 233

Manoeuvre Analysis of Multilegged Walking Robots with Dynamic Constraints
Between Leg Propulsors
A.P. Bessonov, K.S. Tolmachov, N.V. Umnov 241

CHAPTER VIb: **Biomechanical Aspects of Robots and Manipulators**

Biomechanical Aspect of a Mastication Robot
H. Takanobu, A. Takanishi 251

A Strategy for the Finger-Arm Coordinated Motion -- Analysis of Human Motion
in Violin Playing
K. Shibuya, S. Sugano 259

CHAPTER VII: **Control of Motion - I**

High Performance Control of Manipulators Using Base Force/Torque Sensing
G. Morel, S. Dubowsky 269

Control of a Redundant Scara Robot in the Presence of Obstacles
W. Risse, M. Hiller 277

Experiments of Kinematic Control on a Two-Robot System
F. Caccavale, P. Chiacchio, S. Chiaverini, B. Siciliano 285

Optimal Path Planning for Moving Objects
K. Łasiński, M. Galicki, P. Gawłowicz 293

Stiffness Control of an Object Grasped by a Multifingered Hand
K. Tanie, H. Maekawa, Y. Nakamura 301

CHAPTER VIII: **Control of Motion - II**

Practical Stabilisation of Robots Interacting with Dynamic Environment
by Decentralised Control
D. Stokic, M. Vukobratovic 311

Control of Redundant Robots at Singularities in Degenerate Directions
E. Malis, L. Morin, S. Boudet 319

Application of Neural Networks for Control of Robot Manipulators-Simulation
and Implementation
T. Uhl, M. Szymkat, T. Bojko, Z. Korendo, J. Ród 327

CHAPTER IX: **Control of Motion - III**

Active Force Control of Pneumatic Actuators
H. Busch, J. Hewit 337

Using Backpropagation Algorithm for Neural Adaptive Control: Experimental
Validation on an Industrial Mobile Robot
P. Henaff, S. Delaplace 347

Hopfield's Artificial Neural Networks in Multiobjective Optimization Problems
of Resource Allocations Control
J. Balicki, Z. Kitowski 355

CHAPTER X: **Sensing and Machine Intelligence**

Type Synthesis of Contact Sensing Elements for Robotic Fixturing
W.W. Nederbragt, B.Ravani 365

Object-Oriented Approach to Programming a Robot System
C. Zieliński 373

Contact Tasks Realization by Sensing Contact Forces
B. Borovac, L. Nagy, M. Šabli.. 381

CHAPTER XI: **Application and Performance Evaluation**

Application of Force Control in Telerobotics
R. Bicker, D. Glennie, Ow Sin Ming 391

Development and Application of the TELBOT System - A New Tele Robot System
W. Wälischmiller, H.-Y. Lee 399

A Decision-Making System for the Initial Stage Implementation
of Assembly Strategies
C.J. Tsaprounis, N.A. Aspragathos, A. Denstoras 409

APPENDIX A: **Programme and Organising Committee** 421

APPENDIX B: **List of Participants** 427

Opening Lecture

MOBILE MANIPULATOR SYSTEMS

O. Khatib

Stanford University, Stanford, CA, USA

Abstract

Mobile manipulation capabilities are key to many new applications of robotics in space, underwater, construction, and service environments. In these applications, consideration of vehicle/arm dynamics is essential for robot coordination and control. This article discusses the inertial properties of holonomic mobile manipulation systems and presents the basic strategies developed for their dynamic coordination and control. These strategies are based on extensions of the *operational space formulation*, which provides the mathematical models for the description, analysis, and control of robot dynamics with respect to the task behavior.

1 Introduction

A central issue in the development of mobile manipulation systems is vehicle/arm coordination [1,2]. This area of research is relatively new. There is, however, a large body of work that has been devoted to the study of motion coordination in the context of kinematic redundancy. In recent years, these two areas have begun to merge [3], and algorithms developed for redundant manipulators are being extended to mobile manipulation systems. Typical approaches to motion coordination of redundant systems rely on the use of pseudo- or generalized inverses to solve an under-constrained or degenerate system of linear equations, while optimizing some given criterion. These algorithms are essentially driven by kinematic considerations and the dynamic interaction between the end-effector and the manipulator's internal motions are ignored.

Our approach to controlling redundant systems is based on two models: an *end-effector dynamic model* obtained by projecting the mechanism dynamics into the operational space, and a *dynamically consistent force/torque relationship* that provides decoupled control of joint motions in the null space associated with the redundant mechanism. These two models are the basis for the dynamic coordination strategy we are implementing for the mobile platform.

Another important issue in mobile manipulation concerns cooperative operations between multiple vehicle/arm systems. Our study of the dynamics of parallel, multi robot structures reveals an important additive property. The effective mass and inertia of a multi-robot system at some operational point are shown to be given by the sum of the effective masses and inertias associated with the object and each robot. Using this property, the multi-robot system can be treated as a single *augmented object* [5] and controlled by the total operational forces applied by the robots. The control of internal forces is based on the *virtual linkage* [6] which characterizes internal forces.

2   Operational Space Dynamics

The joint space dynamics of a manipulator are described by

$$A(\mathbf{q})\ddot{\mathbf{q}} + \mathbf{b}(\mathbf{q}, \dot{\mathbf{q}}) + \mathbf{g}(\mathbf{q}) = \mathbf{\Gamma}; \tag{1}$$

where $\mathbf{q}$ is the n joint coordinates and $A(\mathbf{q})$ is the $n \times n$ kinetic energy matrix. $\mathbf{b}(\mathbf{q}, \dot{\mathbf{q}})$ is the vector of centrifugal and Coriolis joint-forces and $\mathbf{g}(\mathbf{q})$ is the gravity joint-force vector. $\mathbf{\Gamma}$ is the vector of generalized joint-forces.

The operational space equations of motion of a manipulator are [4]

$$\Lambda(\mathbf{x})\ddot{\mathbf{x}} + \mathbf{\mu}(\mathbf{x}, \dot{\mathbf{x}}) + \mathbf{p}(\mathbf{x}) = \mathbf{F}; \tag{2}$$

where $\mathbf{x}$, is the vector of the m operational coordinates describing the position and orientation of the effector, $\Lambda(\mathbf{x})$ is the $m \times m$ kinetic energy matrix associated with the operational space. $\mathbf{\mu}(\mathbf{x}, \dot{\mathbf{x}})$, $\mathbf{p}(\mathbf{x})$, and $\mathbf{F}$ are respectively the centrifugal and Coriolis force vector, gravity force vector, and generalized force vector acting in operational space.

3   Redundancy

The operational space equations of motion describe the dynamic response of a manipulator to the application of an operational force $\mathbf{F}$ at the end effector. For non-redundant manipulators, the relationship between operational forces, $\mathbf{F}$, and joint forces, $\mathbf{\Gamma}$ is

$$\mathbf{\Gamma} = J^T(\mathbf{q})\mathbf{F}; \tag{3}$$

where $J(\mathbf{q})$ is the Jacobian matrix.

However, this relationship becomes incomplete for redundant systems. We have shown that the relationship between joint torques and operational forces is

$$\mathbf{\Gamma} = J^T(\mathbf{q})\mathbf{F} + \left[I - J^T(\mathbf{q})\bar{J}^T(\mathbf{q})\right]\mathbf{\Gamma}_0; \tag{4}$$

with

$$\bar{J}(\mathbf{q}) = A^{-1}(\mathbf{q})J^T(\mathbf{q})\Lambda(\mathbf{q}); \tag{5}$$

where $\bar{J}(\mathbf{q})$ is the *dynamically consistent generalized inverse* [5] This relationship provides a decomposition of joint forces into two dynamically decoupled control vectors: joint forces corresponding to forces acting at the end effector ($J^T\mathbf{F}$); and joint forces that only affect internal motions, $\left([I - J^T(\mathbf{q})\bar{J}^T(\mathbf{q})]\mathbf{\Gamma}_0\right)$.

Using this decomposition, the end effector can be controlled by operational forces, whereas internal motions can be independently controlled by joint forces that are guaranteed not to alter the end effector's dynamic behavior. This relationship is the basis for implementing the dynamic coordination strategy for a vehicle/arm system.

The end-effector equations of motion for a redundant manipulator are obtained by the projection of the joint-space equations of motion (1), by the *dynamically consistent* generalized inverse $\bar{J}^T(q)$,

$$\bar{J}^T(q)\,[A(q)\ddot{q} + b(q,\dot{q}) + g(q) = \Gamma] \quad \Longrightarrow \quad \Lambda(q)\ddot{x} + \mu(q,\dot{q}) + p(q) = F; \quad (6)$$

The above property also applies to non-redundant manipulators, where the matrix $\bar{J}^T(q)$ reduces to $J^{-T}(q)$.

4 Vehicle/Arm Coordination

In our approach, a mobile manipulator system is viewed as the mechanism resulting from the serial combination of two sub-systems: a "macro" mechanism with coarse, slow, dynamic responses (the mobile base), and a relatively fast and accurate "mini" device (the manipulator).

The mobile base referred to as the *macro structure* is assumed to be holonomic. Let Λ be the *pseudo kinetic energy matrix* associated with the combined macro/mini structures and Λ_{mini} the operational space *kinetic energy matrix* associated with the mini structure alone.

The magnitude of the inertial properties of macro/mini structure in a direction represented by a unit vector w in the m-dimensional space are described by the scalar [5]

$$\sigma_w(\Lambda) = \frac{1}{(w^T\Lambda^{-1}w)};$$

which represents the effective inertial properties in the direction w.

Our study has shown [5] that, *in any direction w, the inertial properties of a macro/mini-manipulator system (see Figure 1) are smaller than or equal to the inertial properties associated with the mini-manipulator in that direction:*

$$\sigma_w(\Lambda) \le \sigma_w(\Lambda_{mini}). \quad (7)$$

A more general statement of this *reduced effective inertial* property is that the inertial properties of a redundant system are bounded above by the inertial properties of the structure formed by the smallest distal set of degrees of freedom that span the operational space.

The reduced effective inertial property shows that the dynamic performance of a combined macro/mini system can be made comparable to (and, in some cases, better than) that of the lightweight mini manipulator. The idea behind our approach for the coordination of macro and mini structures is to treat them as a single redundant system.

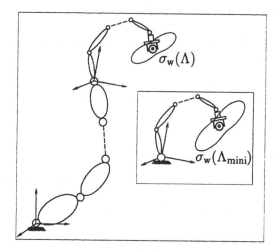

Figure 1: Inertial properties of a macro/mini-manipulator

The *dynamic coordination* we propose is based on combining the operational space control with a minimization of deviation from the midrange joint positions of the mini-manipulator. This minimization is implemented with a joint torques selected from the *dynamically consistent* null space of equation (4) to eliminate any coupling effect on the end-effector.

This is

$$\Gamma_{\text{Null-Space}} = \left[I - J^T(\mathbf{q}) \bar{J}^T(\mathbf{q}) \right] \Gamma_{\text{Coordination}};$$
(8)

5 Cooperative Manipulation

Our research in cooperative manipulation has produced a number of results which provide the basis for the control strategies we are developing for mobile manipulation platforms. Our approach is based on the integration of two basic concepts: The *augmented object* [5] and the *virtual linkage* [6]. The *virtual linkage* characterizes internal forces, while the *augmented object* describes the system's closed-chain dynamics. These models have been successfully used in cooperative manipulation for various compliant motion tasks performed by two and three PUMA 560 manipulators [7].

5.1 Augmented Object

The *augmented object* model provides a description of the dynamics at the operational point for a multi-arm robot system. The simplicity of these equations is the result of an additive property that allows us to obtain the system equations of motion from the equations of motion of the individual mobile manipulators.

The *augmented object* model is

$$\Lambda_{\oplus}(\mathbf{x})\ddot{\mathbf{x}} + \mu_{\oplus}(\mathbf{x}, \dot{\mathbf{x}}) + \mathbf{p}_{\oplus}(\mathbf{x}) = \mathbf{F}_{\oplus};$$
(9)

with

$$\Lambda_{\oplus}(\mathbf{x}) = \Lambda_{\mathcal{L}}(\mathbf{x}) + \sum_{i=1}^{N} \Lambda_i(\mathbf{x}); \tag{10}$$

where $\Lambda_{\mathcal{L}}(\mathbf{x})$ and $\Lambda_i(\mathbf{x})$ are the kinetic energy matrices associated with the object and the ith effector, respectively. The vectors, $\mu_{\oplus}(\mathbf{x}, \dot{\mathbf{x}})$ and $\mathbf{p}_{\oplus}(\mathbf{x})$ also have the additive property.

The generalized operational forces $\mathbf{F}_{\oplus}$ are the resultant of the forces produced by each of the N effectors at the operational point.

$$\mathbf{F}_{\oplus} = \sum_{i=1}^{N} \mathbf{F}_i. \tag{11}$$

The dynamic decoupling and motion control of the augmented object in operational space is achieved by selecting a control structure similar to that of a single manipulator. The dynamic behavior of the augmented object of equation (9) is controlled by the net force $\mathbf{F}_{\oplus}$. Due to the actuator redundancy of multi-effector systems, there is an infinity of joint-torque vectors that correspond to this force.

5.2 Virtual Linkage

Object manipulation requires accurate control of internal forces. Recently, we have proposed the *virtual linkage* [7] as a model of internal forces associated with multi-grasp manipulation. In this model, grasp points are connected by a closed, non-intersecting set of virtual links, as illustrated in Figure 2 for a three-grasp task.

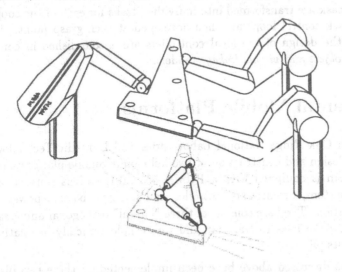

Figure 2: The Virtual Linkage

In the case of an N-grasp manipulation task, a *virtual linkage* model is a $6(N-1)$ degree of freedom mechanism that has $3(N-2)$ linearly actuated members and N spherically actuated joints. Forces and moments applied at the grasp points of this linkage will cause forces and torques at its joints. We can independently specify internal forces in the $3(N-2)$ members, along with $3N$ internal moments at the spherical joints. Internal forces in the object are then characterized by these forces and torques in a physically meaningful way.

The relationship between applied forces, their resultant and internal forces is

$$\begin{bmatrix} \mathbf{F}_{res} \\ \mathbf{F}_{int} \end{bmatrix} = \mathbf{G} \begin{bmatrix} \mathbf{f}_1 \\ \vdots \\ \mathbf{f}_N \end{bmatrix}; \tag{12}$$

where $\mathbf{F}_{res}$ represents the resultant forces at the operational point, $\mathbf{F}_{int}$ the internal forces and $\mathbf{f}_i$ the forces applied at the grasp point i. $\mathbf{G}$ is called the grasp description matrix, and relates forces applied at each grasp to the resultant and internal forces in the object.

5.3 Decentralized Cooperation

For fixed base manipulation, the *augmented object* and *virtual linkage* have been implemented in a multiprocessor system using a centralized control structure. This type of control is not suited for autonomous mobile manipulation platforms.

In a multiple mobile robot system, each robot has real-time access only to its own state information and can only infer information about the other robots' grasp forces through their combined action on the object. Recently, we have developed a new control structure for decentralized cooperative mobile manipulation [8]. In this structure, the object level specifications of the task are transformed into individual tasks for each of the cooperative robots. Local feedback control loops are then developed at each grasp point. The task transformation and the design of the local controllers are accomplished in consistency with the *augmented object* and *virtual linkage* models.

6   Experimental Mobile Platforms

In collaboration with Oak Ridge National Laboratories and Nomadic Technologies, we have completed the design and construction of two holonomic mobile platforms (see Figure 3). Each platform is equipped with a PUMA 560 arm, various sensors, a multiprocessor computer system, a multi-axis controller, and sufficient battery power to allow for autonomous operation. The base consists of three "lateral" orthogonal universal-wheel assemblies which allow the base to translate and rotate holonomically in relatively flat office-like environments [9].

The control strategies discussed above have been implemented on these two platforms. Erasing a whiteboard, cooperating in carrying a basket, and sweeping a desk are examples of tasks demonstrated with the Stanford Mobile Platforms [10]. The dynamic coordination strategy has allowed full use of the relatively high bandwidth of the PUMA. Object motion and force control performance with the Stanford mobile platforms are comparable with the results obtained with fixed base PUMA manipulators.

Figure 3: The Stanford Mobile Platforms

7 Conclusion

We have presented extensions of various operational space methodologies for fixed-base manipulators to mobile manipulation systems. A vehicle/arm platform is treated as a macro/mini structure. This redundant system is controlled using a dynamic coordination strategy, which allows the mini structure's high bandwidth to be fully utilized.

Cooperative operations between multiple platforms rely on the integration of the *augmented object*, which describes the system's closed-chain dynamics, and the *virtual linkage*, which characterizes internal forces. These models are the basis for the decentralized control structure presented in [8].

Vehicle/arm coordination and cooperative operations have been implemented on two mobile manipulator platforms developed at Stanford University.

Acknowledgments

The financial support of Boeing, General Motors, Hitachi Construction Machinery, and NSF (grants IRI-9320017 and CAD-9320419) is gratefully acknowledged. Many thanks to Alain Bowling, Oliver Brock, Arancha Casal, Kyong-Sok Chang, Robert Holmberg, Francois Pin, Diego Ruspini, David Williams, James Slater, John Slater, Stef Sonck, and Kazuhito Yokoi for their valuable contributions to the design and construction of the Stanford Mobile Platforms.

8 References

1 Ullman, M., Cannon, R., Experiments in Global Navigation and Control of a Free-Flying Space Robot. Proc. Winter Annual Meeting, Vol. 15, 1989, pp. 37-43.

2 Umetani, Y., and Yoshida, K., Experimental Study on Two-Dimensional Free-Flying Robot Satellite Model. Proc. NASA Conf. Space Telerobotics, 1989.

3 Papadopoulos, E., Dubowsky, S., Coordinated Manipulator/Spacecraft Motion Control for Space Robotic Systems. Proc. IEEE Int. Conf. Robotics and Automation, 1991, pp. 1696-1701.

4 Khatib, O., A Unified Approach to Motion and Force Control of Robot Manipulators: The Operational Space Formulation. IEEE J. Robotics and Automation, vol. 3, no. 1, 1987, pp. 43-53.

5 Khatib, O., Inertial Properties in Robotics Manipulation: An Object-Level Framework, *Int. J. Robotics Research,* vol. 14, no. 1, February 1995. pp. 19-36.

6 Williams, D. and Khatib, O., The Virtual Linkage: A Model for Internal Forces in Multi-Grasp Manipulation. Proc. IEEE Int. Conf. Robotics and Automation, 1993, pp. 1025-1030.

7 Williams, D. and Khatib, O., Multi-Grasp Manipulation," IEEE Int. Conf. Robotics and Automation Video Proceedings 1995.

8 Khatib, O. Yokoi, K., Chang,K., Ruspini, D., Holmberg, R. Casal, A., Baader., A. "Force Strategies for Cooperative Tasks in Multiple Mobile Manipulation Systems," *Robotics Research 7, The Seventh International Symposium,* G. Giralt and G. Hirzinger, eds., Springer 1996, pp. 333-342.

9 Pin, F. G. and S. M. Killough, "A New Family of Omnidirectional and Holonomic Wheeled Platforms for Mobile Robots", IEEE Trans. on Robotics and Automation, Vol. 10, No. 4, August 1994, pp. 480-489.

10 Khatib, O., K. Yokoi, K. Chang, D. Ruspini, R. Holmberg, A. Casal, and A. Baader. "The Robotic Assistant," *IEEE Int. Conf. Robotics and Automation Video Proceedings,* 1996.

Chapter I
Mechanics I

TOWARDS REDUCING THRUSTER-FLEXIBILITY INTERACTIONS IN SPACE ROBOTS

E. Martin, E. Papadopoulos and J. Angeles

McGill University, Montreal, QUE, Canada

Abstract

Space manipulators mounted on an on-off thruster-controlled base are envisioned to assist in the assembly and maintenance of space structures. When handling large payloads, manipulator joint and link flexibility become important, for they can result in payload-attitude controller fuel-replenishing dynamic interactions. In this paper, the dynamic behavior of a flexible-joint manipulator on a free-flying base is approximated by a single-mode mechanical system, while its parameters are matched with space-manipulator data. Describing functions are used to predict the dynamic performance of three alternative controller-estimator schemes, and to conduct a parametric study on the influence of key system parameters. Design guidelines and a state-estimator are suggested that can minimize such undesirable dynamic interactions as well as thruster fuel consumption.

Introduction

Robotic devices in orbit will play an important role in space exploration and exploitation. The mobility of such devices can be enhanced by mounting them on free-flying bases, controlled by on-off thrusters. Such robots introduce a host of dynamic and control problems not found in terrestrial applications. When handling large payloads, manipulator joint or structural flexibility becomes important and can result in payload-attitude controller fuel-replenishing dynamic interactions. Such interactions may lead to control system instabilities, or manifest themselves as limit cycles[1].

The CANADARM-Space Shuttle system is the only operational space robotic system to date. Its Reaction Control System (RCS), which makes use of on-off thrusters, is designed assuming rigid body motion, and using single-axis, thruster switching logic based on phase-plane techniques. This approach is common in the design of thruster-based control systems. However, the flexible modes of this space robotic system have rather low frequencies, which continuously change with manipulator configuration and payload, and can be excited by the RCS activity. The performance degradation of the RCS due to the deployment of a flexible *payload*, with or without the CANADARM has been studied[2]. A new design for the RCS was developed to reduce the impact of large measurement uncertainties in the rate signal during attitude control, thereby increasing significantly the performance of the RCS for rigid-body motion[3]. However, the flexibility problem was not addressed. Currently, the method for resolving these problems consists of performing extensive simulations. If dynamic interactions occur, corrective actions are taken, which would include adjusting the RCS parameter values, or simply changing the operational procedures[2]. The consequences of such interactions can be problematic, since fuel is an unavailable resource in space; hence, classical attitude controllers must be improved to reduce the possibility of such dynamic interactions.

This problem was studied using a single-mode, linear translational mechanical system to approximate the dynamic behavior of a two-flexible-joint manipulator mounted on a three-degree-of-freedom (dof) base[4]. A state-estimator and design guidelines were suggested to minimize such undesirable dynamic interactions, as well as thruster fuel consumption. In that study, the damping ratio of the system was taken constant and equal to 0.05. However, the damping ratio will in general be a function of the manipulator configuration and its payload. This paper reports results obtained under these more realistic conditions. In Section 2, the required damping ratios for all configurations and payloads are obtained using the dynamics model of a two-dof planar manipulator mounted on a three-dof spacecraft. The control laws and models used are then derived in Section 3. Finally, in Section 4, the describing-function method and simulations are used to study this approximate system, and show the improved performance that can be obtained using the state-estimator model previously developed[4]. Design guidelines are also given.

Modeling

The dynamics model of the two-flexible-joint planar manipulator mounted on a free-floating base of Fig. 1(a), was developed using a Lagrangian formulation under the assumption that all link and joint flexibilities are lumped at the joints[5]. This is reasonable, since joint flexibility is more significant than link flexibility in this kind of system. Each flexible joint is modeled as a torsional spring in parallel with a torsional dashpot. Using linearization techniques and a proper change of variables, the system reduces to two decoupled equations describing the joint flexibility modes, and two equations giving the torque required to brake the joints in a specific configuration. In order to obtain these two decoupled equations, the damping at the joints is assumed proportional to the respective stiffness of these joints, thereby obtaining

$$\ddot{y}_i + \lambda\omega_i^2\dot{y}_i + \omega_i^2 y_i = 0, \qquad i = 1,2 \tag{1}$$

where λ is the proportionality constant and ω_i is the natural frequency, that can be readily derived[5]. Comparing Eq. (1) to the standard second-order form,

$$\ddot{y}_i + 2\zeta_i\omega_i\dot{y}_i + \omega_i^2 y_i = 0, \qquad i = 1,2 \tag{2}$$

we have

$$2\zeta_i\omega_i = \lambda\omega_i^2, \qquad i = 1,2 \tag{3}$$

which implies that $\zeta_2 = \zeta_1(\omega_2/\omega_1)$.

Thus, if $\omega_2 > \omega_1$, we must have $\zeta_2 > \zeta_1$ in order to have proportional damping. This is in agreement with published data for the CANADARM/Space Shuttle system, where it can be observed that the oscillations corresponding to the second and higher modes die out very quickly, and thus, the first mode dominates the response of the system[6]. Therefore, the hypothesis of proportional damping is reasonable.

Considering that two different joints are used, the approximate characteristics of the CANADARM/Space Shuttle system are used together with the frequency expressions

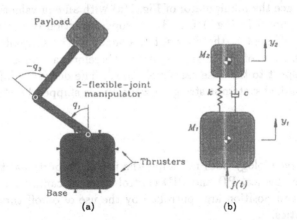

Figure 1: Flexible manipulator replaced by a spring and a dashpot: (a) two-link manip-ulator on a floating base; (b) simplified two-mass system.

ω_i to obtain the first natural frequency and the damping ratio of this system for many configurations and payloads. The results are recorded in Tables 1 and 2 for $q_1 = 45°$, where β is defined as the ratio of the mass of the payload over the mass of the base. In these two tables, varying the payload from 0 to 30% of the spacecraft result in a 12-fold change in natural frequency and damping ratio.

Table 1: First natural frequency evaluation for $q_1 = 45°$ (ω_1) (Hz).

β	$q_3 = -135°$	$q_3 = -90°$	$q_3 = -45°$	$q_3 = 0°$
0.	0.653	0.429	0.343	0.320
0.01	0.254	0.170	0.137	0.128
0.05	0.127	0.091	0.077	0.073
0.1	0.095	0.073	0.064	0.062
0.15	0.082	0.066	0.059	0.057
0.2	0.074	0.062	0.057	0.055
0.25	0.069	0.059	0.055	0.053
0.3	0.065	0.057	0.054	0.052

Table 2: Damping ratio evaluation for $q_1 = 45°$ (ζ_1).

β	$q_3 = -135°$	$q_3 = -90°$	$q_3 = -45°$	$q_3 = 0°$
0.	0.102	0.067	0.054	0.050
0.01	0.040	0.027	0.021	0.020
0.05	0.020	0.014	0.012	0.011
0.1	0.015	0.011	0.010	0.010
0.15	0.013	0.010	0.009	0.009
0.2	0.012	0.010	0.009	0.009
0.25	0.011	0.009	0.009	0.008
0.3	0.010	0.009	0.008	0.008

The dynamics of a simple two-flexible-joint planar manipulator is rather complicated; it is preferable to employ a simplified model to analyze the problem stated in the previous section. We can replace the manipulator of Fig. 1(a) with an equivalent two-mass-spring-dashpot system, as shown in Fig. 1(b). By a proper selection of the spring stiffness k and the damping coefficient c, the resonant frequency of the simplified system can be matched to the lowest one of the original system. Therefore, a similar relative motion of the payload with respect to the base can be obtained. The derivation of the equations of motion for this simplified system is straighforward and is skipped here[4,5].

Control

Control Schemes

Currently available technology does not allow the use of proportional thruster valves in space, and thus, the classical PD and PID control schemes cannot be used. Therefore, spacecraft attitude and position are controlled by the use of on-off thruster valves, that introduce nonlinearities.

The usual scheme to control a spacecraft with on-off thrusters is based on the error phase plane, defined as that with spacecraft attitude error e and error-rate $\dot{e}$ as coordinates. The on-and-off switching is determined by switching lines in the phase plane and can become complex, as is the case in the phase plane controller of the Space Shuttle[2]. To simplify the switching logic, two switching lines with equations $e + \lambda\dot{e} = \pm\delta$ have been used. The deadband limits $[-\delta, \delta]$ are determined by attitude limit requirements, while the slope of the switching lines, by the desired rate of convergence towards equilibrium and by the rate limits. This switching logic can be represented as a relay with a deadband, where the input is $e + \lambda\dot{e}$, the left-hand side of the switching-line equations[4].

To compute the input to the controller, the position and the velocity of the system base are required. Using current space technology, both states can be obtained by sensor readings. However, it can happen that only the attitude is available and then, the velocity must be estimated[4]. In this paper, three models previously studied are considered[4]. For the first model, Case 1, both the position and the velocity of the base are available by sensors and these signals are simply passed through filters to eliminate high-frequency noise. For the other two models, we assume that only the position is available from sensors and state estimators are used to obtain the required velocity. In Case 2, we differentiate the position signal while passing it through a filter to obtain an estimate for the velocity. The position signal is also filtered in this case. Finally, for the model corresponding to Case 3, a classical asymptotic state observer is used to obtain an estimate for the position and the velocity.

In all three cases, a time delay τ has been included to account for the delay between the time a sensor reads a measurement, and the time this measurement is used. Since this delay is more significant than the delay of turning on or off the thrusters, only a sensor time delay is included.

Frequency-Domain Analysis

The attitude controller assumes on-off thrusters, which are nonlinear devices; hence, the system cannot be adequately analyzed through the application of linear analysis methods. This problem is addressed using the describing-function method, which can predict the existence of limit cycles in nonlinear systems[7,8].

In order to use this method, the system under study must be partitioned into a linear and a nonlinear part. Then, the system is transformed into the configuration shown in Fig. 2, whereby $G(j\omega)$ is the frequency response of all the linear elements in the system and $N(A, \omega)$ is the describing function of the nonlinearity, which is tabulated in books[8]. For the three cases mentioned in the previous subsection, it is always possible to reduce the block diagrams in the configuration of Fig. 2. A detailed description of this method can be found in the literature[7,8]. The application of the method to the problem at hand, along with the stability definition used in this paper, are reported elsewhere[4].

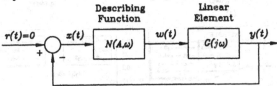

Figure 2: A feedback system whose nonlinear part has been replaced by its corresponding describing function.

Parametric Studies and Results

Using the describing function method, a parametric study was undertaken to investigate the significance of key system parameters. The three cases discussed in Section 3 are analyzed using the fixed parameter values of Table 3, and the range of parameter values of Table 4, both being based on available space manipulator data[5].

Table 3: Fixed-parameter values.

q_1	τ (s)	ω_{se} (rad/s)	ζ_f	ζ_{se}
45°	0.1	0.2513	0.707	0.707

Table 4: Free-parameter values.

β	$0.01 \le \beta \le 0.3$
λ (s)	$0.1 \le \lambda \le 50$
a_0 (m/s^2)	$0.0002 \le a_0 \le 0.02$
δ (m)	$0.001 \le \delta \le 0.1$
q_3	$-135°, -90°, -45°, 0°$
ω_f (rad/s)	$0.2513 \le \omega_f \le 5$

The results of the parametric study for Case 2 are illustrated with the use of stability maps, as depicted in Fig. 3. Figure 3(a) shows the stability boundary for different cutoff frequencies ω_f of the employed second-order filter. The region below such boundary represents a zone where the system is stable, while the region above corresponds to a zone of instability. As shown in the same figure, the stability zone can be increased by increasing the cutoff frequency ω_f. A similar analysis of the graphs of Fig. 3 leads to guidelines for the design of attitude control systems when flexibility is a major concern, namely,

1. The cutoff frequency ω_f for the filters should be chosen as large as possible to avoid instability.

2. The velocity gain λ should be chosen as large as possible to avoid instability.

3. The acceleration of the base a_0 should be kept small for stability. Unstable types of behavior are more likely to occur for large a_0.

4. Deadband limits δ should be chosen as large as possible to avoid instability.

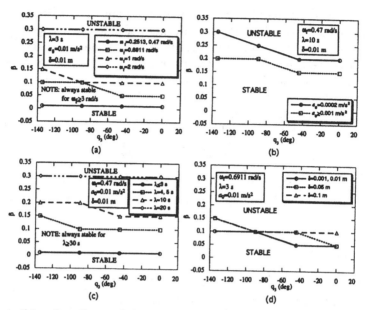

Figure 3: Describing function stability maps for Case 2 showing: (a) the effect of the cutoff frequency ω_f; (b) the effect of the base acceleration a_0; (c) the effect of the velocity gain λ; and (d) the effect of the deadband limit δ.

The upper limits of these parameters are set by design requirements or available hardware.

The same conclusions are drawn when Case 1 is analyzed. However, in general, the performance of Case 1 is worse than that for Case 2. To demonstrate this, the system configuration for Case 1 with parameters given in Tables 3 and 5 was used. Simulation results, for an initial error of 0.05 m, are shown in Fig. 4. Figures 4(b) and (c) show that thrusters are firing continuously, thus resulting in a high total fuel consumption of 491.1 fuel units, and a large rate of fuel consumption. Therefore, the system is classified as unstable. Moreover, the phase-plane trajectories of Fig. 4(a) show that a large limit cycle is reached due to the dynamic interactions.

If the model of Case 2 is simulated with the same parameter values, the results of Fig. 5 are obtained. Examining Figs. 5(b) and (c), we observe that thrusters are firing continuously, and that the total fuel consumption is quite high, namely, 430.8 fuel units. Therefore, this system is also considered unstable. The same conclusion is reached with the describing function method, as illustrated in Fig. 3(c) for $\lambda = 3$ s. Thus, the results corresponding to Cases 1 and 2 are both unstable, but the performance of Case 1 is worse than that for Case 2, since the fuel consumption is higher.

On the other hand, using the system configuration of Case 3, with the same parameters,

Table 5: Free-parameter values used for simulations.

β	λ (s)	a_0 (m/s^2)	δ (m)	ω_n (rad/s)	ζ	ω_f (rad/s)
.3	3	0.01	0.01	$2\pi(0.052)$	0.008	0.47

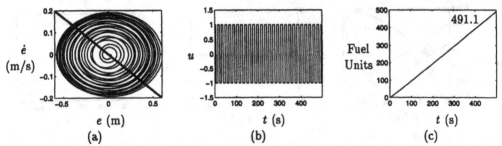

Figure 4: Simulation results using the Case 1 model: (a) Spacecraft error phase plane; (b) Thruster command history; and (c) Fuel consumption.

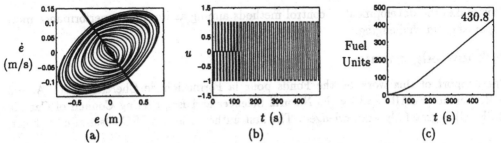

Figure 5: Simulation results using the Case 2 model: (a) Spacecraft error phase plane; (b) Thruster command history; and (c) Fuel consumption.

provides very interesting results, as shown in Fig. 6. From Fig. 6(a), it can be seen that a limit cycle contained between the switching lines is reached, which results in a stable system. One can note a spiral motion due to the relative motion between the two masses, that damps out. Figures. 6(b) and (c) are also typical of a stable system, since the thrusters are not firing continuously and the fuel-consumption curve is flat, thereby resulting in a near-zero rate of fuel consumption, similar to that for a rigid body system. In this case, the total fuel consumption is very small, namely, 7.2 fuel units only. Therefore, it is observed that the use of the proposed state estimator increases the performance of the control system significantly, and extends the operational life of the system. In addition, using the describing function method, it can be shown that this estimator results in a system that is almost always stable for the whole range of parameters, thus resulting in significantly increased stability margins in comparison to Cases 1 and 2. Using simulation, it was shown that the performance of this estimator remains very good in the presence of noise and model uncertainties[5].

Conclusions

This paper reports describing function results when the damping ratio is assumed to vary with the configuration of the manipulator and its payload. These results are almost the same as those obtained using a constant damping ratio of 0.05[4]. In fact, the general trends, or guidelines, are the same, but the particular values are different. Many configurations that were previously ascertained as stable are now found to be unstable. This more accurate analysis shows that the domain of instability is larger than previously tought.

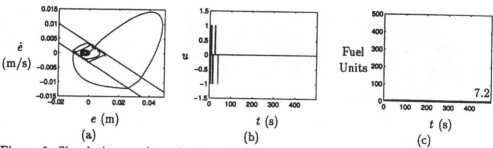

Figure 6: Simulation results using the Case 3 model: (a) Spacecraft error phase plane; (b) Thruster command history; and (c) Fuel consumption.

This makes the development of control methods aiming at improving performance more important and challenging.

Acknowledgements

The support of this work by the Fonds pour la Formation de Chercheurs et l'Aide à la Recherche (FCAR) and by the Natural Sciences and Engineering Council of Canada (NSERC) is gratefully acknowledged. The first author is an NSERC Class-of-67 Scholar.

References

[1] Millar, R. A., and Vigneron, F. R., "Attitude Stability of a Pseudorate Jet-Controlled Flexible Spacecraft," *J. of Guid., Cont., and Dyn.*, Vol. 2, No. 2, 1979, pp. 111–118.

[2] Sackett, L. L., and Kirchwey, C. B., "Dynamic Interaction of the Shuttle On-Orbit Flight Control System with Deployed Flexible Payload," *Proc. of the AIAA Guid. and Cont. Conf.*, San Diego, CA, 1982, pp. 232–245.

[3] Kubiak, E. T., and Martin, M. W., "Minimum Impulse Limit Cycle Design to Compensate for Measurement Uncertainties," *J. of Guid., Cont., and Dyn.*, Vol. 6, No. 6, 1983, pp. 432–435.

[4] Martin, E., Papadopoulos, E., and Angeles, J., "On the Interaction of Flexible Modes and On-off Thrusters in Space Robotic Systems," *Proc. of the 1995 Int. Conf. on Intelligent Robots and Systems, IROS'95*, Vol. 2, Pittsburgh, PA, 1995, pp. 65–70.

[5] Martin, E., "Interaction of Payload and Attitude Controller in Space Robotic Systems," Master Thesis, Dept. of Mech. Eng., McGill University, Montreal, Canada, 1994.

[6] Singer, N. C., "Residual Vibration Reduction in Computer Controlled Machines," Technical Report 1030, MIT Artificial Intelligence Laboratory, Cambridge, MA, 1989.

[7] Slotine, J.-J. E., and Li, W., *Applied Nonlinear Control*, Prentice Hall, Englewood Cliffs, N.J., 1991.

[8] Atherton, D. P., *Nonlinear Control Engineering*, Van Nostrand, New York, 1975.

MODELING AND OPTIMIZATION
OF THE TUBE-CRAWLING ROBOT

F.L. Chernousko and N.N. Bolotnik

Russian Academy of Sciences, Moscow, Russia

The pipe-crawling robot is an eight-legged walking machine that moves inside pipe-lines and can be used for inspection, maintenance, and repair. Optimization of structural parameters and possible gaits of the robot is discussed. The results obtained by computer simulation show a considerable sensitivity of operation characteristics of the robot with respect to its geometrical and kinematical parameters.

1. Introduction

In this paper, we present some investigations concerning an eight-legged pipe-crawling robot intended for motion inside pipes. Such a machine can be used for inspection, maintenance, and repair of pipe-lines. The robot was designed and developed by Prof. F. Pfeiffer and his colleagues in the Institute B of Mechanics at Munich Technical University. The results on optimization described in the paper were obtained in the course of the joint work of research teams from the Institute B of Mechanics, Munich TU, and the Institute for Problems in Mechanics of the Russian Academy of Sciences in Moscow.

To choose structural parameters of the robot (e.g., dimensions of the legs, types of actuators, etc.) and such characteristics of the gait as the length of the step and positions of the legs relative to the robot body during the step, we use the optimization techniques combined with a computer simulation of the robot dynamics. For the simulation, a software has been developed which makes it possible to investigate the behaviour of the robot, depending on the design variables, a control algorithm, and operation conditions (e.g., the angle of crawling). We consider two performance criteria to be maximized: the velocity of motion of the robot along the pipe and the thrusting force produced by the legs. Both the criteria are found to be sensitive to the variation of the lengths of the legs links, the length of the step, and positions of the feet on the pipe surface.

2. Structure and Kinematics of the Robot

The robot consists of a body and eight identical two-link legs attached to it by revolute joints (hip joints). The axes of four of the joints (numbered from 1 to 4) are normal to a plane π_1 and intersect this plane at points A_1, A_2, A_3, and A_4. The axes of the other joints (numbered from 5 to 8) are normal to a plane π_2 and intersect the latter at points A_5, A_6, A_7, and A_8. In what follows, we identify the points A_i ($i = 1, ..., 8$) with the corresponding joints. The planes π_1 and π_2 are orthogonal to each other. We will call the line of their intersection the axis of the robot and assume the mass centre C of the robot body to belong to this axis. We associate with the robot body a reference frame $Cxyz$, with the x-axis directed along the robot axis, the y-axis belonging to the π_1 plane, and the z-axis belonging to the π_2 plane. The points A_i form rectangles $A_1 A_2 A_4 A_3$ and $A_5 A_6 A_8 A_7$, the sides $A_1 A_3$, $A_2 A_4$, $A_5 A_7$, and $A_6 A_8$ are parallel to the axis x. The rectangles have a common symmetry axis x and match each other if we rotate the plane π_2 with respect to π_1 about the x-axis by the angle of $\pi/2$. The positions of the points A_i are specified by three numbers: a and d (the x- and y-coordinates of the point A_2) and b (the x-coordinate of the point A_3). We assume that the robot body is contained inside the cylinder $y^2 + z^2 \leq d^2$.

The links of the legs are rigid rods connected by means of revolute joints (knee joints) B_i ($i = 1, ..., 8$) whose axes are parallel to the axes of the corresponding hip joints A_i. On the end of the second link, there is a foot P_i. Thus, the leg can be considered as a planar two-member linkage $A_i B_i P_i$.

To describe configurations of the legs we introduce reference frames $A_i x_i y_i z_i$ fixed relative to the robot body. The x_i-axis is collinear to the x-axis. The y_i-axis belongs to the plane π_1 or π_2 in which the point A_i is located and is directed towards the robot axis. The z_i-axis completes the right-handed triad of coordinate axes. The configuration of the ith leg is described by the angle q_{1i} between the first link and the negative y-semiaxis and the angle q_{2i} between the links. The coordinates of the foot P_i are given by

$$x_i = l_1 \sin q_{1i} + l_2 \sin(q_{1i} - q_{2i}), \qquad y_i = l_1 \cos q_{1i} + l_2 \cos(q_{1i} - q_{2i}). \qquad (2.1)$$

Here, $l_1 = |A_i B_i|$ ($l_2 = |B_i P_i|$) is the length of the first (second) links of the legs.

The robot is controlled by driving torques M_{1i} and M_{2i} applied to the axes of joints A_i and B_i, respectively.

We consider a regular motion of the robot inside a circular cylindrical pipe, in which the robot's body travels translationally at a constant speed v, with the x-axis coinciding with the pipe axis. The pipe axis is inclined at a certain angle δ to the horizontal plane while the robot body is oriented in such a way that the y-axis forms an angle α with the vertical plane containing the pipe axis.

3. Equations of Motion

The first group of equations describes the motion of the robot as a whole and is given by

$$M\mathbf{g} + \sum_{i=1}^{8} \mathbf{F}_i = M\ddot{\mathbf{R}}_C, \qquad M\mathbf{R}_C \times \mathbf{g} + \sum_{i=1}^{8} \mathbf{R}_i^P \times \mathbf{F}_i = \dot{\mathbf{L}}_C. \qquad (3.1)$$

Here, $M = m + 8(m_1 + m_2)$ is the total mass of the robot (m is the mass of the robot body, while m_1 and m_2 are masses of the leg links), $\mathbf{R}_C$ is the position vector of its mass centre with respect to the point C, $\mathbf{R}_i^P$ is the position vector of the point P_i with respect to the point C, $\mathbf{g}$ is the gravity acceleration vector, $\mathbf{L}_C$ is the angular momentum of the robot with respect to the point C, and $\mathbf{F}_i$ is the contact force exerted on the foot P_i by the pipe surface.

Two vector equations (3.1) can be represented as six scalar differential equations of the second order with respect to q_{1i} and q_{2i}.

Denote by F_i, N_i, and Φ_i the x_i-, y_i-, and z_i-components of the force $\mathbf{F}_i$ in the $A_i x_i y_i z_i$ reference frame. The component N_i is the normal reaction applied to the foot P_i by the pipe surface, while F_i and Φ_i are the components of the friction force. We assume that there is no slippage between the feet of support legs and the pipe and that only dry friction is present here. Therefore, the inequalities

$$N_i \geq 0, \qquad F_i^2 + \Phi_i^2 \leq \mu N_i^2, \qquad (3.2)$$

where μ is the friction coefficient, must hold.

The other group of equations includes 16 Lagrangian equations describing the motion of the legs of the robot relative to the body. These equations are

$$\frac{d}{dt}\frac{\partial K_i}{\partial \dot{q}_{1i}} - \frac{\partial K_i}{\partial q_{1i}} + \frac{\partial \Pi_i}{\partial q_{1i}} = -M_{1i} + F_i\left[l_1 \cos q_{1i} + l_2 \cos(q_{1i} - q_{2i})\right]$$
$$+ N_i\left[l_1 \sin q_{1i} + l_2 \sin(q_{1i} - q_{2i})\right], \qquad (3.3)$$

$$\frac{d}{dt}\frac{\partial K_i}{\partial \dot{q}_{2i}} - \frac{\partial K_i}{\partial q_{2i}} + \frac{\partial \Pi_i}{\partial q_{2i}} = M_{2i} - F_i l_2 \cos(q_{1i} - q_{2i}) - N_i l_2 \sin(q_{1i} - q_{2i}).$$

Here, $K_i = K_i(q_{2i}, \dot{q}_{1i}, \dot{q}_{2i})$ is the kinetic energy of the ith leg and $\Pi_i = \Pi_i(q_{1i}, q_{2i})$ is the potential energy due to gravity. Besides, the kinetic and potential energies depend on geometrical and inertial parameters of the robot.

4. Gaits

We confine ourselves to the gaits satisfying the following conditions.

1. At each time instant, all legs belonging to one of the planes π_1 or π_2 are support ones, while the legs belonging to the other plane are in the transfer phase.

2. All support legs and all transferred legs move synchronously and have identical configurations at each time instant, i.e.,

$$q_{s1}(t) = q_{s2}(t) = q_{s3}(t) = q_{s4}(t); \quad q_{s5}(t) = q_{s6}(t) = q_{s7}(t) = q_{s8}(t), \quad s = 1, 2.$$

3. The gait is periodic, i.e.,

$$q_{si}(t + T) = q_{si}(t), \qquad s = 1, 2, \qquad i = 1, ..., 8. \qquad (4.1)$$

with some positive T for any t. The minimal T satisfying (4.1) is called the period of the gait.

4. The legs belonging to the planes π_1 and π_2 repeat the motions of each other with a half-period time shift.

5. The transfer phase of each leg includes two time intervals τ during which the foot stays on the pipe surface but exerts no pressure on it. One of the intervals immediately follows the support phase of a current step, while the other precedes the support phase of the next step. These *safety intervals* are introduced to avoid the loss of contact between the robot and the pipe when changing support legs.

The gaits can be described as time histories either of the angles q_{1i} or q_{2i} or of the feet coordinates x_i and y_i, see (2.1). In the latter case, we assume the knees to be bent in the direction of motion.

To describe a gait satisfying conditions 1 - 4 it is sufficient to specify the motion of one leg during one step $(0 \le t \le T)$. Denote by x the position of the foot (related to the corresponding $A_i x_i y_i z_i$ frame) at the beginning of the support phase $(t = 0)$ and by $s = vT$, the length of the step. For time periods $[0, T/2 + \tau]$ and $[T - \tau, T]$ during which the foot stays on the pipe surface, we have

$$x_i(t) = x - vt, \quad t \in [0, T/2 + \tau]; \quad x_i(t) = x - v(t - T), \quad t \in [T - \tau, T];$$
$$y_i(t) \equiv -h = -(\rho - d), \tag{4.2}$$

where ρ is the radius of the pipe. Let the motion of the foot in the transfer phase between the instants of contact be governed by the equation

$$\ddot{x}_i = \frac{s}{(T/4 - \tau)^2} \text{sign}\,(3T/4 - t), \quad y_i(t) = -h + \frac{\Delta}{2}\left[1 - \cos\frac{\pi(t - T/2 - \tau)}{T/4 - \tau}\right], \tag{4.3}$$
$$t \in [T/2 + \tau, T - \tau],$$

where Δ is the minimum lift of the foot from the pipe surface.

5. Determination of control torques

Using (2.1) where $x_i = x_i(t)$ and $y_i = y_i(t)$ we find the functions $q_{1i}(t)$ and $q_{2i}(t)$ and substitute them into (3.1) to obtain the contact forces $\mathbf{F}_i$. Note that 12 components F_i, N_i, and Φ_i of these forces for four support feet are not uniquely determined from six equations (3.1). To avoid the nonuniqueness we impose the additional conditions

$$\Phi_{2j-1} = -\Phi_{2j} = \xi_j P \cos \delta \sin \alpha_j, \quad F_{2j-1} = F_{2j} = (P \sin \delta + D(t))/4,$$

$$N_{2j-1} - N_{2j} = \nu_j P \cos \delta \cos \alpha_j,$$

$$\alpha_j = \alpha \text{ for } j = 1, 2; \quad \alpha_j = \pi/2 - \alpha \text{ for } j = 3, 4; \quad j = 1, 2, 3, 4. \quad . \tag{5.1}$$

Here, P is the weight of the robot, $D(t)$ is the inertial force acting along the x-axis, and ξ_j and ν_j are unknown coefficients; $j = 1, 2$ (3, 4) when the legs belonging to the plane π_1 (π_2) are support legs. The function $D(t)$ is expressed through $q_{1i}(t)$ and $q_{2i}(t)$ for a chosen gait. Relations (5.1) are valid only for support legs; for transferred legs, $F_i = N_i = \Phi_i = 0$.

Substituting (5.1) into (3.1) we uniquely determine ξ_i and ν_i. Thus, the problem is reduced to choosing N_1 and N_3 for any instant t. We determine N_1 and N_3 by minimizing the maximum normal reaction $N = \max_i N_i$ for support feet under constraints (3.2) and (5.1).

On determining the contact forces $\mathbf{F}_i$, the control torques can be found from (3.3).

6. Optimization of Parameters

We consider optimization with respect to two performance criteria: the speed v of motion of the robot body and the thrusting force F_i in the direction of motion (in the latter case, the constancy of speed is not supposed). The parameters to be varied are l_2, s, and x. Other parameters of the robot and the pipe are fixed. For examples described below, we take

$$a = 0.4\,\text{m}, \quad b = -0.4\,\text{m}, \quad d = 0.113\,\text{m}, \quad l_1 = 0.15\,m, \quad \rho = 0.375\,\text{m}, \quad \mu = 1,$$

$$m = 5\,\text{kg}, \quad m_1 = m_2 = 1\,\text{kg}, \quad M_1^* = 68\,\text{N} \cdot \text{m}, \quad M_2^* = 35\,\text{N} \cdot \text{m},$$

$$\dot{q}_{01} = 1.363\,\text{rad/s}, \quad \dot{q}_{02} = 2.726\,\text{rad/s}.$$

Here, M_1^* and M_2^* are maximum allowable torques at the joints A_i and B_i, respectively; and $\dot{q}_{01}$ and $\dot{q}_{02}$ are maximum allowable angular velocities of the corresponding links.

The variable parameters l_2, s, and x are subject to a number of constraints implying the implementability of the gait. These *parametric constraints* ensure that the links of the legs do not touch the robot body and the pipe. Besides, certain constraints are imposed on control torques and angular velocities of the legs links. These *drive constraints* are determined by the chosen electric motors and gears. Both the parametric and drive constraints are not presented here.

Problem 1. Find the length l_2 of the second links of the legs, and the gait parameters s and x maximizing the speed v of motion of the robot under the parametric and drive constraints.

Problem 1 was solved by direct computer simulation and variation of parameters within the constraints imposed. Investigations reveal a considerable sensitivity of the maximum velocity v to variation of s and l_2.

E x a m p l e 1. We set $\alpha = 0$ and $\delta = 0$.

Let us fix $l_2 = 0.15\,\text{m}$ (the links of the legs are identical). When varying s from 0.01 m up to 0.46 m, for $x = 0.11\,\text{m}$, the maximum allowable velocity v changes from 0.004 m/s to 0.085 m/s. When changing l_2 from 0.15 m to 0.25 m for $x = 0.11\,\text{m}$ and $s = 0.24\,\text{m}$, the maximum allowable velocity v changes from 0.087 m/s to 0.196 m/s.

Let us illustrate the optimization efficiency. Investigation show that the dependence of v on the angles α and δ is weak, and hence the data given below are representative.

First, we fix $l_2 = 0.15\,\text{m}$, $s = 0.30\,\text{m}$, and $x = 0.08\,\text{m}$. In this case, the maximal reachable velocity is $v = v_1 = 0.079\,\text{m/s}$. If we fix $l_2 = 0.15\,\text{m}$ and vary s and x, we find that the optimal velocity $v = v_2 = 0.085\,\text{m/s}$ is reached at $s = 0.26\,\text{m}$ and $x = 0.11\,\text{m}$. By varying all the three parameters l_2, s, and x, we achieve the maximum velocity $v = v_3 = 0.245\,\text{m/s}$ at $l_2 = 0.25\,\text{m}$, $s = 0.12\,\text{m}$, and $x = 0.15\,\text{m}$. The comparison of ratios $v_2/v_1 = 1.08$, $v_3/v_1 = 3.1$, and $v_3/v_2 = 2.88$ shows a considerable sensitivity of the maximum velocity to the variation of l_2, s and x (especially, of l_2). Therefore the adjustment of these parameters when designing and operating the robot is advisable.

Consider now the problem of maximizing the thrusting force of the robot. For simplicity, we confine ourselves to the case where the mass of the legs is much less than that of the body, and hence the methods of statics are applicable.

Problem 2. Given s, find x and l_2 maximizing the guaranteed maximum of the thrusting force during the support phase

$$J(x, l_2, s) = \min_{x_i \in [x - s/2, x]} \max_{M_{1i}, M_{2i}} \sum_i F_i$$

under the constraints $| M_{1i} | \leq M_1^*$, $| M_{2i} | \leq M_2^*$, (3.2), and the parametric constraints.

The solution of Problem 2 confirms the advisability of the optimization.

E x a m p l e 2. We set $M_1^* = 68 \, \text{N} \cdot \text{m}$, $M_2^* = 27 \, \text{N} \cdot \text{m}$, $s = 0.56 \, \text{m}$, $l_1 = 0.15 \, \text{m}$, $h = 0.24 \, \text{m}$, and $\mu = 1$.

Let us fix $x = 0.14 \, \text{m}$ and $l_2 = 0.15 \, \text{m}$. Then we have $J = J_1 = 642 \, \text{N}$. The variation of only one parameter x gives $J = J_2 = 786 \, \text{N}$ reached at $x = 0.105 \, \text{m}$. The variation of both x and l_2 gives the maximum thrusting force $J = J_3 = 1047 \, \text{N}$ at $x = 0.02 \, \text{m}$ and $l_2 = 0.21 \, \text{m}$. Comparison of the ratios $J_2/J_1 = 1.22$, $J_3/J_1 = 1.63$, and $J_3/J_2 = 1.33$ indicates high sensitivity of the thrusting force to the choice of positions of the support feet and the length of the second links of the legs.

7. Conclusion

The results of calculations show that the characteristics of the pipe-crawling robot depend significantly on the length of the second link of its legs: the longer the link, the greater are the speed and the thrusting force. These characteristics depend also on the position of the feet at the beginning and at the end of the support phase. Thus, optimization of parameters is recommended when designing the structure of the robot and planning its gaits.

This work is supported by the Russian Foundation for Basic Research (grant No. 96-01-01142).

AN INVESTIGATION OF A QUALITY INDEX FOR THE STABILITY OF IN-PARALLEL PLANAR PLATFORM DEVICES

J. Lee and J. Duffy

University of Florida, Gainesville, FL, USA

M. Keler

FH Munich, Germany

Abstract

The paper investigates primarily the geometrical meaning of the determinant of the Jacobian (det j) of the three connector lines of a planar in-parallel platform device using reciprocity. A remarkably simple result is deduced : *The maximum value of det j namely, det j_m is simply one-half of the sum of the lengths of the sides of the moving triangular platform. Further, this result is shown to be independent of the location of the fixed pivots in the base.*

A dimensionless ratio $\lambda = |\det j| / \det j_m$ is defined as the quality index ($0 \leq \lambda \leq 1$) and it is proposed here to use it to measure "closeness" to a singularity.

An example which determines the optimal design by comparing different shaped moving platforms having the same det j_m is given and demonstrates that the optimal shape is in fact an equilateral triangle.

Introduction

The geometry of the singularities of serial manipulators is well known[1,2,3]. The conditions for determining singularities are obtained from the vanishing of the determinant of the 6x6 Jacobian (det j) relating the instantaneous joint motions to the instant motion of the end effector. For virtually all industrial manipulators with special geometry for which pairs of joint axes are parallel or intersect at right angles, det j can be expanded as a product of factors. The algebra can be simplified considerably by choosing the reference point on the third or fourth joint axes counting the grounded joint as the first joint in the chain. It has been established that the first and sixth joint parameters of a six joint manipulator cannot produce a singular configuration.

There were several methods to determine how close a manipulator is to a singularity as the end effector moves from one location to another and this is reported in detail in [4] where it is concluded that no one method can be judged as superior to another. In this paper the determinant of j is used to measure a quality of a system. However a problem is that det j has dimensions(length^3) for spatial serial manipulators. Recently Keler[5] made an important inroad into this problem by determining expressions for the maximum value of det j, namely det j_m, for a number of industrial manipulators. In this way he has able to define a quality index λ where

$$\lambda = \frac{|\det j|}{\det j_m}$$

and for which $0 \le \lambda \le 1$.

Even less is known about the singularities of the parallel mechanisms[6]. In general there is no judicious choice for the reference point to simplify the determinant of the 6x6 Jacobian matrix, det j, of the lines of the six connectors joining the moving platform to the fixed base. Here, singularity configurations must be defined by the linear dependence of systems of lines (screws of zero pitch). Now all the special screw systems have been identified and investigated in great detail by Hunt[7] and later by Hunt-Gibson[8] and Rico-Duffy[9]. However, examining special configurations of a particular in-parallel mechanism and relating them to special screw systems is not a simple task.

At the outset, it appears that the problem of how to measure 'closeness' to a singularity may be solved by determining a quality index λ. However, Angeles and López-Cajún point out that these may be problems with such a quantity when the system is ill-conditioned which will magnify errors. They also discuss inhomogeneous problems in employing condition numbers[10]. At least this is a starting point. When $\lambda = 0$ the device is in singular configuration and when $\lambda = 1$ the device is in its best configuration to sustain load.

This paper investigates the geometrical meaning of det j for in-parallel planar manipulators. A definition of det j_m is proposed which enables a definition for a quality index for in-parallel planar manipulators.

Line Coordinates

In this paper the ray and axial coordinates of lines $\$_k$ and $\$_{ij}$ are defined by $\hat{s}_k = \{\underline{S}_k; \underline{S}_{0k}\}$ and $\hat{s}_{ij} = \{\underline{S}_{0ij}; \underline{S}_{ij}\}$ where $\underline{S}$ is the unit vector defining the direction of the line and $\underline{S}_0 = \underline{r} \times \underline{S}$ is the moment of the line about a reference point O. Clearly $\underline{S} \cdot \underline{S}_0 = \underline{S} \cdot (\underline{r} \times \underline{S}) = 0$ and $\underline{S} \cdot \underline{S} = 1$.

Employing matrix notation, the mutual moment (see Hunt[7] and Duffy[11]) of a pair of skew lines $\$_{ij}$ and $\$_k$ with ray $\hat{s}_k = \begin{bmatrix} S_k \\ \underline{S}_{0k} \end{bmatrix}$ and axis coordinates $\hat{S}_{ij} = \begin{bmatrix} \underline{S}_{0ij} \\ \underline{S}_{ij} \end{bmatrix}$ can be expressed as

$$\hat{s}_k^T \hat{S}_{ij} = \hat{S}_{ij}^T \hat{s}_k = -q_k \sin\alpha \tag{1}$$

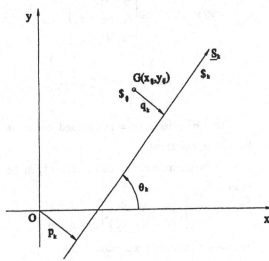

Figure 1 : Mutual Moment of Lines $\$_{ij}$ and $\$_k$

where α is the angle between them and q_k is the mutual perpendicular distance.

Here, the basis for $\hat{s}_k$ is $\{ i , j , 0k ;$

$0\epsilon i^*$, $0\epsilon j$, ϵk $\}$ or simply $\{ i , j ; \epsilon k \}$ whereas the basis for $\hat{s}_{ij}$ is $\{ \epsilon i , \epsilon j , 0\epsilon k; 0i$

, $0j , k \}$ or simply $\{ \epsilon i , \epsilon j ; k \}$ (see Figure-1). The moments of the line $\$_{ij}$ about the x and y axes are respectively $(y_{ij}, -x_{ij})$ and its direction is k.

From here on, lines such as $\$_k$ in the x-y plane will be designated in ray coordinates (see Figure 1) and

$$\hat{s}_k = \begin{bmatrix} c_k \\ s_k \\ p_k \end{bmatrix} \quad (2)$$

where the abbreviates $\cos \theta_k = c_k$ and $\sin \theta_k = s_k$ have been introduced and p_k is the moment of the line about the origin O.

Lines such as $\$_{ij}$ passing through the penetration point G of the x-y plane parallel to the z-axis will be designated in axis coordinates as

$$\hat{S}_{ij} = \begin{bmatrix} y_{ij} \\ -x_{ij} \\ 1 \end{bmatrix} \quad (3)$$

It is simple to deduce from (1) or Figure 1 that the mutual moment of the lines $\$_k$ and $\$_{ij}$ which are themselves mutually perpendicular is given by

$$\left| \hat{s}_k{}^T \hat{S}_{ij} \right| = \left| \hat{S}_{ij}{}^T \hat{s}_k \right| = q_k \quad (4)$$

where q_k is the absolute magnitude of the moment of $\underline{S}_k$ and $\underline{S}_{ij}$.

The Geometrical Meaning of det j and det j_m for Planar In-Parallel Platforms with RPR Connectors

Figure-2 illustrates an in-parallel platform with three serial RPR connector chains. An applied force is related to the resultant leg force by

$$\hat{w} = j\underline{f} \quad (5)$$

$^*\epsilon$ is a dual operator for which $\epsilon^2 = 0$.

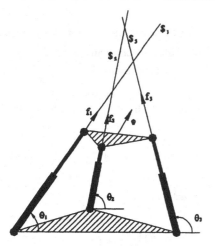

**Figure 2 : An In-Parallel Platform
with General Geometry**

where

$$j = \begin{bmatrix} c_1 & c_2 & c_3 \\ s_1 & s_2 & s_3 \\ p_1 & p_2 & p_3 \end{bmatrix} \quad (6)$$

$$\underline{f} = \begin{bmatrix} f_1 \\ f_2 \\ f_3 \end{bmatrix} \quad \text{and} \quad \hat{w} = \begin{bmatrix} f_x \\ f_y \\ m_z \end{bmatrix}$$

The abbreviations $c_i = \cos \theta_i$ and $s_i = \sin \theta_i$ have been introduced.

Arranging the pairs of the columns of j in the form

$$\begin{bmatrix} c_1 & s_1 & p_1 \\ c_2 & s_2 & p_2 \end{bmatrix}, \begin{bmatrix} c_2 & s_2 & p_2 \\ c_3 & s_3 & p_3 \end{bmatrix}, \begin{bmatrix} c_3 & s_3 & p_3 \\ c_1 & s_1 & p_1 \end{bmatrix} \quad (7)$$

and using Grassman's expansion yields

$$y_{12}:-x_{12}:1 = (s_1 p_2 - s_2 p_1):-(c_1 p_2 - c_2 p_1):(c_1 s_2 - c_2 s_1) \quad (8)$$

$$y_{23}:-x_{23}:1 = (s_2 p_3 - s_3 p_2):-(c_2 p_3 - c_3 p_2):(c_2 s_3 - c_3 s_2) \quad (9)$$

$$y_{31}:-x_{31}:1 = (s_3 p_1 - s_1 p_3):-(c_3 p_1 - c_1 p_3):(c_3 s_1 - c_1 s_3) \quad (10)$$

where (x_{12}, y_{12}), (x_{23}, y_{23}) and (x_{31}, y_{31}) are the coordinates for the points of intersection of the pairs of lines $(\$_1, \$_2)$, $(\$_2, \$_3)$ and $(\$_3, \$_1)$.

Now

$$\hat{S}_{12} = \begin{bmatrix} y_{12} \\ -x_{12} \\ 1 \end{bmatrix}, \quad \hat{S}_{23} = \begin{bmatrix} y_{23} \\ -x_{23} \\ 1 \end{bmatrix}, \quad \hat{S}_{31} = \begin{bmatrix} y_{31} \\ -x_{31} \\ 1 \end{bmatrix} \quad (11)$$

are respectively the unitized coordinates of the lines $\$_{12}$, $\$_{23}$ and $\$_{31}$. Expanding the determinant of j from the first column of (6) yields

$$\det j = \begin{bmatrix} c_1 & s_1 & p_1 \end{bmatrix} \begin{bmatrix} (s_2 p_3 - s_3 p_2) \\ -(c_2 p_3 - c_3 p_2) \\ (c_2 s_3 - c_3 s_2) \end{bmatrix} \quad (12)$$

Therefore

$$\det j = \hat{s}_1^T s_{3-2} \hat{S}_{23} = s_{3-2} \hat{S}_{23}^T \hat{s}_1 \quad (13)$$

where the abbreviation $s_{3-2} = s_3 c_2 - c_3 s_2 = \sin(\theta_3 - \theta_2)$ has been introduced. Analogously expanding from the second and third columns of (6)

$$\det j = \hat{s}_2^T s_{1-3} \hat{S}_{31} = s_{1-3} \hat{S}_{31}^T \hat{s}_2 \quad \text{and} \quad \det j = \hat{s}_3^T s_{2-1} \hat{S}_{12} = s_{2-1} \hat{S}_{12}^T \hat{s}_3 \quad (14,15)$$

The determinant of j has thus been expressed in three alternative forms. Now it is apparent from Figure-3 that

$$\hat{S}_{23}^T \hat{s}_1 = q_1, \quad \hat{S}_{31}^T \hat{s}_2 = -q_2, \quad \hat{S}_{12}^T \hat{s}_3 = q_3 \quad (16)$$

The determinant of j can thus be expressed in three alternative forms as scalar multiples of the three mutual moments. The scalar multiples are respectively $\sin(\theta_3-\theta_2)$, $\sin(\theta_1-\theta_3)$, $\sin(\theta_2-\theta_1)$ and therefore

$$\det j = s_{3-2}q_1 \quad (17\text{-}1) \qquad \det j = -s_{3-1}q_2 = s_{1-3}q_2 \quad (17\text{-}2) \qquad \det j = s_{2-1}q_3 \quad (17\text{-}3)$$

The geometrical meaning for the alternative expressions (see Figure-3) is clear. However q_1, q_2 and q_3 are variables and it is not immediately clear how useful these expressions for $\det j$ are in determining a quality index.

It is interesting to examine the expressions for $\det j$ when any pair of the connectors become parallel. Consider for example that the connector lines $\$_1$ and $\$_2$ are parallel (or they could be anti-parallel) as illustrated by Figure-4.

For the parallel case $\theta_1 = \theta_2$ and $s_{2-1} = 0$. The third expression for $\det j$ (17-3) fails since $q_3 \to \infty$ and hence $\det j = 0 \cdot \infty$. However the other expressions (17-1) and (17-2) yield equivalent values for $\det j$. Analogously expressions (17-1) and (17-2) fail when $s_{3-2}=0$ and $s_{1-3}=0$.

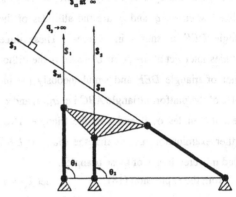

Figure 4 : Connectors $\$_1$ and $\$_2$ Parallel

It is clear (see Figure-3) that the three connectors $\$_1$, $\$_2$ and $\$_3$ form a triangle and through the vertices pass the lines $\$_{12}$, $\$_{23}$ and $\$_{31}$. This construction will now be used to determine $\det j_m$ for a general planar in-parallel mechanism (a mechanism consisting of a fixed triangular base jointed to a moving triangular platform by three RPR connectors in parallel). In Figure-5, the vertices of the triangle are labeled D, E and F. The base of the mechanism has fixed pivots at points G, H and I which lie somewhere along the lines $\$_1$, $\$_2$ and $\$_3$.

The moving platform has pivots A, B and C which also lie on the lines $\$_1$, $\$_2$ and $\$_3$. (Any pair of fixed and moving pivots A-H, B-I and C-G must not of course coincide.)   For the optimal configuration the $\det j_m$ can be expressed in three ways and

$$\det j_m = q_1 \sin\beta' = q_2 \sin\gamma' = q_3 \sin\alpha' \quad (18)$$

When the angles α', β' and γ' are known it is required that q_1, q_2 and q_3 are all maximum subject to the condition expressed by (18).

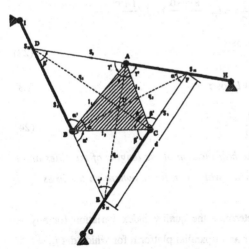

Figure-5 : General In-parallel Mechanism
in Optimal Configuration

For example, for a given value of β' which is an angle between the lines $\$_3$ and $\$_2$ Figure-6 illustrates that q_1 is maximum when it joins D to the joint at C which is on the line $\$_1$. Hence q_1 is an altitude of the triangle DEF and q_1' and q_1'' are both less than q_1.

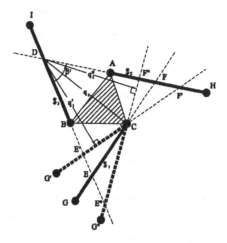

Figure 6 : Geometry of the maximum q_1

Further, the optimal configuration for det j_m is realized when q_1, q_2 and q_3 are the altitudes of the triangle DEF as shown in Figure-5. These three altitudes intersect at the point U which is the ortho-center of triangle DEF and simultaneously the in-center of the platform triangle ABC i.e. q_1, q_2 and q_3 bisect the angles α, β and γ of the triangle ABC.

Further examination reveals that the triangles DBA, CFA, CBE are similar and the triangle DEF is a scaled mirrored image of these triangles.

From the expression (18)
$$\det j_m = q_2 \sin\gamma' \tag{19}$$

and from Figure-5
$$q_2 = d\sin\alpha' \quad \text{where} \quad d = \frac{l_3\sin\alpha'}{\sin(\alpha'+\beta')} + \frac{l_2\sin\gamma'}{\sin(\beta'+\gamma')} \tag{20}$$

Therefore
$$\det j_m = \frac{l_3\sin^2\alpha'\sin\gamma'}{\sin(\alpha'+\beta')} + \frac{l_2\sin\alpha'\sin^2\gamma'}{\sin(\beta'+\gamma')} \tag{21}$$

Now $\alpha' = (\pi-\alpha)/2$, $\beta' = (\alpha+\gamma)/2$ and $\gamma = (\pi-\gamma)/2$ thus

$$\det j_m = l_3\cos^2(\frac{\alpha}{2}) + l_2\cos^2(\frac{\gamma}{2}) = l_3(\frac{1+\cos\alpha}{2}) + l_2(\frac{1+\cos\gamma}{2})$$

$$= \frac{l_2 + l_3 + l_2\cos\gamma + l_3\cos\alpha}{2} \tag{22}$$

From Figure-5
$$l_1 = l_2\cos\gamma + l_3\cos\alpha \tag{23}$$

Therefore from (22)
$$\det j_m = \frac{l_1 + l_2 + l_3}{2} = \frac{L}{2} \tag{24}$$

Hence for a general planar platform det j_m is one-half the sum of the lengths of the sides and is independent of the location of the fixed pivots G, H and I which can lie anywhere on the lines $\$_1$, $\$_2$ and $\$_3$.

It is now possible to employ these results to determine the quality index variation for any in-parallel planar platform. It is proposed to use det j_m for an in-parallel platform for which det $j_m = 0.5L$ where L is the sum of the lengths of three sides of the moving platform (the quality index will be defined by $\lambda = \frac{|\det j|}{0.5L}$).

Figure-7 illustrates a general triangular platform which has interior angles of 50°, 60° and 70° and its base pivots coincide with the intersection points of the connector action lines $\$_1$, $\$_2$ and $\$_3$. The platform is in its optimal configuration and $\lambda = 1$ (det j_m=7.4241(m)).

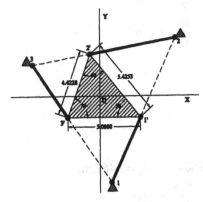

Figure-8 shows the quality index λ as a function of a rotation γ, about the z-axis through the mass center Q of the moving platform. It is interesting to note that the quality index $\lambda=0$ when $\gamma=120°$ and $\gamma=-60°$ and this is because q_1, q_2 and q_3 are all zero. Clearly $\lambda=1$ when $\gamma=0°$.

Figure-9 shows the quality index λ as a function of linear displacements parallel to x-y plane. The set of singular configurations in the x-y plane lie on an elliptic contour. It is important to note that when the moving joint 1' coincides with the base joint 1 the center of the platform Q coincides with the point J (see Figure-9). The three expressions for det j

Figure 7 : Optimal configuration of a general triangular platform

are undefined because it is possible to rotate the moving platform about the axis through the coincident joints. Analogously this also happens when the moving joints 2' and 3' coincide with base joints 2 and 3. However, they are impossible to reach in a real mechanism because connector lengths cannot be zero and circles within which the platform cannot reach can be drawn at the center J, K and L with radii of minimum connector lengths.

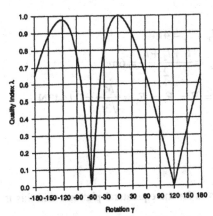

Figure 8 : Variation of λ with rotation γ about Q

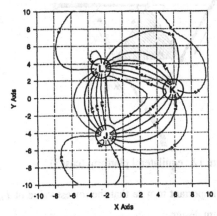

Figure 9 : Contour of constant λ in the x-y plane

Determination Of An Absolute Quality Index λ_{abs}

Figures 10 through 12 illustrate three different mechanisms which have same value of det j_m = 6m and they are in their optimal configurations. A moment $M = 1$ Nm (Newton-meter) is applied to these platforms in order to determine the maximum leg force of each platform.

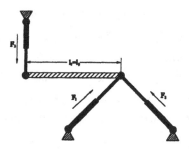

$$l_1 = l_3 = 6\text{m} \qquad l_2 = 0\text{m}$$

$$\det j_m = \frac{l_1 + l_2 + l_3}{2} = 6\text{m}$$

$$F_{\text{max}} = F_2 = \frac{M}{l_1} = 0.1667\text{N}$$

Figure 10 : Line segment platform (case1)

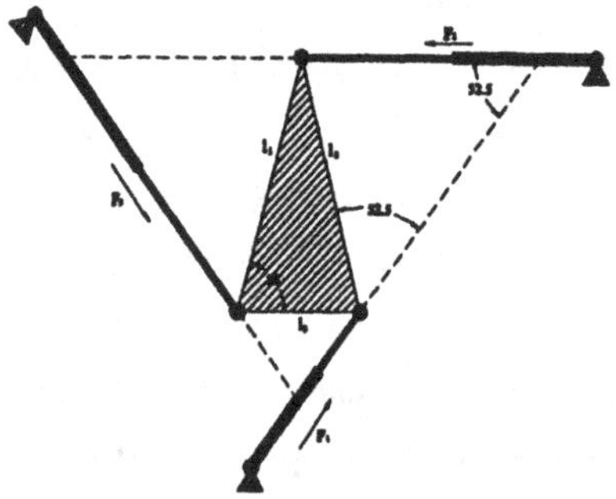

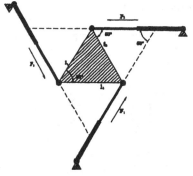

$$.\; l_1 = l_2 = 4.7664\text{m} \qquad l_3 = 2.4672\text{m}$$

$$\det j_m = \frac{l_1 + l_2 + l_3}{2} = 6\text{m}$$

$$F_{\text{max}} = F_2 = \frac{M}{l_2 \tan 52.5} = 0.1610\text{N}$$

Figure 11 : isosceles platform (case2)

$$l_1 = l_2 = l_3 = 4\text{m}$$

$$\det j_m = \frac{l_1 + l_2 + l_3}{2} = 6\text{m}$$

$$F_{\text{max}} = F_1 = F_2 = F_3 = \frac{M}{\sqrt{3}l_1} = 0.1443\text{N}$$

Figure 12 : equilateral platform (case3)

It is important to note that the lowest value of F_{max} occurs for the equilateral platform and this is the optimum design.

Further, from these results it is possible to determine an absolute quality index λ_{abs} using an equilateral platform as the basis for comparison. In other words for an arbitrary platform it is necessary to define λ_{abs} using the longest side. For example for case 1

$$\lambda_{\text{abs}} = \frac{6}{1.5 l_{\text{max}}} = \frac{6}{1.5 \times 6} = 0.6667$$

for case 2

$$\lambda_{abs} = \frac{6}{1.5 l_{max}} = \frac{6}{1.5 \times 4.7664} = 0.8392$$

and for case 3

$$\lambda_{abs} = \frac{6}{1.5 l_{max}} = \frac{6}{1.5 \times 4} = 1.0$$

It is now clear that in general $0 \leq \lambda_{abs} < 1$ unless of course the moving platform is an equilateral triangle for which $0 \leq \lambda_{abs} \leq 1$.

References

[1] K. J. Waldron, S. L. Wang and S. J. Bolin, "*A Study of the Jacobian Matrix of Serial Manipulators,*" ASME Journal of Mechanism, Transmission and Automation in Design, Vol. 107, No. 2, p. 230-238 (1984)

[2] H. Lipkin and E. Pohl, "*Enumeration of Singular Configuration for Robotic Manipulators,*" Journal of Mechanical Design, Vol. 113, p. 272-279 (1991)

[3] G. Beni and S. Hackwood, "*Recent Advances in Robotics,*" Chapter 5, Willy-Interscience Publication (1985)

[4] C. Klein and B. Blaho, "*Dexterity Measures for the Design and Control of Kinematically Redundant Manipulators*", The International Journal of Robotics Research, vol 6, No.2, p. 72-83 (1987)

[5] M. Keler, "*Kriterien der Führbarkeitseinschränkung von Effektoren in Robotern,*" Maschinenbautechnik, Berlin 40, p.155-158 (1991)

[6] H. Mohammadi Daniali, P. Zsombor-Murray and J Angeles, "*Singularity Analysis of Planar Parallel Manipulators*", Mech. Mach. Theory, Vol 3, No.5, p. 665-678 (1995)

[7] K. H. Hunt, "*Kinematic Geometry of Mechanisms*", Oxford University Press, Oxford (1978).

[8] G. Gibson and K. H. Hunt, "*Geometry of the screw systems - 1. Screws: genesis and geometry 2. Classification of screw systems,*" Mechanism and Machine Theory Vol.25, p. 1-10, 11-27 (1990).

[9] M. Rico Martinez and J. Duffy, "*Classification of screw systems - I. One- and two-systems II. Three-systems,*" Mechanism and Machine Theory Vol. 27, p. 451-458, 471-490 (1992).

[10] J. Angeles and C. López-Cajún, "*Kinematic Isotrophy and the Conditioning Index of Serial Robotic Manipulators*", The International Journal of Robotics Research, vol 11, No. 6, p. 560-571 (1992)

[11] J. Duffy, "*Statics and Kinematics with Applications to Robotics,*" Cambridge University Press (April, 1996)

WORKSPACES OF PLANAR PARALLEL MANIPULATORS

J-P. Merlet

INRIA, Sophia-Antipolis, France

C.M. Gosselin

Laval University, Ste-Foy, QUE, Canada

N. Mouly

INRIA, Grenoble, France

Abstract: This paper presents geometrical algorithms for the determination of various workspaces of planar parallel manipulators. Workspaces are defined as regions which can be reached by a reference point C located on the mobile platform. First, the *maximal* workspace is determined as the region which can be reached by point C with at least one orientation. From the above regions, the *inclusive* workspace, i.e., the region which can be attained by point C with at least one orientation in a given range, can be obtained. Then, the *total orientation* workspace, i.e., the region which can be reached by point C with every orientation of the platform in a given range, is determined. Three types of planar parallel manipulators are described and one of them is used to illustrate the algorithms.

1 Introduction

Parallel manipulators have been proposed as mechanical architectures which can overcome the limitations of serial robots [5]. Parallel manipulators lead to complex kinematic equations and the determination of their workspace is a challenging problem. Some researchers have addressed the problem of the determination of the workspace of parallel manipulators ([2]; [8]; [11]; [12]), especially for computing the workspace of the robot when its orientation is fixed.

In this paper, the problem of the determination of the workspaces of planar 3 d.o.f parallel manipulators is addressed. Algorithms are proposed for the determination of the maximal workspace, a problem which has been elusive to previous analyses.

Planar 3 d.o.f. parallel manipulators are composed of three kinematic chains connecting a mobile platform to a fixed base. The manipulator of particular interest in this study is referred to as the $3 - RPR$ manipulator. In this manipulator, the mobile platform is connected to the base via three identical chains consisting of a revolute joint attached to the ground followed by an actuated prismatic joint which is connected to the platform by a revolute joint (figure 1). Henceforth, the center of the joint connecting the ith chain to the ground will be denoted A_i and the center of the joint connecting the ith chain to the platform will be referred to as B_i. Others types of planar parallel manipulators are: the $3 - RRR$ robot ([1];[9];[5]) in which the joints attached to the ground are the only actuated joints and the $3 - PRR$ robot in which the prismatic joints are actuated.

A fixed reference frame is defined on the base and a moving reference frame is attached to the platform with its origin at a reference point C. The position of the moving platform is defined by the coordinates of point C in the fixed reference frame and its orientation is given by the angle θ between one axis of the fixed reference frame and the corresponding axis of the moving frame.

For the $3 - RPR$ manipulator, the workspace limitations are due to the limitations of the prismatic actuators. The maximum and minimum lengths of the prismatic actuator of the jth chain are denoted ρ^j_{max}, ρ^j_{min}. These values will be referred to as *extreme values* of the joint coordinates.

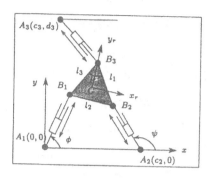

Figure 1: The $3 - RPR$ parallel manipulator.

Furthermore, an *annular region* $\mathcal{E}$ is defined as the region which lies between two concentric circles with different radii. The circle $\mathcal{E}^e$ with the largest radius will be referred to as the *external circle* and the smaller circle $\mathcal{E}^i$ (if it exists) will be referred to as the *internal circle*. The dimensions of the manipulators which are used in the examples are given in the appendix. In what follows, the presentation of the various workspaces will focus on the $3 - RPR$ manipulator.

2 Maximal workspace

The *maximal workspace* is defined as the region which the reference point C can reach with at least one orientation. It shall be noted that the maximal workspace will depend upon the choice of the reference point. One of the objectives of the present work is to determine geometrically the boundary of the maximal workspace.

The determination of the maximal workspace has been addressed by Kassner [7] who pointed out that the boundary of this workspace is composed of circular arcs and of portions of sextic curves, but was only able to compute them with a discretization method. The same observation was made by Kumar [8] but his method, based on screw analysis, cannot be used for a manipulator with prismatic actuators.

2.1 Determining if a point is in the maximal workspace

First, a simple algorithm is derived to determine if a location of the reference point is in the maximal workspace, this being equivalent to determining if there is at least one possible orientation of the platform for this location.

For a given position of C, point B_1 can move on a circle C_B^1 with center C and radius $\|CB_1\|$. We first verify if C_B^1 is completely inside or outside the annular region $\mathcal{E}_1$, corresponding to the constraint for leg 1, by checking the distance between the centers of C_B^1 and $\mathcal{E}_1$ with respect to their radii. If C_B^1 is inside $\mathcal{E}_1$, then any orientation is allowed for the platform, with respect to the constraints on leg 1. If C_B^1 is outside $\mathcal{E}_1$ then no orientation is allowed for the platform and C is outside the maximal workspace.

If the preceding test fails it may be assumed that there are intersection points between C_B^1 and $\mathcal{E}_1^e$, $\mathcal{E}_1^i$. For each of these intersection points there is a unique orientation angle possible for the platform. These angles are ordered in the interval $[0, 2\pi]$ in order to obtain a set of consecutive intervals. Then, in order to determine which intervals define valid orientations for the platform, the middle value of each interval is used as the orientation of the platform and the constraints on leg 1 are tested for the corresponding configuration. A similar procedure is performed for the legs 2 and 3. For leg i, n various intervals are obtained and the set I_n^i of possible orientations of the platform with respect to the constraint on the leg can be determined.

The intersection I_n of these lists is then determined as the intersection of all the sets of three intervals $\{I_1 \in I_{n_1}^1, I_2 \in I_{n_2}^2, I_3 \in I_{n_3}^3\}$. If I_n is not empty then C belongs to the maximal workspace and I_n defines the possible orientation for the moving platform at this point.

2.2 Determination of the boundary of the maximal workspace

For purposes of simplification, it is first assumed that the reference point on the platform is chosen as one of the B_i's, for example point B_3. The general case will be presented later on as a generalization. If a location of B_3 belongs to the boundary of the maximal workspace then at least one of the legs is at an extreme value (otherwise B_3 will be capable of moving in any direction which is in contradiction with B_3 being on the boundary of the workspace). Note that the configuration with three legs in an extreme extension defines only isolated points of the boundary since they are solutions of the direct kinematics of the manipulator, a problem which admits at most 6 different solutions [3].

2.2.1 Boundary points with one extreme leg length

In order to geometrically determine the points of the boundary for which one leg length of the manipulator is at an extreme value, the kinematic chain $A_iB_iB_3$ is considered as a planar serial 2 d.o.f. manipulator whose joint at A_i is fixed to the ground. It is well known that the positions of B_3 belonging to the boundary of the workspace are such that A_i, B_i, B_3 lie on the same line. For instance, consider leg 1: two types of alignment are possible. Either $A_1B_1B_3$ or $A_1B_3B_1$ (or $B_3A_1B_1$) are aligned in this order. Consequently, point B_3 lies on a circle C_{B_3} centered at A_i. As B_3 moves on C_{B_3}, points B_1, B_2 will move on circles denoted C_{B_1}, C_{B_2}. Valid positions of B_3 on its circle are such that the corresponding positions of B_1, B_2, B_3 respectively belong to the annular regions $\mathcal{E}_1$, $\mathcal{E}_2$, $\mathcal{E}_3$.

Let α denote the rotation angle of leg 1 around A_1. The intersection points of circle C_{B_i} with the annular region $\mathcal{E}_i$ are then computed. All the intersection points define specific values for the angle α and the orientation of the platform. These values are ordered in a list leading to a set of intervals I^i. It is then possible to determine which intervals are components of the boundary of the workspace by taking the middle point of the arc and verifying if the corresponding pose of the platform belongs to the workspace.

As mentioned previously various types of alignment with various extreme values of the leg length are possible and some of them will lead to a component of the boundary. The arcs which are obtained after studying these different cases are placed in an appropriate structure and will be denoted *phase 1 arcs*.

2.2.2 Boundary points with two extreme leg lengths

The case for which the reference point lies on the boundary of the workspace while two leg lengths of the manipulator are in an extreme extension is now investigated. Since the reference point is point B_3, only the cases where the legs with extreme lengths are legs 1 and 2 need to be considered.

When legs 1 and 2 have a fixed length, the trajectory of point B_3 is the coupler curve of a four-bar mechanism. This mechanism has been well studied [4] and it is well known that the coupler curve is a sextic. Consequently it can be deduced that the boundary of the maximal workspace will be constituted of circular arcs and of portions of sextics.

Four sextics will play an important role in this study. They are the coupler curves of the four-bar mechanisms with leg lengths corresponding to the various combinations of extreme lengths of legs 1 and 2, i.e., $(\rho_{max}^1, \rho_{max}^2)$, $(\rho_{max}^1, \rho_{min}^2)$, $(\rho_{min}^1, \rho_{min}^2)$, $(\rho_{min}^1, \rho_{max}^2)$.

Some particular points, referred to as the *critical points* will determine the circular arcs and the portions of sextic which define the boundary of the maximal workspace. The critical points can be of five different types, thereby defining five sets of such points.

The first set consists of the intersection points of the sextics and the annular region $\mathcal{E}_3$: in this case the three leg lengths are at an extreme value. Therefore, these points are solutions of the *direct kinematics* and can be found numerically. The second set of critical points consists of the intersection points of the sextics with the phase 1 arcs. In this case the length of legs 1 and 2 are defined by the sextics and the length of leg 3 is the radius of the arc. These points will also be critical points for the arcs. A third set of critical points are the multiple points of the sextics. Indeed we may have only one critical point on a circuit on the sextic: therefore introducing the multiple points as critical points enables to define two arcs of sextic on the circuit, one of them being a member of the boundary of the workspace. Finding these multiple points is a well known problem [4]. The fourth set of critical points for the sextics will be the limit points of the coupler curve. Indeed, for some value of the leg lengths the four-bar mechanism may not be a crank, i.e., the angle ϕ is restricted to belong to some intervals. Consequently the sextic is not continuous and each position of B_3 corresponding to one of the bounds of the intervals is a critical point. The last set of critical points for the sextics consists of the set of intersection points between the sextics. Recently Innocenti has proposed an algorithm to solve this problem [6].

2.2.3 Determination of the portions of sextic belonging to the boundary

Any portion of sextic belonging to the boundary must lie between two critical points. For each critical point T_i the unique pair of angles ϕ_i, ψ_i corresponding to T_i is determined (note that for the critical points which are multiple points of the coupler curve although the coupler point is identical, the angles ϕ_i, ψ_i are different). For a given value of ϕ, there are in general two possible solutions for ψ which are obtained by solving a second order equation in the tangent of the half-angle of ψ. Consequently ψ is determined using one of the two expressions of the tangent of the half-angle.

First, the T_i's are sorted according to the expression which is used for determining the corresponding angle ψ_i, thereby giving rise to two sets of T_i's. Each of these sets is then sorted according to an increasing value of the angle ϕ. Consequently, the sextics are split into *arcs* of sextics, some of which are components of the boundary of the maximal workspace.

A component of the boundary will be such that for any point on the arc a motion along one of the normals to the sextic will lead to a violation of the constraints while a motion along the other normal will lead to feasible values for the link lengths. Any other combination implies that the arc is not a component of the boundary of the maximal workspace. In order to perform this test, the inverse jacobian matrix for a point on the arc (for example the middle point i.e. the coupler point obtained for ψ as the middle value between the angles ψ of the extreme points of the arc) is computed as well as the unit normal vectors n_1, n_2 of the sextic at this point. Then the joint velocities are calculated for a cartesian velocity directed along n_1, n_2. The sign of the joint velocities obtained indicates whether or not the arc is a component of the boundary. A similar procedure is used to identify the circular arcs which are components of the boundary. To this end, the phase 1 arcs, for which the critical points — the intersection points with the sextics and the extreme points of the arcs — have been determined, are considered. Each of the arcs between two critical points is examined to determine if the arc is a component of the boundary by using the same test as for the arcs of sextic. The boundary of the maximal workspace is finally obtained as a list of circular arcs and portions of sextics. The maximal workspace of the manipulators described in the appendix are shown in figure 2.

The computation time of the boundary of the maximal workspace is heavily dependent on the result. On a SUN 4-60 workstation this time may vary from 1500 to 15000 ms. The most expensive part of the procedure is the calculation of the intersection of the sextics.

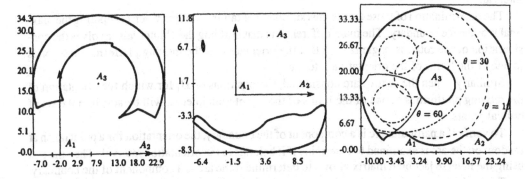

Figure 2: Left: the maximal workspace for manipulator 3 with $\rho_1 \in [8, 12]$, $\rho_2 \in [5, 15]$, $\rho_3 \in [10, 17]$. Middle: the maximal workspace for manipulator 4, $\rho_1 \in [8, 12]$, $\rho_2 \in [5, 15]$, $\rho_3 \in [10, 17]$. Right: the area within the thick lines is the maximal workspace of manipulator 3 with $\rho_1 \in [5, 20]$, $\rho_2 \in [5, 20]$, $\rho_3 \in [5, 20]$. The dashed and thin lines represent the constant orientation workspace for various orientations of the platform.

2.2.4 Maximal workspace for any reference point

To compute the maximal workspace for a reference point different from B_3, a similar algorithm can be used. Basically, only the complexity of the algorithm will be increased. Indeed, not only the eight circles of type C_1, C_2 have to be considered but also the four circles centered at A_3 which correspond to the case where the length of link 3 has an extreme value. Similarly, the twelve sextics which can be obtained from all the possible values for the extreme lengths of links 1, 2, 3 must now be considered.

3 Inclusive maximal workspace

The *inclusive maximal workspace* (denoted IMW) is defined as the set of all the positions which can be reached by the reference point with at least one orientation of the platform in a given interval referred to as the *orientation interval*. Hence, the maximal workspace is simply a particular case of IMW for which the prescribed orientation interval is $[0, 2\pi]$. In what follows, it is assumed that the orientation of the moving platform is defined by the angle between the x axis and the line B_3B_1. Moreover, it is also assumed that the reference point of the moving platform is B_3.

The computation of the boundary of the IMW is similar to the computation of the boundary of the maximal workspace. First, it is recalled that it is simple to determine if a point belongs to the IMW since one can compute the possible orientations of the moving platform at this point. It is also clear that a point lies on the boundary if and only if at least one of the link lengths is at an extreme value.

Consider first the circles described by B_3 when points A_i, B_i, B_3 lie on the same line. For each position of B_3 on the circles, the orientation of the moving platform is uniquely defined. The valid circular arcs must satisfy the following constraints: the points B_1, B_2, B_3 lie inside the annular regions $\mathcal{E}_1, \mathcal{E}_2, \mathcal{E}_3$ and the orientation of the moving platform belongs to the orientation interval.

The determination of these arcs is thus similar to obtaining the arcs when computing the maximal workspace boundary. The main difference is that building the I^i intervals involves the consideration of the rotation angle α such that the orientation of the moving platform corresponds to one of the limits of the orientation interval.

Similarly, when the sextics are considered, the positions of B_3 for which the orientation of the moving platform is at one of the limits of the orientation interval will be added in the set of critical points.

To verify if a particular arc is a component of the boundary, the orientation for a point taken at random on the arc is examined to determine if it belongs to the orientation interval. Then the test using the inverse jacobian matrix allows to determine if the arc is a component of the boundary.

Typically the computation time for an IMW is about 1000 to 20000 ms on a SUN 4-60 workstation. Figure 3 presents some IMW for various orientation intervals.

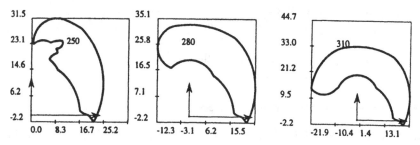

Figure 3: IMW of manipulator 1 for various orientation intervals (the orientation intervals always begin at 0). The limits are $\rho_1 \in [2, 8]$, $\rho_2 \in [5, 25]$, $\rho_3 \in [10, 25]$.

4   Total orientation workspace

This section addresses the problem of determining the region reachable by the reference point with all the orientations in a given set $[\theta_i, \theta_j]$ which will be referred to as the orientation interval. This workspace will be denoted as TOW.

It is relatively easy to determine if a point belongs to this workspace since it is possible to compute the possible orientations for any position of the reference point. For a point belonging to the boundary of the TOW, one leg will be at an extreme value. Indeed, two legs cannot be at an extreme value since in this case the orientation of the moving platform is unique and consequently the point cannot belong to the TOW.

Assume that for a point on the boundary the orientation of the moving platform is one of the bounds of the interval, i.e., θ_i or θ_j while the length of leg i is at an extreme value. As B_i moves on the circle of the annular region $\mathcal{E}_i$ corresponding to the value of the leg length, point B_3 moves on a circle C_w^i with the same radius whose center is obtained by translating the center of $\mathcal{E}_i$ by the vector $\mathbf{B_iB_3}$, which is fixed since the orientation of the moving platform is known. Any point in the TOW must lie within the circle C_w^i. Therefore if the bounds θ_i, θ_j and all the possible B_i's are considered, any point of the TOW must be inside the 12 circles with center and radii $(A_3, \rho_{max}^3), (A_3, \rho_{min}^3), (A_1 + \mathbf{B_1B_3}, \rho_{max}^1), (A_1 + \mathbf{B_1B_3}, \rho_{min}^1), (A_2 + \mathbf{B_2B_3}, \rho_{max}^2), (A_2 + \mathbf{B_2B_3}, \rho_{min}^2)$.

Assume now that a point on the boundary is reached with an orientation different from θ_i, θ_j and that the length of link 1 has an extreme value, say ρ_{max}^1. When the orientation of the moving platform lies in the orientation interval, B_1 belongs to a circular arc defined by its center B_3, its radius $\|\mathbf{B_3B_1}\|$ and the angles θ_i, θ_j. As the point belongs to the TOW the arc must lie inside the annular region $\mathcal{E}_1$. Furthermore this arc is tangent at some point to the external circle of $\mathcal{E}_1$ since B_3 lies on the boundary of the TOW. This tangency implies that point B_3 lies on a circle of center A_1 and radius $\rho_{max}^1 - \|\mathbf{B_1B_3}\|$. Any point within the TOW must be inside this circle. Four such circles may exist, whose center and radii are $(A_1, \rho_{max}^1 - \|\mathbf{B_1B_3}\|), (A_1, \rho_{min}^1 - \|\mathbf{B_1B_3}\|), (A_2, \rho_{max}^2 - \|\mathbf{B_2B_3}\|), (A_2, \rho_{min}^2 - \|\mathbf{B_2B_3}\|)$.

If a point B_3 belongs to the TOW it is necessary that the point is included in the 16 circles which have been determined. Consequently the boundary of the TOW is the intersection of these sixteen circles. Note that a particular case of TOW is the *dextrous workspace*, which is the region which can be reached by the reference point with any orientation [8],[12],[11].

5   Conclusion

Geometrical algorithms for the determination of the boundary of various workspaces for planar parallel manipulators have been described. Basically the presented algorithms can be extended without any difficulties to other types of planar parallel robots [10]. The authors want to acknowledge that this work has been supported in part by the France-Canada collaboration contract n° 070191.

Appendix

The dimensions of the manipulators used in the examples of this paper are defined in figure 1 and their numerical values are presented in the following table.

Manipulator	Type	l_1	l_2	l_3	c_2	c_3	d_3	μ
1	RPR	25	25	25	20	0	10	60
2	RPR	20.839	17.045	16.54	15.91	0	10	52.74
3	RPR	25	25	25	20	10	17.32	60
4	RPR	2	2	2	10	5	8.66	60

References

[1] Gosselin C. *Kinematic analysis optimization and programming of parallel robotic manipulators.* Ph.D. Thesis, McGill University, Montréal, June, 15, 1988.

[2] Gosselin C. and Angeles J. The optimum kinematic design of a planar three-degree-of-freedom parallel manipulator. *J. of Mechanisms, Transmissions and Automation in Design,* 110(1):35–41, March 1988.

[3] Gosselin C., Sefrioui J., and Richard M.J. Solution polynomiale au problème de la cinématique directe des manipulateurs parallèles plans à 3 degrés de liberté. *Mechanism and Machine Theory,* 27(2):107–119, March 1992.

[4] Hunt K.H. *Kinematic geometry of mechanisms.* Clarendon Press, 1978.

[5] Hunt K.H. Structural kinematics of in parallel actuated robot arms. *J. of Mechanisms, Transmissions and Automation in Design,* 105:705–712, March 1983.

[6] Innocenti C. Analytical determination of the intersection of two coupler-point curves generated by two four-bar linkages. In J. Angeles P. Kovacs, G. Hommel, editor, *Computational Kinematics,* pages 251–262. Kluwer, 1993.

[7] Kassner D.J. Kinematics analysis of a planar three-degree-of-freedom platform-type robot manipulator. Master's thesis, Purdue University, Purdue, December 1990.

[8] Kumar V. Characterization of workspaces of parallel manipulators. *ASME J. of Mechanical Design,* 114:368–375, September 1992.

[9] Ma O. and Angeles J. Direct kinematics and dynamics of a planar three-dof parallel manipulator. In *ASME Design and Automation Conf.,* volume 3, pages 313–320, Montréal, September, 17-20, 1989.

[10] Merlet J-P. and Mouly N. Espaces de travail et planification de trajectoire des robots parallèles plans. Research Report 2291, INRIA, Fevrier September 1994.

[11] Pennock G.R. and Kassner D.J. The workspace of a general geometry planar three degree of freedom platform manipulator. *ASME J. of Mechanical Design,* 115:269–276, June 1993.

[12] Williams II R.L. and Reinholtz C.F. Closed-form workspace determination and optimization for parallel robot mechanisms. In *ASME Proc. of the the 20th Biennial Mechanisms Conf.,* pages 341–351, Kissimmee, Orlando, September, 25-27, 1988.

NONLINEAR CONTROL OF A PARALLEL ROBOT INCLUDING MOTOR DYNAMICS

L. Beji
CEMIF, Evry, France

A. Abichou
INAT, Cité Mahrajène, Tunis, Tunisia

P. Joli and M. Pascal
CEMIF, Evry, France

Abstract. The tracking control problem of a parallel robot including the electrical actuator dynamics is addressed in this paper. For the electrically actuated robots we design a nonlinear control law in armatures' input voltages. The model obtained is in standard form to allow the application of the singular perturbation method. To validate the proposed corrective controller, the passivity concept and singular perturbation techniques are considered. Simulation tests are presented that confirm the efficiency of the proposed nonlinear control law.

1. Introduction

Various control methods for robot motion control are designed at torque input level and the actuator dynamics are excluded (more details can be found in Stepanenco et al[1]). However, actuator dynamics constitute an important part of the complete robot dynamics. Furthermore the actuators cannot be controlled directly in forces or torques. In this context, for a Rigid-Link Electrically-Driven (RLED) robot, Tarn et al[2] develop a feedback linearizing control.

The RLED manipulator is transformed to a third-order dynamic model that requires acceleration measurements. Dawson et al[3] use the assumption of exact model knowledge and propose a corrective tracking controller for the RLED robots. In the presence of unknown parameters of a RLED robot, recently an adaptive controller scheme was designed by Stepanenco et al[1]. For a hydraulically actuated serial robot[4,5], the tracking control problem is addressed using singular perturbation method where the hydraulic part is the fast subsystem. The focus of this work is the tracking control of a 6-DOF parallel Space Manipulator (SM) which is provided to perform assemblage tasks with high dexterity. The formulated model of the parallel SM leads to a singularly perturbed model where the input is the armatures' input voltages (see section 2). In section 3, a corrective control law is designed and its validity is achieved. As a result, an exponential stability of the closed loop system is obtained.

2. Dynamic model of the global system

In accordance with the modelisation theories of a closed kinematic chains mechanism[6], the 6-DOF parallel SM exhibits the following dynamic model of the mechanical part[7]

$$A(q)\ddot{q} + C(q,\dot{q})\dot{q} + G(q) = \tau \tag{1}$$

where $q \in R^n$ denotes the controlled variables (n denotes the end effector mobility). $\tau \in R^n$ denotes the vector of driving torques and forces. $A(q) \in R^{n \times n}$ denotes the inertia matrix, which is a symmetric positive definite matrix. $C(q,\dot{q})\dot{q} \in R^n$ represents the vector of centrifugal and Coriolis forces. $G(q) \in R^n$ is the vector of gravitational forces. The active joints of the parallel SM are driven by a DC-motors which are modelled by[11]

$$L_{dc}\dot{\tau} + R_{dc}\tau + K_{dc}\dot{q} = u \tag{2}$$

where $R_{dc} = R(NK_t)^{-1}$; $K_{dc} = K_b N$; $L_{dc} = L(NK_t)^{-1}$. R and L are diagonal positive matrices of the armature resistances and inductances respectively. K_b is a diagonal matrix of the back emf constant of the motors. u is the armature' input voltages. $N \in R^{n \times n}$ is a diagonal matrix of the gear ratios ($N > 0$). $K_t \in R^{n \times n}$ ($K_t > 0$) is a diagonal matrix of the motor torque constants. We consider the following decomposition: $N = N_M N'$, $K_t = K_t^M K_t'$ and $L = L_m L'$ where for $j = 1$ to n, $N_M = \max(N_j)$, $L_m = \min(L_j)$, $K_t^M = \max(K_{ij})$.

In accordance with the relations (1) and (2), the dynamic behaviour of a rigid-link electrically-driven robot such that the parallel SM is governed by the following pair of equations :

$$\begin{cases} A(q)\ddot{q} + C(q,\dot{q})\dot{q} + G(q) = \tau \\ \varepsilon\dot{\tau} = \alpha_1\dot{q} + \alpha_2\tau + \alpha_3 u \end{cases} \tag{3}$$

with $\varepsilon = L_m / (N_M K_t^M)$; $\alpha_1 = -N' K_t' K_b N(L')^{-1}$; $\alpha_2 = -(N_M K_t^M)^{-1} R(L')^{-1}$; $\alpha_3 = N' K_t'(L')^{-1}$.

α_1, α_2 are constants and diagonal matrices. α_3 is a constant and an invertible diagonal matrix. ε is assumed a "small" scalar parameter ($\varepsilon \approx 2.10^{-4}$ in the case of the parallel SM).

3. Singular perturbation method

Before studying the case of our robot, we consider the following nonlinear system:

$$\begin{cases} \dot{x} = f(x,y,\varepsilon,t) & x \in B_x \subset R^n, x(0) = x_0 \\ \varepsilon\dot{y} = g(x,y,\varepsilon,t) & y \in B_y \subset R^m, y(0) = y_0 \end{cases} \tag{4}$$

Further, for all $[t,x,y,\varepsilon] \in [0, \infty) \times B_x \times B_y \times [0,\varepsilon_0]$ we consider the following assumption:

A1) The functions f and g are smooth enough with $f(0,0,\varepsilon,t) = 0$ and $g(0,0,\varepsilon,t) = 0$. Moreover, the equation $g(x,y,0,t) = 0$ has a unique real root $y = h(x,t)$ with $h(0,t) = 0$. A2) f, g and their first partial derivatives with respect to x, y and ε are continuous and bounded. A3) The function h and $\partial g(x,y,0,t)/\partial y$ have continuous first partial derivatives, $\partial f(x,h(x,t),0,t)/\partial x$ has bounded first partial derivatives with respect to x. A4) The initial conditions x_0 and y_0 are regular functions of ε. A5) The origin of the reduced system; $\dot{x} = f(\bar{x},h(\bar{x},t),0,t)$ is exponentially stable. A6) The origin of the following *boundary layer model* is exponentially stable uniformly with respect to (x,t); $d\hat{y}/d\varsigma = g(x,\hat{y}(\varsigma) + h(\bar{x},t),0,t)$ where $\varsigma = t/\varepsilon$.

Theorem[10] **1.** *We suppose that Assumption A1-6 hold, then the singular perturbation problem (4) has a unique solution $x(t,\varepsilon)$, $y(t,\varepsilon)$ defined for all $t \geq t_0 \geq 0$ with $x(t,\varepsilon) - \bar{x}(t) = O(\varepsilon)$ and $y(t,\varepsilon) - h(t,\bar{x}(t)) - \hat{y}(\varsigma) = O(\varepsilon)$ hold uniformly for $t \in [t_0, \infty)$, where $\bar{x}(t)$ and $\hat{y}(\varsigma)$ are the solutions of the reduced and boundary layer problems.*

3. Nonlinear control scheme for a RLED parallel robot

We recall the dynamic model of the mechanical part

$$A(q)\ddot{q} + C(q,\dot{q})\dot{q} + G(q) = \tau_d - (\tau_d - \tau) = \tau_d - \tilde{\tau} \tag{5}$$

where $\tilde{\tau} = \tau_d - \tau$ can be viewed as a disturbance signal for the mechanical dynamics. τ_d represents a reference torque vector. The stability of system (5) can be achieved by the rejection of the disturbance $\tilde{\tau}$ (i.e. $\tilde{\tau} \to 0$) and by the design of an adequate τ_d. We consider now for the singularly perturbed system (3) the following first feedback in armature voltages:

$$u = \alpha_3^{-1}\tau_d - \alpha_3^{-1}(I_d + \alpha_2)\tau - \alpha_3^{-1}\alpha_1\dot{q} \tag{6}$$

I_d denotes the identity matrix. In electric drives for a DC-motors[11], τ is proportional to the armature current vector i which is assumed to be accessible. Based on the passivity approach[8], we propose the following auxiliary law in torques and forces:

$$\tau_d = A(q)\ddot{q}_r + C(q,\dot{q})\dot{q}_r + G(q) - K_d s \tag{7}$$

where $e = q - q_d$; $s = \dot{e} + \Lambda e$; $\dot{q}_r = \dot{q}_d - \Lambda e$. $K_d = K_d' > 0$, $\Lambda = \Lambda' > 0$ and are constant and diagonal matrices. q_d is the vector of the reference trajectory.

Theorem 2. *For the system* (3) *we consider the feedback in armature voltages* (6) *and the passivity-based controller* (7). *Under the conditions* $K_{d,m} > \dfrac{1}{2}$ *and* $\Lambda_m > \dfrac{1}{2}$, *the tracking error* e *tends exponentially to zero.* $K_{d,m}$ *(resp.* Λ_m*) is the smallest eigenvalue of* K_d *(resp.* Λ*).*

Proof. Substituting (6) into (3), we get the following closed-loop global system

$$\begin{cases} A(q)\ddot{q} + C(q,\dot{q})\dot{q} + G(q) = \tau_d - \tilde{\tau} \\ \varepsilon\dot{\tau} = \tilde{\tau} \end{cases} \tag{8}$$

Into the coordinate $x = (e', s')'$, the system (8) is transformed to (' denotes the transposition):

$$\begin{cases} \dot{x} = f(x,\tau,t) \\ \varepsilon\dot{\tau} = g(x,\tau,t) \end{cases} \tag{9}$$

with $f(x,\tau,t) = (s - \Lambda e, A^{-1}(e + q_d)(\tau - C(e + q_d, \dot{e} + \dot{q}_d)(\dot{e} + \dot{q}_d) - G(e + q_d)) + \Lambda\dot{e} - \ddot{q}_d)'$ and $g(x,\tau,t) = \tilde{\tau}$.

Further, we assume that $(x = 0, \tau = 0)$ is an equilibrium point of system (9); $f(0,0,t) = 0$ and $g(0,0,t) = 0$. Our aim is to prove that this equilibrium position is exponentially stable. Let us now verify the conditions of Theorem 1. The roots of the algebraic equation $g(\bar{x},\bar{\tau},t) = 0$ are given by

$$g(\bar{x},\bar{\tau},t) = 0 \Leftrightarrow \tau_d - \bar{\tau} = 0 \Leftrightarrow \bar{\tau} = \tau_d = h(\bar{x},t) \tag{10}$$

As $\bar{\tau}$ is an isolated root for (10), therefore model (9) is in *standard form*[10]. Further the function f, g are considered smooth enough, thus, assumptions A1-4 are satisfied. The slow subsystem is given by

$$\dot{x} = f(\bar{x},h(\bar{x},t),t) \tag{11}$$

which gives the dynamic of the mechanical part controlled by τ_d (for simplicity's sake we note the slow variables as x in (9)). Under (7) the dynamic behaviour of the closed-loop slow subsystem is given by

$$\begin{cases} \dot{e} = -\Lambda e + s \\ \dot{s} = A^{-1}(e+q_d)(-C(e+q_d,\dot{e}+\dot{q}_d)s - K_d s) \end{cases} \tag{12}$$

where we consider the following Lyapunov function

$$V(x) = \frac{1}{2}(s^t A(q)s + e^t e) \tag{13}$$

It is an easy computation to see using the passivity property[8]: $\xi^t(\dot{A}(q) - 2C(q,\dot{q}))\xi = 0$ ($\forall \xi$) and the conditions given by Theorem 2 that:

$$\dot{V}(x) = -s^t(K_d - \frac{1}{2})s - e^t(\Lambda - \frac{1}{2})e - \frac{1}{2}(e-s)^t(e-s) \le 0.$$

Then A5 is satisfied. Now, the boundary layer system is given by

$$d\eta / d\varsigma = g(x,\eta + h(x,t),t) = \tau_d - (\eta + \tau_d) = -\eta \tag{14}$$

Then A6 is trivially satisfied. We consider for (14) the Lyapunov function $W(\eta) = \frac{1}{2}\eta^t\eta$.

Remark 4 *By Theorem 1 we can assert that the tracking errors remain in a small neighbourhood of zero which leads to a practical stability result. To achieve a complete stability result, it remains to prove that the real system is exponentially stable.*

We consider the composite Lyapunov function candidate: $L(x,\eta) = V(x) + W(\eta)$. The time-derivative of L along the trajectories of the standard model (9) is given by

$$\dot{L}(x,\eta) = \frac{\partial V}{\partial x} f(x,\eta + h(x,t),t) + \frac{1}{\varepsilon}\frac{\partial W}{\partial \eta} g(x,\eta + h(x,t),t)$$

$$- \frac{\partial W}{\partial \eta}\frac{\partial h}{\partial x} f(x,\eta + h(x,t),t) - \frac{\partial W}{\partial \eta}\frac{\partial h}{\partial t}(x,t) \tag{15}$$

Remark 5 *An upper bound for all quantities in (15) can be easily achieved using assumption A1-6, except the term* $\|(\partial W / \partial \eta)(\partial h / \partial t)\| \le k\|\eta\|\|\partial h / \partial t\|$ *(k is a constant). The biais is due to the presence of the unbounded term* $\partial h(x,t) / \partial t$.

Lemma 1 *The term $h(x,t)$ satisfies the following inequality*

$$\|\partial h(x,t) / \partial t\| \le k_1 \|x\| \tag{16}$$

Proof. We recall the expression of $h(x,t)$: $h(x,t) = A(q)\ddot{q}_r + C(q,\dot{q})\dot{q}_r + G(q) - K_d s$.

Since we have $\dot{q}_r = \dot{q}_d - \Lambda e$, $q = e + q_d$, the partial derivative of $h(x,t)$ can be written as:

$$\frac{\partial h(x,t)}{\partial t} = h_1(x,t) + h_2(x,t) \tag{17}$$

where

$$h_2(x,t) = -\frac{\partial A}{\partial q}\dot{q}_d\Lambda\dot{e} - \left[\frac{\partial C}{\partial q}\dot{q}_d + \frac{\partial C}{\partial \dot{q}}\ddot{q}_d\right]\Lambda e \tag{18}$$

$$h_1(x,t) = \frac{\partial A}{\partial q}\dot{q}_d\ddot{q}_d + A(q)\dddot{q}_d + \frac{\partial C}{\partial q}\dot{q}_d^2 + \frac{\partial C}{\partial \dot{q}}\ddot{q}_d\dot{q}_d + C(q,\dot{q})\ddot{q}_d + \frac{\partial G}{\partial q}\dot{q}_d \tag{19}$$

We assume that the dynamic coefficients A, C and G are C^∞. Or $\|e\| \le k_2\|x\|$ and $\|\dot{e}\| \le k_3\|x\|$.

By the fact the reference trajectories are considered bounded, the term $h_2(x,t)$ satisfies the following inequality:

$$\|\partial h_2(x,t) / \partial t\| \le k_4\|x\| \tag{20}$$

Let us now examine the term $h_1(x,t)$. By assumption A1 we have $A(q_d)\ddot{q}_d + C(q_d,\dot{q}_d)\dot{q}_d + G(q_d) = 0$. In fact this means that the trajectory tracking objectives are achieved (i.e. $x = 0, \tau = 0$). Using this result and (19), after some arrangements $h_1(x,t)$ can be written as:

$$h_1(x,t) = \left[\frac{\partial A(e+q_d)}{\partial q} - \frac{\partial A(q_d)}{\partial q}\right]\dot{q}_d\ddot{q}_d + \left[A(e+q_d) - A(q_d)\right]\dddot{q}_d$$

$$+ \left[\frac{\partial C(e+q_d, \dot{e}+\dot{q}_d)}{\partial q} - \frac{\partial C(q_d,\dot{q}_d)}{\partial q}\right]\dot{q}_d^2 + \left[\frac{\partial C(e+q_d, \dot{e}+\dot{q}_d)}{\partial \dot{q}} - \frac{\partial C(q_d,\dot{q}_d)}{\partial \dot{q}}\right]\ddot{q}_d\dot{q}_d$$

$$+ \left[C(e+q_d, \dot{e}+\dot{q}_d) - C(q_d,\dot{q}_d)\right]\ddot{q}_d + \left[\frac{\partial G(e+q_d)}{\partial q} - \frac{\partial G(q_d)}{\partial q}\right]\dot{q}_d \tag{21}$$

Now, using the growth finite theorem an ultimate bound of (21) can be easily found in the sens of Lemma 1. As an example we take the first term of (21) which has the following bound

$$\left\| \frac{\partial A(e + q_d)}{\partial q} - \frac{\partial A(q_d)}{\partial q} \right\| \leq \left\| \frac{\partial^2 A(q_d)}{\partial q^2} \right\| \|e\| \leq k_s \|e\| \leq k_6 \|x\|, \text{ etc.} \qquad \square$$

Therefore by Remark 5 and Lemma 1 there is ε^{**} such that $\dot{L} \leq 0$. Then the tracking error e tends exponentially to zero. This complete the proof of the theorem. $\qquad \square$

4. Simulation results

Simulations tests are performed on a electrically-driven 6-DOF parallel robot. It is constituted by a top plate, a fixed plate and three limbs. A complete geometric description of this robot can be found in Beji *et al*[7]. The desired trajectory was specified in the joint space by: $q_d(t) = A_M \cos(t)$, where the amplitude $A_M = 0.1m$ for the prismatic joints and $A_M = -0.1rd$ for the revolute active joints. The simulations are given for initial conditions of tracking errors not equal to 0. The parameters of the DC-motors are as follow. For the RS110M motors (prismatic actuated joints); $K_t = 0.08368Nm/A$, $R = 4.8\Omega$, $K_b = 0.0242V/rd/s$, $L = 1.6mH$ and $N = 46.8$. For the RS240B motors (revolute actuated joints); $K_t = 0.068Nm/A$, $R = 0.64\Omega$, $K_b = 0.0449V/rd/s$, $L = 0.45mH$ and $N = 100$.

The controller gains $\Lambda = diag\{30,30,30,25,25,25\}$; $K_p = diag\{10,10,10,10,10,10\}$ and $K_d = diag\{15,15,15,10,10,10\}$. The simulations are performed with the numerical language ACSL. Figure 1 shows the trajectory tracking errors of the SM actuated joints. It is shows the performance of the proposed controller with a good trajectories tracking. Furthermore we have found that the disturbance errors $\tilde{\tau}$ converge rapidly to zero and this ensure the stability of the mechanical part.

4. Conclusion

In this paper, a control law at the voltage input, has been derived that incorporates the robot manipulator dynamic as well as the dynamic of the actuators. The present corrective controller requires the measurement of joint positions, velocities and the motor armature currents. Passivity concept and singular perturbation techniques are used to validate this controller. The performance of the proposed nonlinear control law is supported by a simulation on a 6-DOF RLED parallel SM. We point out that the problem encountered in Remark 5 can be surrounded by a static state feedback law (see d'Andréa-Novel[5] for a hydraulically robot case) which depends on ε (ε denotes the fluid compressibility). However, in a electrically or hydraulically robot case ε is not known with sufficient accuracy. In this paper we overcome this problem. Without ε-dependency of the controller a complete stability result is obtained.

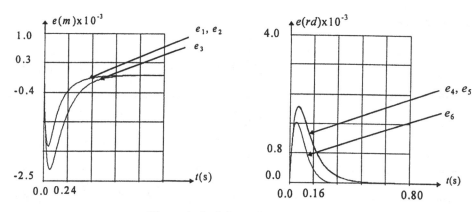

Figure 1a, b. Joint trajectory errors.

References

[1] Y. Stepanenco, C.-Y. Su, "Adaptative Motion Control of Rigid-Link Electrically-Driven Robot Manipulators", *IEEE Conference on Decision and Control*, 1994, pp. 1050-4729.

[2] T.J Tarn, A.K. Bejczy, X. Yun, Z. Li, "Effect of motor dynamics on nonlinear feedback robot arm control", *IEEE Trans. on Robotics and Automation*, Vol.7, 1991, pp. 114-122.

[3] D.M. Dawson, Z. Qu., J.J. Carroll, "Tracking control of rigid-link electrically-driven robot manipulators", *International Journal of Control*, Vol.56, N°5, 1992, pp. 991-1006.

[4] A. Abichou, "Stabilisation de systèmes mécaniques avec bifurcation fourchette. Commande non linéaire d'un robot hydraulique", *Thesis in Applied Mathematics and Automatic*, Ecole des Mines de Paris, Dec. 1993.

[5] B. d'Andréa-Novel, M.A. Garnero; A. Abichou, "Nonlinear control of a hydraulic robot using singular perturbations", *IEEE-SMC, San Diego*, 1994, pp. 1932-1937.

[6] E. Dombre, W. Khalil, "Modélisation et commande des robots", *Ed. Hermès*, Paris, 1989.

[7] L. Beji, P. Joli, M. Pascal, "Dynamic Study of a 6 DOF and Three limbs parallel Manipulator", *Proc. of the IEEE-SMC Conf.*, July, 1996, Lille, France.

[8] H. Berguis, H. Nijmeijer, "A Passivity Approach to Controller-observer Design for robot", *Internal report* No. 92R109, 1992, University of Twente.

[9] J.J.-E. Slotine, W. Li, "On the Adaptive Control of Robot Manipulators", *International Journal of Robotics Research*, Vol. 6, 1987, pp. 49-59.

[10] H.K Khalil, "Nonlinear Systems", *Macmillan Publishing Company*, NewYork, 1992.

[11] M.S Mahmoud, "Robust control of robot arms including motor dynamics", *International Journal of Control*, Vol.58, N°4, 1993, pp.853-873.

AN INVERSE FORCE ANALYSIS OF THE SPHERICAL 3-DOF PARALLEL MANIPULATOR WITH THREE LINEAR ACTUATORS CONSIDERED AS SPRING SYSTEM

J. Knapczyk and G. Tora
Cracow University of Technology Cracow, Poland

Introduction

The subject of our investigation is the spherical 3-DOF parallel manipulator with three linear actuators, any of them is attached to the base with a universal joint and to the platform with a spherical joint (see Fig.1). The fourth spherical joint between the platform and the base permits the spherical motion of the platform. Similar mechanical structures are widely used as mounting or supporting structures for solar panels, telescopes, radar and satellite antennas, mirrors for laser beams etc.

Proper force and position control of the compliant parallel manipulator requires the knowledge of all possible equilibrium configurations for a given external load acting on the platform. This problem, related to the inverse force analysis, has not been tackled yet for the general case of the considered spherical manipulator. Recently, P.Dietmaier [1] has studied the inverse force analysis problem for a tetrahedral three-spring system, which can be considered as a particular case of the spherical parallel manipulator where three spherical joints are linked by a common pivot in the platform. The actuator drive systems contain compliant elements, which can be considered as springs with characteristics assumed to be linearly elastic. The task was to compute the stable equilibrium orientation of the platform under a given external load.

In general, spherical displacement of the platform driven by three in-parallel linear actuators with known elasticities can be decomposed into two components: - the main displacement caused by actuator displacements as the input variables; - an additional displacement resulting from elastic deflections of actuators. The first component can be obtained as the solution of the direct kinematic problem. The second component is dependent on the external load acting on the platform and its actual orientation.

Direct kinematic analysis by using vector method

A mathematical description of the manipulator uses the following notation: A_i - the *i*th base point (the centre of *i*th joint connecting the actuator leg with the base); B_i-the platform point (i=1,2,3). The vectors listed below are described in the base coordinate system $Ox_by_bz_b$ (Fig.1):

$a_i = OA_i = \text{const.}, \quad b_i = OB_i = \text{var.}, \quad a_{ij} = a_j - a_i, \quad b_{ij} = b_j - b_i, \quad d_i = a_i - b_i, \quad a_i^{\circ} = a_i/a_i, \quad b_i^{\circ} = b_i/b_i.$

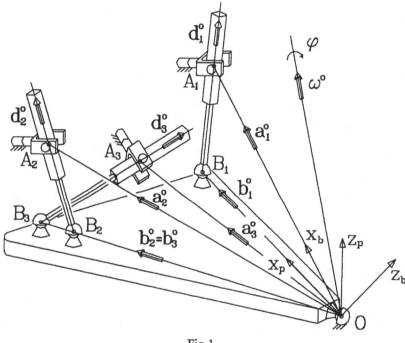

Fig.1.

The geometry of the considered manipulator is given. In particular, the positions of points A_i (i=1, 2, 3) are given in an arbitrary reference system fixed to the base; the positions of points B_i are given in an arbitrary reference system fixed to the platform; the leg lengths $A_iB_i = d_i$ are known. Without loss of generality, the reference systems are chosen with origins coincident with point O - the center of the spherical joint connecting the platform with the base, x_b- axis directed from point O to point A_1, x_p- axis directed to point B_1, y_b - axis in the plane OA_1A_2 and y_p - axis in the plane OB_1B_2.

The direct position analysis for the general fully-parallel 3-DOF spherical manipulator results in two-equation system in echelon form; the first equation is 8th order and the remaining one is linear. As a consequence, when a set of actuator displacements is given, eight configuration of the manipulator are possible [5].The forward position analysis of the general spherical manipulator can be solved in a compact, closed form when three platform points O, B_2 and B_3 are collinear so the unit orientation vectors of points B_2 and B_3 are equal ($b_2^o = b_3^o$).

Two unknown unit vectors b_1^o and b_2^o can be found directly, using twice the general formula for finding one of three unit vectors [6] in the following way:

$$b_2^o = \frac{[(a_2^o \cdot b_2^o) - (a_2^o \cdot a_3^o)(a_3^o \cdot b_3^o)]a_2^o + [(a_3^o \cdot b_3^o) - (a_2^o \cdot a_3^o)(a_2^o \cdot b_2^o)]a_3^o \pm \sqrt{D_2}(a_2^o \times a_3^o)}{1 - (a_2^o \cdot a_3^o)^2} \quad (1)$$

where: $D_2 = 1 - (a_2^o \cdot b_2^o)^2 - (a_2^o \cdot a_3^o)^2 - (a_3^o \cdot b_3^o)^2 + 2(a_2^o \cdot b_2^o)(a_2^o \cdot a_3^o)(a_3^o \cdot b_3^o)$

(1a)

$$b_1^o = \frac{[(a_1^o \cdot b_1^o) - (a_1^o \cdot b_2^o)(b_1^o \cdot b_2^o)]a_1^o + [(b_1^o \cdot b_2^o) - (a_1^o \cdot b_2^o)(a_1^o \cdot b_1^o)]b_2^o \pm \sqrt{D_1}(a_1^o \times b_2^o)}{1 - (a_1^o \cdot b_2^o)^2} \quad (2)$$

where: $D_1 = 1 - (a_1^o \cdot b_1^o)^2 - (a_1^o \cdot b_2^o)^2 - (b_1^o \cdot b_2^o)^2 + 2(a_1^o \cdot b_1^o)(a_1^o \cdot b_2^o)(b_1^o \cdot b_2^o)$ (2a)

The dot products of the corresponding unit vectors can be obtained as follows:

$$a_i^o \cdot a_j^o = \frac{a_i^2 + a_j^2 - a_{ij}^2}{2a_i a_j}; \qquad b_i^o \cdot b_j^o = \frac{b_i^2 + b_j^2 - b_{ij}^2}{2b_i b_j}; \qquad a_i^o \cdot b_i^o = \frac{a_i^2 + b_i^2 - d_i^2}{2a_i b_i}; \quad (3)$$

In order to solve the direct velocity problem, the vector closure equation for ith leg $b_i + d_i = a_i$ was differentiated with respect to time. Taking dot products of both sides of the resulting equation by d_i one can obtain:

$$\dot{b}_i \cdot a_i = -d_i \dot{d}_i \quad (4)$$

On the other hand:

$$\dot{b}_i = \omega \times b_i \quad (5)$$

where: ω – the platform angular velocity.

Substituting (5) into (4):

$$a_i \times b_i \cdot \omega = d_i \dot{d}_i \quad (6)$$

Vector equation (6) can be expressed in matrix form:

$$G\omega = D\dot{d} \tag{7}$$

where:

$$G = [a_1 \times b_1 \quad a_2 \times b_2 \quad a_3 \times b_3]^T, \qquad D = Id, \qquad I\text{=3x3 unity matrix,}$$

$$d = [d_1 \quad d_2 \quad d_3]^T, \qquad \dot{d} = [\dot{d}_1 \quad \dot{d}_2 \quad \dot{d}_3]^T.$$

Inverse force analysis by using compliance matrix and iterative procedure

The manipulator, shown in Fig.1. is in static equilibrium position. Moment equilibrium equation of the isolated platform can be written with respect to point O in the form:

$$M_O + \sum_{i=1}^{3} b_i \times F_i = 0 \tag{8}$$

where: M_O - is the vector of external torque acting on the platform,

F_i - the vector of reaction force of ith actuator on the platform.

$$F_i \approx F_i \frac{b_i - a_i}{d_i} \tag{9}$$

Substituting (9) into (8):

$$\sum_{i=1}^{3} (a_i \times b_i) \frac{F_i}{d_i} = -M_O \tag{10}$$

Vector equation (10) can be described in matrix form:

$$G^T D^{-1} F = -M_O \tag{11}$$

where:

$F = [F_1 \quad F_2 \quad F_3]^T$, the matrices G and D are defined above (see (7)).

Assuming linear elasticity of actuators, the compliance matrix is used to relate the small changes in the actuator reaction forces ΔF to the corresponding deflections Δd by the formula:

$$\Delta d = C(-\Delta F) \tag{12}$$

where:

$$C = \begin{bmatrix} c_1 & 0 & 0 \\ 0 & c_2 & 0 \\ 0 & 0 & c_3 \end{bmatrix}, \qquad \Delta d = \begin{bmatrix} \Delta d_1 \\ \Delta d_2 \\ \Delta d_3 \end{bmatrix}, \qquad \Delta d_i = d_{i0} - d_i$$

d_{i0} - the actuator (spring) free length, d_i - the actuator actual length, c_i - the compliance constant of the ith actuator.

Assuming small angular displacement of the platform according to equation (7) we can write:

$$G\Delta\varphi = D\Delta d \tag{13}$$

where: $\Delta\varphi$ - the platform angular displacement corresponding to Δd - the vector of actuator deflections

From equation (13) using (12) and (11) one can obtain:

$$\Delta\varphi = K_\varphi \Delta M \tag{14}$$

where:

$$K_\varphi = G^{-1}DC(G^{-1}D)^T \tag{15}$$

K_φ - the angular compliance matrix, ΔM - an increment of the vector of the external moment acting on the platform

The equilibrium position of the manipulator can be obtained by using an iterative procedure. After computing the compliance matrix (15) in the first step of iteration, the computations are repeated for the next positions. The convergence of the iterative procedure causes the difference between the actuator lengths computed in subsequent iteration steps to decrease. Simultaneously, the potential energy is evaluated in each iteration step and its value is monitored during the iteration process, in order to confirm the suitability of the procedure. In the equilibrium position this function should reach its global minimum. This algorithm requires a compliant manipulator to remain near the unloaded configuration.

Inverse force analysis by solving the system of equilibrium equations

The force equilibrium equation for the ith actuator can be written in the form:

$$\frac{1}{c_i}(d_{i0} - d_i)\frac{a_i - b_i}{d_i} + \Delta F_i = 0 \tag{16}$$

Substituting (16) into (8)

$$\Delta M - \sum_{i=1}^{3}(a_i \times b_i)\frac{1}{c_i}\left(1 - \frac{d_{i0}}{d_i}\right) = 0 \tag{17}$$

where: $d_i = \sqrt{a_i^2 + b_i^2 - 2a_i \cdot b_i}$, $b_i = R(\alpha,\beta,\gamma)b_{i,p}$, R - the rotation matrix for the transformation from the platform system to the base system [5,7], α, β, γ - the angular coordinates of the platform in relation to the base system.

The finite angular displacement between two platform orientations: - obtained for the unloaded manipulator or with rigid actuators (denoted by upper script ^{r}) and - obtained for the loaded manipulator with flexible actuators (denoted by upper script ^{f}) can be determined by using following formula:

$$\omega^{\circ} tg(\Delta\varphi) = [(^{f}b_{jk} - ^{r}b_{jk}) \times (^{f}b_{ji} - ^{r}b_{ji})] / [(^{f}b_{jk} - ^{r}b_{jk}) \cdot (^{f}b_{ji} - ^{r}b_{ji})] \qquad (18)$$

where: $b_{jk} = b_k - b_j$, $b_{ji} = b_i - b_j$, $i \neq j$, $i \neq k$, $i, j, k \in \{1, 2, 3\}$;

ω° – the unit vector of the finite rotation axis called the compliance axis.

The angular displacements of the platform caused by actuator compliances and obtained by using compliance matrix and solving equilibrium equations are presented in Fig.2. as functions of the multiplication factor of compliance constants.

Sensitivity analysis

The algorithm described earlier permits a computer simulation of the influence of design (geometrical and stiffness) parameters on the elastokinematical characteristics. Such simulation provide information about effects caused by changes of particular parameters and permits the selection of the set of most influencing ones.

The angular displacement of the platform corresponding to different values of selected parameters can be used to evaluate the results of sensitivity analysis. The changes of the parameter values are denoted by their relative values (the actual values were divided by the corresponding initial values). the results of the sensitivity analysis are presented in Fig.3.

Numerical example

The input data for the considered manipulator were determined by measurements of a road-building machine (Baukema SHM4 - 120A). The coordinates of the joint centers are described in the base system (for the base joints) and in the platform systems (for the platform joints). Input data are given as follows:

$a_1 = [2.47 \quad 0 \quad 0]^{T}$ [m], $a_2 = [2.25 \quad 1.02 \quad 0]^{T}$ [m], $a_3 = [2.72 \quad 0.13 \quad -0.46]^{T}$ [m],

$b_1 = [2.40 \quad 0 \quad 0]^{T}$ [m], $b_2 = [2.18 \quad 1.01 \quad 0]^{T}$ [m], $b_i = [2.41 \quad 1.07 \quad -0.03]^{T}$ [m],

$d_1 = 0.60$ [m], $d_2 = 0.65$ [m], $d_3 = 1.33$ [m], $M_0 = [6.51 \quad -2.39 \quad -2.33]10^{3}$ [Nm]

$c_1 = 2.394 \ 10^{7}$ [N/m], $c_2 = 3.223 \ 10^{7}$ [N/m], $c_3 = 1.343 \ 10^{7}$ [N/m],

The calculation results are presented as the platform angular displacement vectors obtained:

- by using compliance matrix: $\Delta\varphi = [4.005 \quad 1.071 \quad -0.841] \ 10^{-4}$ [rd],

- by solving the equilibrium equations: $\Delta\varphi = [4.008 \quad 1.072 \quad -0.843] \ 10^{-4}$ [rd].

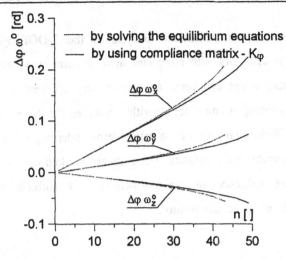

Fig.2. The relations between: $\Delta\varphi\,\omega°$ – the platform angular displacement and n - multiplication factor of actuators compliances.

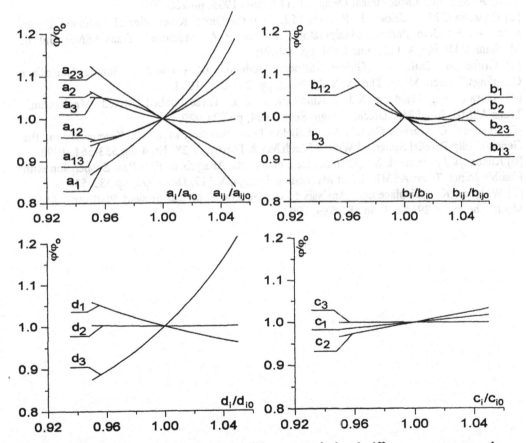

Fig.3. The results of sensitivity analysis of the geometrical and stiffness parameters on the compliant angular displacement of the platform loaded by the same external moment.

Conclusion

A method for the elastokinematic analysis of the 3-DOF spherical manipulator containing three linear actuators with compliant drive systems was developed. The vector method used for displacement and force analysis is very efficient, relatively concise and convenient for programming. A computer algorithm prepared for this analysis allows the user to determine the equilibrium position of the manipulator under a given external load and to calculate its elastokinematic characteristics. The software worked out for PC computers was used to evaluate the influence of the geometrical and stiffness parameters on the elastokinematic properties of the manipulator.

References

[1] Dietmaier P.: „An Inverse Force Analysis of a Tetrahedral Three-Spring System." Trans.ASME, Jnl of Mechanical Design, V.117, June 1995, pp.286-291.
[2] Gosselin C.M., Sefriouri J., Richard M.J.: „On the Direct Kinematics of Spherical Three-Degree-of-Freedom Parallel Manipulators of General Architecture." Trans.ASME, Jnl of Mechanical Design, V.116, June 1994, pp.594-598.
[3] Griffis M., Duffy J.: „Global Stiffness Modelling of a Class of Simple Compliant Couplings." Mech. Mach. Theory. V.28, No.2, pp. 207-224, 1993.
[4] Husain M., Waldron K.J.: „Kinematics of a Three-Limbed Mixed Mechanism." Trans.ASME, Jnl of Mechanical Design, Sept. 1994, pp.924-929.
[5] Innocenti C., Parenti-Castelli V.: „Echelon Form Solution of Direct Kinematics for the General Fully-Parellel Spherical Wrist." Mech.Mach.Theory, V.28, No.4, pp.553-561, 1993.
[6] Knapczyk J., Dzierżek S.: „Displacement and Force Analysis of Five-Rod Suspension with Flexible Joints. Trans.ASME, Jnl of Mechanical Design, V.117, Dec.1995, pp.532-538.
[7] Wohlhart K.: „Displacement Analysis of the General Spherical Stewart Platform." Mech. Mach. Theory, V.29, No.4, pp.581-589, 1994.

Chapter II
Mechanics II

EXPERIMENTAL RESEARCH OF DYNAMICS OF AN ELEPHANT TRUNK TYPE ELASTIC MANIPULATOR

R. Cieslak

Technical University of Wroclaw, Wroclaw, Poland

A. Morecki

Warsaw University of Technology, Warsaw, Poland

1. Introduction

Elastic manipulator is an arm, which can operate in a work-space with obstacles. Such properties are characteristic of the following manipulators:

- built of a number of rigid links >3 connected with each other by kinematic joints of third, fourth and five class and controlled by a cable drive or a direct drive.
- built as a continuos joint structure,
- built of elastic links connected to each other by the joints or rigidly.

The elephant trunk type elastic manipulator is built of elastic links (coil springs) connected rigidly and operated by cable manipulation in work spaces with obstacles.

The prototype was designed in 1990 in Warsaw University of Technology (team headed by A. Morecki; Fig. 1a). In 1991 a new project started at the Chair of Design and Automation of Off-Road and Heavy Machines of Technical University of Wrocław (R. Cieślak and A. Morecki). The new model of an experimental manipulator is based on the results obtained by the theoretical and experimental investigations. The manipulator is driven by hydraulic actuators (Fig. 1b). Some new results concerning the static and dynamic properties of this manipulator have been obtained. The main goal of these experiments was the experimental verification of the proposed theoretical model.

a)   b)

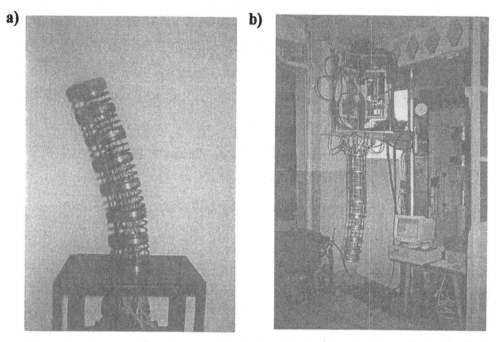

Fig. 1. Elastic manipulator of an elephant trunk type: a) the prototype, b) experimental manipulator with hydraulic drive.

2. Construction of an experimental manipulator of an elephant trunk type.

The arm of the elastic manipulator is built of eight coil springs connected with each other by rigid separators. In the arm we can distinguish three segments: the base consist of four springs, the middle one consists of three springs and the terminal one consists of one spring. Each of the segments is driven by two pairs of cables which act independently. The action of each pair of cables is coupled i.e. the shortening of one cable causes the lengthening of the other one by the some length.

In each segment the cables are located at 90 degree. Each pair is controlled by one reciprocal actuator. The arm is controlled by a set of six hydraulic actuators and six proportional hydraulic distributors. The diagram of the arm and the hydraulic supply system is shown on Fig. 2.

a)

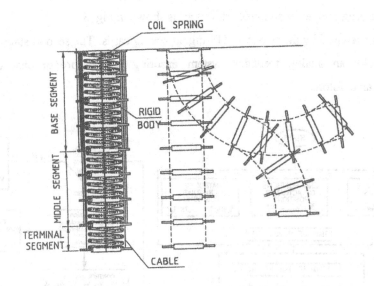

b)

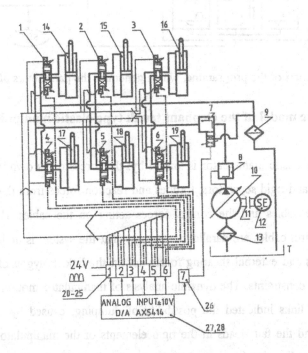

Fig. 2. The diagram of the arm of the elastic manipulator: a) the diagram, b) the control system: 1-6-proportional distributors, 7-proportional valve, 8-safety valve, 9,13- filters, 10-pump, 11-coupling, 12-electric engine, 14-19-hydraulic actuators, 20-26 - digital control cards, 27-28 A/D control cards

Two stage system with an open architecture is used to control the manipulator. The first stage of the control system is realised as an open digital - analog system. The motion parameters are generated by a programme. Its simplifies flowchart is shown in Fig. 3.

The control is exercised by the two AX5414 digital analog cards. The second stage of control is performed by an analog feedback system ensuring precise proportional control of proportional valve pistons.

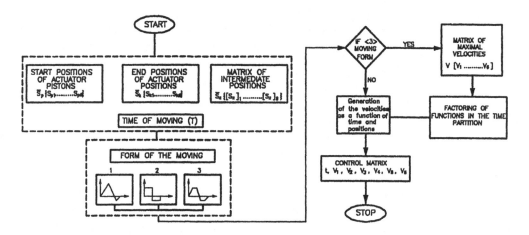

Fig 3. Flow chart of the programme, which generates the parameters of actuator motion.

3. The dynamic model of the elephant trunk type elastic manipulator.

Each link of the elastic manipulator is loaded by: the forces of its own weight i.e. weight of elastic elements and rigid separators; torques and reaction forces from the neighbouring links, pre-tensin in the cables and the control forces acting on the cables. Because in the initial position the control cables are parallel we can say that the system is under constrained when torsional torques and external shearing forces act on the link. Polygon of forces is closed by elastic nonlinear constraints. The dynamic analysis of manipulator motion taking into account vibration of the links indicated the possibility of dumping, caused by the friction between control cables and the fair -leads in the rigid elements of the manipulator. For mathematical description of the direct and inverse dynamics problem, physical model shown in Fig. 4 was assumed. Figure 4 presents the graphical interpretation of the model. Coefficients of the rigidity in this model represents the equivalent values of real rigidity for real links.

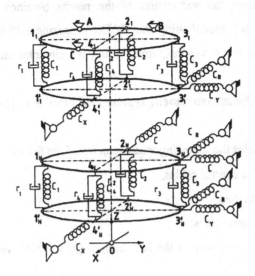

Fig. 4. The model of the elastic manipulator

4. Problem description

The elephant trunk type elastic manipulator is built of elastic elements characterised by high geometrical nonlinearities. Its structure is non homogeneous i.e. the segments are not ideally similar to each other. Additionally, strong influence of the reaction forces caused by the neighbouring links was observed. Besides that, as it was mentioned above, we cannot neglect the influence of dumping forces caused by dry friction and hydraulic actuators.

In particular the elastic phenomena and damping of the hydraulic system (the frequency of free vibrations of the proportional distributors and elasticity of oil and elastic pipe) should be taken into account.

The analysis of technical aspects of the construction of the elephant trunk type elastic manipulator and the theoretical discussion presented by the authors in [2,3,4,5] lead to conclusion that full description requires static and dynamic experiments with the arm which would enable quantitative description of the phenomena taking place in the manipulator arm.

This description will allow the verification of the results obtained during the computer simulation of the multibody model using the DADS programme (from CADSi) and if the results are comparable the model will be „turned" using experimental data.

5. The description of the measurement system and results of the experiments

The experimental rig enables the following measurements of the forces:

- the pre-tension in the control cables,
- the forces in the actuators,
- the forces in the control cables.

The acquisition of the data realised by the A/D system (PCL818H) with sampling frequency 1000Hz.

The dynamical investigations had been proceed by the static of the dry friction of the control cables in the fair leads as a function of bending of a single element. The experimental rig for these measurements is shown in Fig. 5.

a) b)

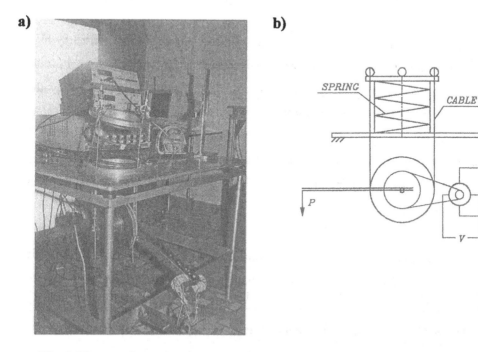

Fig. 5. The experimental rig for the statical measurements of an element of the elastic manipulator: a) photography, b) diagram

6. The results of dynamical investigations of manipulator arm

As a result of statical investigations of a single element the relationship between changes of the forces in the control cables as a function of bending angle was determined. The results are presented in Fig. 6.

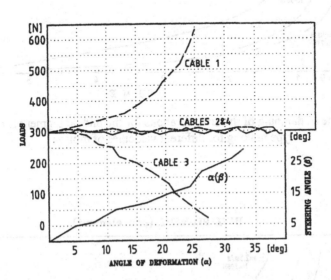

Fig. 6. The changes in the control cables as a function of the bending angle

Figure 7 presents the time of damping of manipulators vibrations caused by step function type reaction as a function of different bending angles of the links in a segment for all three segments.

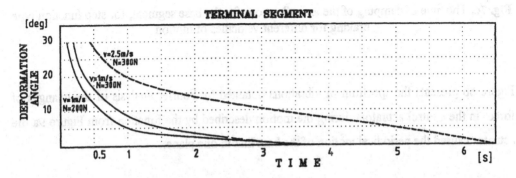

Fig. 7a. The time of damping of the arm vibrations for the terminal segment, for step function type braking for different velocities of motion

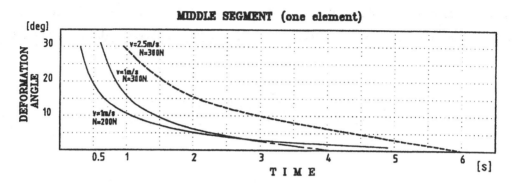

Fig. 7b. The time of damping of the arm vibrations for the middle segment for step function type braking for different velocities of motion

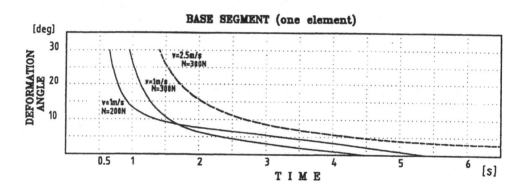

Fig. 7c. The time of damping of the arm vibrations for the base segment, for step function type braking for different velocities of motion

Figure 8b presents the spectrum of vibrations obtained by Fourier analysis of the changes of forces in the control actuators during the motion described by the function from Figure 9a the actuator caused the periodical motion of the base link in one plane.

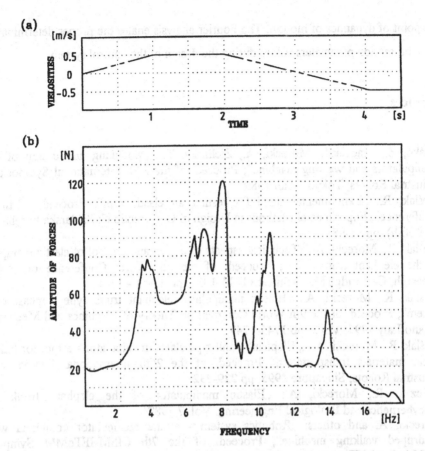

Fig. 8. Fourier analysis of the arm vibrations: (a) the excitation function, (b) the analysis of the response

7. The analysis of the experimental data and comparison with the analytical results

The initial investigations prove the assumption concerning the nonlinear function of elasticity during the bending of static elements.

The obtained results enable the determination of the coefficients of elasticity and damping using an statical methods.

The friction has strong influence on a damping of the free vibrations of the arm. These results enable us to conclude that although the load - carrying ability of the arm decreases with the deflection from the initial position, but the accuracy of positioning and orientation increase from the point of dynamics of motion. The Fourier analysis enales the proper determination of the maximum motion parameters values during the design of the control system.

8. References

[1]   Buśko, Z., Frączek,J., Morecki, A., Zielińska, T.: „Modelling and design of elastic manipulators and walking machine"; Proceed. of the 20th International Symposium on Industrial Robots; Tokyo Japan 1989

[2]   Cieślak, R.: „Experimental stand for testing an elastic arm"; Proceed. of the 2nd Conference: „Experimental Methods in Design of Construction"; Szklarska Poręba 1995; (in Polish), pp 91-98.

[3]   Cieślak, R., Morecki, A.: „Technical aspects of design and control of elastic manipulator of the elephant trunk type"; Proceed. of the 5th World Conference on Robotics Research, Cambridge Massachusetts USA 1994; pp 13-51 - 13-64

[4]   Cieślak, R., Morecki, A.: „Elastic manipulator elephant trunk type - measurements system"; Proceed. of the 9th World Congress on Theory of Machines and Mechanisms; Milano Italy 1995; Vol 3, pp 1745-1747

[5]   Cieślak, R., Morecki, A.: „ Elephant trunk type elastic manipulator - a tool for bulk and liquid materials transportation" Proceed. of the 26th International Symposium on Industrial Robots; Singapore 1995, pp 229-232

[6]   Malczyk, R., Morecki, A.: „Elastic manipulator of the elephant trunk type" Biocybernetics and Biological Engineering. Vol. 7 1987

[7]   Morecki, A. and others: „Robotics system - elastic manipulator combined with a quadruped walking machine"; Proceed. of the 7th CISM-IFToMM Symposium Ro.Man.Sy '86 Hermes 1988.

VIBRATION SUPPRESSING
FOR HYDRAULICALLY DRIVEN LARGE
REDUNDANT MANIPULATORS

M. Schneider and M. Hiller
Gerhard-Mercator University, Duisburg, Germany

Abstract

The application of large manipulators in servicing and building environments is a field of growing research interest. Typical tasks performed by the manipulators are pumping of wet concrete, handling of loads under difficult conditions and cleaning of aircraft. One of the main problems is the occurence of vibrations induced by hydraulic pressure oscillation and elasticity effects. In order to suppress these vibrations and to coordinate the motion for tracking given trajectories, a nonlinear position control concept is required. In this paper, a particular method is developed for this purpose, taking into account effects of mechanical transmission elements, hydraulic actuators and boom flexibility. As an application, the control of a large-scale manipulator carrying out wet concrete pumping tasks is presented.

1 Introduction

In contrast to conventional industrial robots, which are driven by electric motors with a large reducing gear ratio, large-scale manipulators are driven by hydraulic actuators by means of transmission mechanisms leading to closed kinematic loops [1]. Consequently, nonlinear dynamic coupling effects between the arms of the manipulator, which are usually neglected in elastic robotics, play an important role here, and must be modelled correspondingly. The most important problems one has to deal with in this respect are the complexity of the system with a highly nonlinear kinematic and dynamic behaviour, the nonlinear friction and elasticity of the hydraulic drives and the redundant structure of the arm package.

Fig. 1 shows a manipulator with four arm elements. Although this manipulator is used as an example in this paper, all modeling techniques and control concepts are suitable for a whole class of large redundant manipulators, e. g. manipulators with three, five or six arm elements and a total reach between 20 and 100 meters.

The vibrations which occur at the free end of the pumping device in the case of wet concrete pumping have different reasons. On the one hand, the elasticity of the arm elements leads to small deflections, and on the other hand, the compressibility of the oil

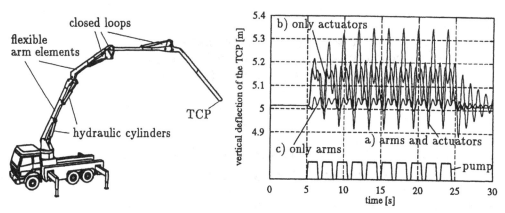

Figure 1: Large-scale manipulator   Figure 2: Vertical deflection during pumping

in the hydraulic cylinders and pipes makes possible small movements in the cylinders and therefore in the joints connecting the arm elements. Fig. 2 shows the vertical deflection of the tool center point (TCP) during a pumping process of the unmoved arm package. The following three simulations are considered: a) elasticity in the arm elements and in the hydraulic actuators, b) elasticity only in the hydraulic actuators, c) elasticity only in the arm elements. Fig. 2 shows clearly that the influence of the oil compressibility is much higher than the influence of the flexibility of the arm elements, although the manipulator for pumping wet concrete which is regarded here has a considerable higher elasticity in the arm elements than manipulators designed for handling loads.

This paper considers only the dominating deflections arising from the oil compressibility. In order to suppress the vibrations and to coordinate the motion for tracking given trajectories, a nonlinear position control concept is adapted.

2 Model of the Manipulator

2.1 Equations of Motion

The constraint equations arising from the seven closed kinematic loops are solved explicitly at the kinematic level. This allows one to explicitly formulate the mathematical model using a minimal set of generalized coordinates. According to the number of degrees of freedom $f = 5$, the following set of minimal coordinates q is introduced:

$$q = [\beta_0, s_1, s_2, s_3, s_4]^T, \tag{1}$$

where β_0 describes the rotation about the vertical axis and s_1 to s_4 are the displacements of the hydraulic actuators. By this, the equations of motion are represented by a coupled system of nonlinear differential equations:

$$M(q)\ddot{q} + g^c(q, \dot{q}) = g^a(q, \dot{q}) + u \quad \text{with} \quad u = F_R = [\tau_0, F_{R1}, F_{R2}, F_{R3}, F_{R4}]^T. \tag{2}$$

Here, $M \in \mathbb{R}^{f \times f}$ is the symmetric, positive-definite inertia matrix, $g^c \in \mathbb{R}^f$ is the vector of the generalized centrifugal forces and $g^a \in \mathbb{R}^f$ denotes the vector of the generalized applied forces *not* including the $p = 5$ generalized driving forces $u = F_R$. τ_0 is the driving torque at the basis joint and F_{Ri} is the resulting driving force of the ith hydraulic drive $(i = 1, \ldots, 4)$. With given value for τ_0 and known rolling circle radius r_{rc} of the slewing track ring, the force F_{R0} in the linear cylinder of the slewing unit can be calculated.

Even though the system is described in minimal coordinates, all constraint forces can be calculated directly using the *kinetostatic method* described in [2]. In particular, the inverse dynamics can be formulated very efficiently by the help of the *force transmission* without formulating the inertia matrix explicitly, providing *residual forces* $\bar{Q}$ at the joints of the generalized coordinates (input joints). These residual forces represent generalized forces and can be interpreted as driving forces at the input joints:

$$- \bar{Q} = u \quad \text{with} \quad u = M(q)\,\ddot{q} + g^c(q, \dot{q}) - g^a(q, \dot{q}). \tag{3}$$

A more detailed description of a simulation model with all relevant coupling effects is given in [3], taking into account also small deformations of the flexible arm elements.

2.2 Hydraulic Subsystems

In order to be able to model the mechanical-hydraulic multibody system, one has to slightly modify the models of hydraulic components found in the pertaining literature. This modification includes the separation of the involved state variables into a hydraulic part (pressure p_A in the cylinder chamber A and pressure p_B in the cylinder chamber B) modelled *within* the hydraulic component, and a mechanical part (mechanical displacement x, $\dot{x}$ and mass properties of the cylinder parts) which are considered within the multibody system.

The resulting force F_R of the hydraulic cylinder acting on the mechanical system as shown in Fig. 3 is

$$F_R = (p_A - \alpha\, p_B)\,A - F_F - F_E. \tag{4}$$

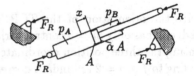

In Eq. (4), F_E denotes the limit stop forces and F_F the frictional forces, composed of the viscous friction, the Coulomb friction and the static friction.

Figure 3: Hydraulic cylinder

The differential equations describing the dynamics of the pressure variables p_A and p_B are given in detail in [4]. The dynamics of the servo valve can be neglected as its eigenfrequency is considerably higher than the eigenfrequency of the controlled system.

2.3 Coupled Mechanical-Hydraulic System

Eq. (6) shows the overall system of 20 strongly coupled (nonlinear) stiff differential equations:

$$z_{mech} = \left[q^T,\, \dot{q}^T\right]^T, \quad z_{hydr} = \left[p_A^T,\, p_B^T\right]^T = [p_{A0}, \ldots, p_{A4}, p_{B0}, \ldots, p_{B4}]^T, \tag{5}$$

$$\dot{z} = \left[\begin{array}{c} \dot{z}_{mech} \\ \dot{z}_{hydr} \end{array}\right] = \left[\begin{array}{c} \dot{q} \\ M^{-1}(q)\,[g^a(q, \dot{q}) - g^c(q, \dot{q}) + F_R(q, \dot{q}, p_A, p_B)] \\ \dot{p}_A(q, \dot{q}, p_A, p_B, y) \\ \dot{p}_B(q, \dot{q}, p_A, p_B, y) \end{array}\right] \tag{6}$$

3   Hierarchical Control Concept

Due to the strong kinematic and dynamic nonlinearities of the investigated manipulators, pure linear controllers do not provide adequate results. In the last decade, however, significant advances have been achieved for new nonlinear control concepts based on differential-geometric methods (e. g. [5]). In [4], two control concepts have been tested which make use of one of these methods – the *exact input-output-linearization*. The first concept uses an exact linearization for calculating the *forces* that are required to obtain a desired motion. Decentralized *force controllers* are responsible to provide these forces via hydraulic actuators. The second concept uses *motion rate control* for calculating the desired values of the joint *positions* and employs decentralized *position controllers* for the hydraulic actuators.

The simulations in [4] for tracking prescribed trajectories with simultaneous collision avoidance have shown that the first concept with force controlled actuators provides much better results due to the fact that this concept linearizes not only the kinematic nonlinearities of the mechanical system but also the dynamic nonlinearities. Therefore this concept will be used in the following.

From the literature it is known that linearization techniques are not applicable if the arm elasticities are significant, as they lead to unstable behaviour. In this case, the method of exact linearization has to be combined with other control concepts. This is currently under investigation.

3.1   Control Task and Control Variables

Since the orientation of the TCP is not considered, one has only $m = 3$ control variables y representing the three position coordinates of the TCP, leading to $f - m = 2$ redundant degrees of freedom:

$$y = r_{tcp}(q) = [r_{xtcp}, r_{ytcp}, r_{ztcp}]^T. \quad (7)$$

The derivatives with respect to time – the velocity v_{tcp} and the acceleration a_{tcp} – can be expressed as

$$\dot{y} = \dot{r}_{tcp} = v_{tcp} = \frac{\partial r_{tcp}}{\partial q}\dot{q} = J_{tcp}\dot{q}, \quad (8)$$

$$\ddot{y} = \ddot{r}_{tcp} = a_{tcp} = J_{tcp}\ddot{q} + \underbrace{\dot{J}_{tcp}\dot{q}}_{\bar{a}_{tcp}}. \quad (9)$$

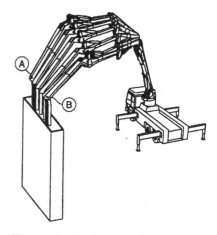

Figure 4: Manipulator filling a concrete form

The Jacobian J_{tcp} and the drift vector $\bar{a}_{tcp}$ can be obtained easily by using the method of *kinematic differentials* [2]. For the simulations in section 4, the manipulator has to track a straight line between point A and point B, while pumping wet concrete, e. g. to fill a concrete form (Fig. 4).

3.2   Decoupling Feedback

Solving the equations of motion (2) with respect to the accelerations $\ddot{q}$ and inserting $\ddot{q}$ in Eq. (9) yields the linear input-output equation for the manipulator:

$$\ddot{y} = \underbrace{-J_{tcp}M^{-1}g + \bar{a}_{tcp}}_{d(q,\dot{q})} + \underbrace{J_{tcp}M^{-1}}_{D(q)}u, \qquad g = g^c - g^a. \tag{10}$$

The solution of Eq. (10) is not unique, as for $m \neq f$ the *decoupling matrix* $D \in \mathbb{R}^{m \times f}$ is rectangular. Therefore, an arbitrary generalized right-inverse $D^R \in \mathbb{R}^{f \times m}$ with $DD^R = I$ is used to obtain the well-known linearizing state feedback (the control outputs $\ddot{y}$ are now replaced by new inputs $w \in \mathbb{R}^m$):

$$u = D^R w - D^R d + (I - D^R D)\,w^*, \tag{11}$$

If $f = p$, which is fulfilled here, the matrices D and D^R do not have to be computed explicitly. In this case the decoupling problem can be devided up into two parts: the inverse kinematics and the inverse dynamics.

The inverse kinematics can be described by a set of linear equations

$$\begin{bmatrix} P & J_{tcp}^T \\ J_{tcp} & 0 \end{bmatrix} \begin{bmatrix} \ddot{q}^* \\ \lambda \end{bmatrix} = \begin{bmatrix} p \\ -\bar{a}_{tcp} + w \end{bmatrix}, \tag{12}$$

arising from an optimization problem for the control of the $f - m = 2$ redundant degrees of freedom:

$$Z(\ddot{q}^*) = \frac{1}{2}\ddot{q}^{*T}P\ddot{q}^* - p^T\ddot{q}^* \overset{!}{=} \min, \qquad P \in \mathbb{R}^{f \times f}, \quad p \in \mathbb{R}^f, \tag{13}$$

which is constrained by Eq. (9). In the case $P = I$ and $p = 0$, the MOORE-PENROSE inverse is obtained. For controlling large-scale manipulators, p is set to artificial forces which are used to enable repeatable motions, to avoid collisions, to take care of joint limitations and to stabilize the zero dynamics. For more details refer to [6] and [4].

By solving Eq. (12) with respect to the desired accelerations $\ddot{q}^*$ and by again using the inverse dynamics (3), the desired driving forces $u^* = F_R^*$ can be obtained in one step without formulating the inertia matrix. Fig. 5 shows the overall control structure, where block I represents the controller model and block II the real system (which has been replaced by a simulation model that takes into account all coupling effects and nonlinearities).

3.3   Stabilizing PID-Controllers

If the parameter sets p_I of the controller model (block I) and p_{II} of the real system (block II), respectively, match exactly, the output variables y and the desired values $\hat{y}$ are identical (Fig. 5). However, due to unavoidable deviations of physical parameters in a real system, a discrepancy between y and $\hat{y}$ occurs in practice that even may become unstable. By additional feedback for the m linear input-output channels, these discrepancies are

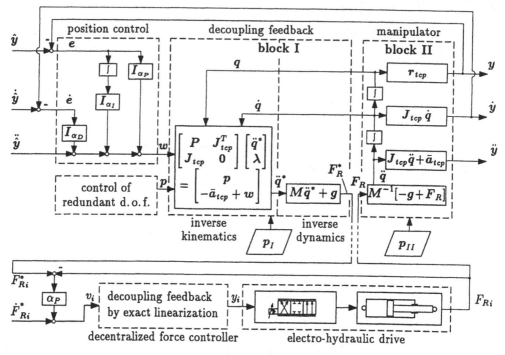

Figure 5: Overall control structure

easily eliminated. In the present model, we employ for this purpose PID-controllers (for the simulations in section 4, only PD-controllers were used):

$$w_i = \ddot{\hat{y}}_i + \alpha_{Di} (\dot{\hat{y}}_i - \dot{y}_i) + \alpha_{Pi} (\hat{y}_i - y_i) + \alpha_{Ii} \int (\hat{y}_i - y_i)\, dt, \quad i = 1, \ldots, m. \tag{14}$$

3.4 Decentralized Force Controllers

The desired resulting driving forces F_R^* acting on the mechanical system, which have been calculated in section 3.2, are provided by decentralized force controllers as shown in Fig. 5. Here the method of exact linearization is employed again to compute the input voltage of the servo valve [4]. The resulting ith input-output channel v_i is asymptotically stabilized by a simple P-feedback with superimposition of the ith reference input $\dot{F}_{Ri}^*$:

$$v_i = \dot{F}_{Ri}^* + \alpha_{Pi} (F_{Ri}^* - F_{Ri}). \tag{15}$$

Normally, the reference input $\dot{F}_{Ri}^*$ in Eq. (15), i.e. the time derivative of the desired driving force, is not available. In order to circumvent this difficulty, one can compute it by numerical differentiation, or – as done in section 4 – simply neglect it. This worsens the follow-up behaviour, but does not influence the stability.

4 Simulation of the Concrete Pumping Manipulator

The following simulations show the behaviour of the manipulator during wet concrete pumping as described in section 3.1. The manipulator has to track a straight line from point A to point B (length: 4 m, see Fig. 4) in a simulation time of 30 s. The pumping process begins at $t_1 = 5$ s and ends at $t_2 = 25$ s (refer to Fig. 2).

Due to the high pressure in the concrete pipes (ca. 100 bar), large forces and torques are induced at the TCP and at the joints connecting the arm elements. These forces and torques depend for the most part on the flexibility of the pipe mountings.

4.1 Uncontrolled Manipulator

In order to provide the input voltages of all servo valves corresponding to the desired motion, a reference run is performed without pumping forces. Now these values are used as input for the uncontrolled system. Fig. 6 shows the spatial TCP-trajectory of the uncontrolled manipulator together with the reference trajectory in a simple 3D-representation. Fig. 7 shows the trajectory in more detail as a projection on the x-y-plane.

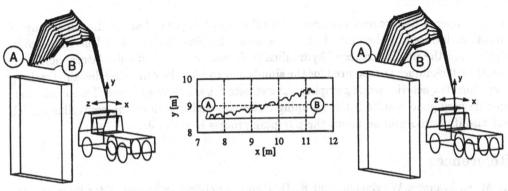

Figure 6: Uncontrolled   Figure 7: Trajectories   Figure 8: Controlled

4.2 Explicit Incorporation of Pumping Force Terms in the Controller Model

If the mathematical model of the manipulator which is used in the decoupling feedback contains a model and the appropriate parameters of the pumping process, very good results can be obtained (see Fig. 8). In the diagram of Fig. 7, the trajectory of the controlled manipulator cannot be distinguished from the reference trajectory. Fig. 9 shows the time response of the pressure variables in the hydraulic cylinder of the first arm element (cylinder A).

4.3 Unknown Pumping Forces as Disturbances in the Controller Model

If no model of the pumping process is available for the controller model, the follow-up behaviour worsens. The simulation results are nevertheless good, so that no difference between the TCP-trajectory and the reference trajectory can be seen in the resolution of Fig. 7.

As there is no model information about the pumping forces, the only way to control the manipulator is to change the input vector w for the decoupling feedback. For this change to be computed, the PD-controller must be fed with an error $e = \hat{y} - y = r_{tcp,desired} - r_{tcp,actual}$ of finite magnitude. This leads to unavoidable errors e during operation, as shown in Fig. 10.

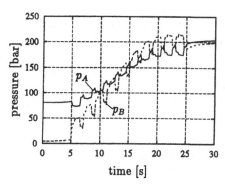

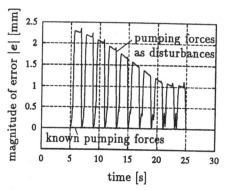

Figure 9: Pressure in cylinder A   Figure 10: Magnitude of the TCP-error

5   Conclusions

In this paper a hierarchical nonlinear control concept is presented for the supressing of vibrations induced by the forces in the process of pumping wet concrete. Due to the complexity and the nonlinearities of hydraulically driven large-scale manipulators, nonlinear modeling techniques are required for the simulation and for the controller design. Tracking prescribed trajectories while pumping wet concrete is used as an example. The simulations provide very good results if the pumping forces are well-known to the controller model, but also if no information about the pumping process is available.

References

[1] M. C. Wanner, W. Engeln, and K. D. Rupp. Enabling technologies for large manipulators – ESPRIT II project LAMA. In *8th International Symposium on Automation and Robotics in Construction ISARC*, volume 1, pages 439–446, 1991.

[2] A. Kecskeméthy. *Objektorientierte Modellierung der Dynamik von Mehrkörpersystemen mit Hilfe von Übertragungselementen.* Fortschrittberichte VDI, Reihe 20, Nr. 88. VDI Verlag, Düsseldorf, 1993.

[3] M. Schneider, M. Hiller, and S. Wagner. Dynamic simulation, nonlinear control and collision avoidance of hydraulically driven large redundant manipulators. In E. Budny, A. McCrea, and K. Szymański, editors, *Automation and Robotics in Construction XII*, pages 341–348, 12th ISARC, Warszawa, Poland, 30 May – 1 June 1995. Instytut Mechanizacji Budownictwa i Górnictwa Skalnego (IMBiGS), Warszawa, Poland.

[4] M. Schneider and M. Hiller. Nonlinear motion control of hydraulically driven large redundant manipulators. In *Proceedings of the IFAC-Workshop Motion Control*, pages 269–278, Munich, Germany, 9–11 October 1995.

[5] A. Isidori. *Nonlinear Control Systems: An Introduction.* Springer-Verlag, 2nd edition, 1989.

[6] C. Woernle. *Regelung von Mehrkörpersystemen durch externe Linearisierung.* Fortschrittberichte VDI, Reihe 8, Nr. 517. VDI Verlag, Düsseldorf, 1995. Habilitationsschrift.

NONLINEAR DYNAMICS
OF COMPUTER CONTROLLED MACHINES

G. Stépán

Technical University of Budapest, Budapest, Hungary

E. Enikov

University of Illinois at Chicago, Chicago, IL, USA

T. Müller

Technical University of Budapest, Budapest, Hungary

Abstract

Position and force-control is a standard mechanical controlling problem, for example in robotics. When unstable equilibria of mechanical systems are stabilized by means of some control, the task can be considered as a position or force control of a system with negative stiffness. Analytical investigation of simple models with 1-2 DOF have a central role in understanding technical phenomena and in forming the common sense in design work. Basic text books often call the attention for the destabilizing effect of sampling in these systems. In this paper, simple closed form formulae for stability limits are presented in basic parameter space of the sampling time, control gains and mechanical parameters. However, the experimentally verified performances of these systems are often worse than expected after the stability analysis of the digitally controlled system. Another digital effect, the quantization have a strong effect on the system behavior. This nonlinearity is very strong locally, and often results in some small amplitude stochastic oscillation of the robot around its desired end position. The paper will show the chaotic nature of these oscillations and some ways to characterize them.

Introduction

The vibration phenomena discussed in this paper by means of simple mechanical models are considered here because position and/or force controlled robots as well as robots carrying out tasks of balancing a mechanical system often have stability problems. These systems are usually controlled digitally, and basic text books [1] and papers [2,3,4] often call the attention for the destabilizing effect of sampling in these systems via simple mechanical models.

All the vibration phenomena in question can be explained with a simple model of a one degree-of-freedom vibratory system controlled digitally in the presence of Coulomb friction (see Figure 1). The mathematical model assumes the form

$$m\ddot{q}(t) + C \operatorname{sign} \dot{q}(t) + sq(t) = Q(t) \tag{1}$$

where q is the scalar general coordinate, Q is the control force and C is the constant Coulomb friction force. Three different cases can easily be modelled by means of different parameters: position control is described by $m > 0$, $s = 0$, while force control is presented if $m > 0$, $s > 0$. When the uncontrolled system is unstable, that is $s < 0$, then the control force Q may still stabilize the linear system. This may describe the case of balancing an inverted pendulum. If the 1 DOF system were unstable because of some negative damping (which is not considered in this study), the system could still be stabilized. This happens when stick-and-slip motion is stabilized by a suitable digital control force.

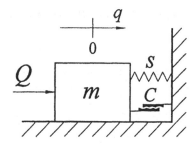

Figure 1. Mechanical model

The simplest PD controller

$$Q = -K_p(q - q_0) - K_d\dot{q} \tag{2}$$

with appropriate control gains K_p and K_d and desired position q_0 could give a perfect result with an asymptotically stable $q \equiv q_0$ position if there are no digital effects in the control and there is no Coulomb friction. Apart of its contribution to nonlinear vibrations, the Coulomb friction decreases the accuracy of the robot. This accuracy can be improved by increasing the control gains. Great values of gains, however, result stability problems when digital effects are not negligible.

There are two kinds of digital effects to be analyzed in this paper. The quantization in the time domain, that is the sampling, is a linear effect, it increases the dimension of the state space, though. This may become important in case of the classical position and force control tasks. The quantization in the state space, which takes place at the analog-digital-analog (ADA) converters, is a strong nonlinear effect, and together with the sampling, it causes small amplitude stochastic vibrations around the desired equilibrium. This becomes essential in the above mentioned tasks of balancing a system at a desired equilibrium.

In the subsequent sections the analytical results, experimental observations and simulation results are given for the three cases of position control, force control, and balancing.

Position control

When the stiffness is $s = 0$ in the model in Figure 1, the simplest case of position control is obtained. As it is well-known, the position error of the robot can be characterized by

$$\Delta = \frac{C}{K_p},$$

that is, we are interested in applying as great proportional gain K_p as possible.

The control force is based on the sampled values of the position and velocity at the time instants $j\tau$, $j = 0, 1, 2, \ldots$ where τ is the sampling time:

$$Q(t) = -K_p q(t_j - \tau) - K_d \dot{q}(t_j - \tau), \quad t \in [t_j, t_j + \tau),$$

that is, the control force is piece-wise constant and the desired equilibrium is assumed to be $q_0 = 0$. As shown in [5], the equation of motion (1) can be solved in closed form for each sampling interval, and the stability of the desired position can be analyzed without the Coulomb friction force C with the help of a 3D discrete mapping. The result of this stability analysis for the parameters

$$p = \frac{K_p}{m}\tau^2, \quad d = \frac{K_d}{m}\tau$$

shows that $q \equiv 0$ is asymptotically stable if and only if

$$0 < p < 2d - 3 + \sqrt{9 - 8d}.$$

Simple maximum calculation shows that the greatest proportional gain within the stability limit, and the smallest position error which can possibly be achieved are

$$p_{max} = \frac{1}{4} \quad \Rightarrow \quad \Delta_{min} = 4\tau^2 \frac{C}{m}.$$

More details of this calculation is given below for the case of force control.

Force control

The control force Q is considered in the form of a PD controller according to [1]:

$$Q = -K_p(F_s - F_0) - K_d \dot{F}_s + F_0. \tag{3}$$

The contact force F_s is sensed by the spring deformation q and the constant desired force is $F_0 = sq_0$. It can be seen from (1),(3) that for any equilibrium q^* the force error satisfies

$$|F_s^* - F_0| = s|q^* - q_0| \leq \frac{C}{|1 + K_p|}. \tag{4}$$

Again, we are interested in applying as great proportional gain K_p as possible without losing stability.

The stability of the desired equilibrium q_0 is investigated by dropping the Coulomb friction from the model and introducing the new dimensionless gains $p = K_p$ and $d = \alpha K_d$ and the dimensionless time $T = \alpha t$, where $\alpha = \sqrt{s/m}$ is the natural angular frequency of the uncontrolled system. For the new convenient state vector

$$\mathbf{x} = \operatorname{col}(x_1\ x_2\ x_3) = \operatorname{col}(m\alpha\dot{q}\ m\alpha^2(q - q_0)\ Q - sq_0),$$

the equation of motion (1) and the control force definition (3) can be written as

$$\frac{d}{dT}\begin{pmatrix} x_1 \\ x_2 \\ x_3 \end{pmatrix} = \begin{pmatrix} 0 & -1 & 1 \\ 1 & 0 & 0 \\ 0 & 0 & 0 \end{pmatrix}\begin{pmatrix} x_1 \\ x_2 \\ x_3 \end{pmatrix}, \quad T \in [T_j, T_{j+1}), \tag{5}$$

$$x_3(T) = -px_2(T_{j-1}) - dx_1(T_{j-1}), \quad T \in [T_j, T_{j+1}). \tag{6}$$

This control force expression considers the sampling effect using the sampling moments $T_j\ j = 0, 1, \ldots$ and the constant dimensionless sampling time $\tau = T_{j+1} - T_j$ for this case. By means of the solution of (5) and with the control force definition (6), we can construct a 3 dimensional discrete mapping in the form

$$\mathbf{x}_{j+1} = \begin{pmatrix} \cos\tau & -\sin\tau & \sin\tau \\ \sin\tau & \cos\tau & 1 - \cos\tau \\ -d & -p & 0 \end{pmatrix} \mathbf{x}_j. \tag{7}$$

The asymptotic stability of the trivial solution is investigated by means of the characteristic polynomial of the principal matrix in (7):

$$-\mu^3 + 2\mu^2 \cos\tau + \mu(p(\cos\tau - 1) - d\sin\tau - 1) + p(\cos\tau - 1) + d\sin\tau.$$

For the characteristic multipliers, the stability condition $|\mu_{1,2,3}| < 1$ can be checked by Jury's criterion [5] in the parameter space p, d, τ. For some sampling times, the stability domains in the p, d plane are shaded in Figure 2.

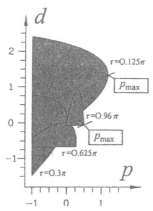

Figure 2. Stability chart for force control parameters

In accordance with (4), the maximum values of the proportional gain $p = K_p$ is calculated for each sampling time τ:

$$\max p = \begin{cases} \frac{(2\cos\tau - 1)^2}{8(1 - \cos\tau)}, & \tau \in \cup_k[2k\pi + \frac{2}{3}\pi, 2k\pi + \frac{4}{3}\pi]; \\ -\frac{\cos\tau}{1 - \cos\tau}, & \tau \notin \cup_k[2k\pi + \frac{2}{3}\pi, 2k\pi + \frac{4}{3}\pi]. \end{cases}$$

These maximum values are denoted in Figure 2 for the cases of two sampling times which represent the way of calculations in two different cases. The above results are also summarized in Figure 3 with respect to the maximum proportional gain and the inversely related minimum force error. For most of the sampling times, the force error varies between C and $2C/3$. In some narrow regions, the proportional gain can be as great as we want, but it is difficult to tune these sensitive parameters in real robotic structures [6]. Since the force accuracy $2C/3$ can easily be achived without the differential gain K_d, this result may explain why differential gain is not used commonly in force control strategies.

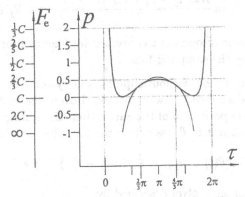

Figure 3. Minimum force error

Balancing

In order to represent a simple mechanical system with negative stiffness in (1), consider the experimental device in Figure 4 which consists of a pendulum hinged to a mobile cart. The balancing of the inverted pendulum is an old and challanging task for researchers (see e.g., [7,8]). The cart consists of an electric motor with a gear-box connected to a pair of wheels by a chain.

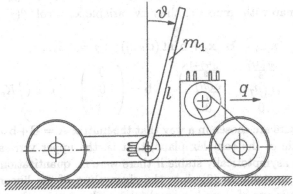

Figure 4. Inverted pendulum on a cart

Here, m_2 includes the mass of the whole cart, the mass moments of inertia of its wheels, and the rotor of the electric motor, while m_1 and l are the mass and the length of the homogeneous pendulum, respectively. The bar is attached to the cart with a ball bearing, so the viscous friction is negligible along θ.

The control force Q is applied via the electric motor and the Coulomb friction at the contact points of the wheels and the ground. It is considered in the form of a simple PD controller with gains K_p and K_d as above. However, when quantization both in time and space are modelled, the control force now has the form

$$Q(t) = h \operatorname{int} \frac{K_p \theta(t_j - \tau) + K_d \dot\theta(t_j - \tau)}{h}, \quad t \in [t_j, t_j + \tau)$$

where h denotes the value of one digit converted into control force, while the sampling effect is characterized by the sampling time τ.

If the wheels roll on the ground without slipping, our system has two degrees of freedom and its motion can be described by two generalized coordinates: the angle θ of the pendulum and the horizontal position q of the cart. After eliminating the cyclic coordinate q and linearization with respect to θ, we obtain the following simple equation of motion:

$$\ddot\theta(t) - \beta^2 \theta(t) = u_j, \quad t \in [j\tau, (j+1)\tau), \quad j = 1, 2, \ldots \tag{8}$$

with the piece-wise constant control described by

$$u_j = -\gamma h \operatorname{int} \frac{K_p \theta((j-1)\tau) + K_d \dot\theta((j-1)\tau)}{h}$$

and with the new parameters introduced as

$$\beta = \sqrt{\frac{6(m_1 + m_2)g}{(m_1 + 4m_2)l}}, \quad \gamma = \frac{6}{(m_1 + 4m_2)l}.$$

By integrating (8) for each sampling time interval $[j\tau, (j+1)\tau)$, and using the Poincaré sections of the trajectories at the time instants $j\tau$, $j = 1, 2, \ldots$, we construct a three dimensional linear map with respect to the new variable $x_j = \operatorname{col}\left(\theta(j\tau) \; \dot\theta(j\tau) \; u_j\right) \in \mathbb{R}^3$ in the form:

$$x_{j+1} = B \cdot x_j + b \operatorname{int}(c \cdot x_j), \quad j = 0, 1, \ldots, \tag{9}$$

$$B = \begin{pmatrix} \operatorname{ch}(\beta\tau) & \frac{\operatorname{sh}(\beta\tau)}{\beta} & \frac{\operatorname{ch}(\beta\tau)-1}{\beta^2} \\ \beta \operatorname{sh}(\beta\tau) & \operatorname{ch}(\beta\tau) & \frac{\operatorname{sh}(\beta\tau)}{\beta} \\ 0 & 0 & 0 \end{pmatrix}, \quad b = \begin{pmatrix} 0 \\ 0 \\ -h\gamma \end{pmatrix}, \quad c = \left(\tfrac{1}{h}K_p \; \tfrac{1}{h}K_d \; 0\right).$$

The control parameters are chosen in a way that the matrix $A = B + b \circ c$ has eigenvalues within the unit circle of the complex plane, that is, the upper vertical position of the pendulum would be asymptotically stable if there were no quantization effect. However, the zero solution is unstable in (9) since the eigenvalues of the linear part are $\lambda_{1,2} = e^{\pm\beta\tau}$, $\lambda_3 = 0$, that is one of them has an absolute value greater than 1.

In the case of this so-called micro-chaos map, an invariant and globally attractive set can be found in the form

$$A = \left\{ x \in \mathbb{R}^n \mid \|x\| \le \frac{\|b\|}{1 - \rho} \right\} \tag{10}$$

where we suppose that

$$\rho = \sigma(B + b \circ c) < 1 \tag{11}$$

is true for the spectral radius of A which can be achieved with the appropriate choice of the gains.

However, there is no stable equilibrium or stable peridoc motion within this attractor. As it is proved for the micro-chaos map in [9], this A is a chaotic attractor and the digitally controlled system will have small amplitude stochastic oscillations around the desired but still unstable equilibrium position. This theoretical result is clearly confirmed by our experiments. Figure 5 shows the "stationary" angle signal θ which also presents the reason of this chaotic behaviour: the finite resolution of the digital converter which introduces the strong local nonlinearity into the system.

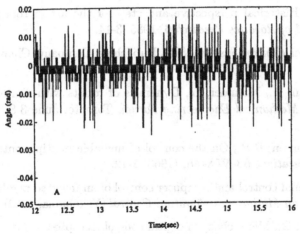

Figure 5. Angle signal of inverted pendulum

Conclusion

Two types of digital effects have been considered in case of controlled mechanical systems. Sampling, i.e. quantization in time, is a linear effect and causes limitations for the maximum proportional gains and the minimum position and/or force errors in case of position and force control. The quantization in space coordinates is a strongly nonlinear effect. This results in small amplitude chaotic oscillations when unstable equilibria of mechanical systems are to be stabilized by means of digital control. This affects the positional accuracy of these systems, and the error can be estimated by the maximum vibration amplitude, that is, by the radius of an attractive set in the state space.

Acknowledgement

This research has been supported by the Hungarian Scientific Research Foundation OTKA No. 4-041 and by the U.S.–Hungarian Science & Technology Program JFID No. 336.

References

[1] Raibert, M. H., Craig, J. J., Hybrid Position/Force Control for Computer Controlled Manipulators, *ASME J. Dynamic Systems, Measurement and Control*, **102** (1981) 125-133.

[2] Craig, J. J., *Introduction to Robotics Mechanics and Control*, Addison-Wisley, Reading, 1986.

[3] Whitney, D. E., Force feedback control of manipulator fine motions, *ASME Journal of Dynamic Systems, Measurement and Control* **98**, 1977, 91-97.

[4] Whitney, D. E., Historical perspective and state of the art in robot force control, *International Journal of Robotics Research* **6**, 1985, 3-14.

[5] Kuo, B. C., *Digital Control Systems*, SRL Publishing Company, Champaign, Illinois, 1977.

[6] Stépán, G., Steven, A., Maunder, L., Dynamics of robots with digital force control. In: *Proc. of CSME Mechanical Engineering Forum*, Toronto, June 3-8, 1990, Vol. III, 355-360.

[7] Higdon, D.T., Cannon, R.H., On the control of unstable multiple-output mechanical systems. *ASME Publications* **63-WA-48**, (1963) 1-12.

[8] Kawazoe, Y., Manual control and computer control of an inverted pendulum on a cart. *Proc. 1st Int. Conf. on Motion and Vibration Control* (Yokohama, 1992), 930-935.

[9] Haller, G., Stépán, G., Micro-chaos in digital control, accepted in *Int. J. of Nonlinear Science* (1996).

IDENTIFICATION AND COMPENSATION
OF GEAR FRICTION FOR MODELING OF ROBOTS

M. Daemi and B. Heimann
University of Hannover, Hannover, Germany

Abstract - *The identification of the dynamics of a standard industrial robot is solved by the application of the multivariable least square method. In order to eliminate the dominant influence of friction in gears and joints a nonlinear friction model is adapted to measured friction characteristics. Its influence is compensated in the identification step. The base parameter vector is grouped and optimal trajectories are used to identify each group. The quality of the identified model is verified by comparison of measured trajectories and torques predicted by the model.*

Introduction

A common approach in robotics for the identification of the equation of motion is the use of multivariable least square methods. It allows a relatively fast identification of a set of linear base parameters of the robot without the need of disassembling it. A main concern in practical applications of this method is the large influence of the friction in gears and joints and their variation with changing operating conditions.

In this paper the identification of the 6-dof dynamic model of a standard industrial robot is presented, which emphasizes on the prediction of a precise friction model and the compensation of its influence on the identification process. The measurements rely only on the robots internal incremental encoders and current sensors for torque measurements. Although this work is adapted to the characteristics of the manutec-r15, the approach can be used for most industrial robots.

A common method for smaller systems (dof $\leq$ 3) to avoid the influence of the variation of friction parameters is to include them into the identification process by modeling a term of dry friction and a term for viscous damping (e.g.[1]). This approach would enlarge the parameter vector for a 6-dof robot by another 12 elements and would lead to problems in the identification process. In this paper effects of the friction model for each link are evaluated separately and their influence on the measured torques are compensated before starting the identification. In order to enable the excitation of all parameters of the dynamic model, an optimization scheme is proposed for defining trajectories that yield good identification results.

Modeling of gear friction

The torques exerted by the drives of each joint of the robot consist of a number of effects which are: acceleration of the moments of inertia, centrifugal- and Coriolis forces,

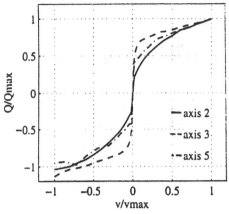 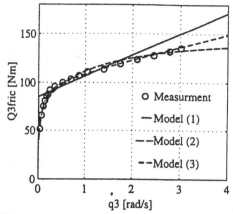

Fig. 1 Normalized friction characteristics   Fig. 2 Adaption of friction models to mea-
for different axes of an industrial robot.   sured friction characteristics of axis 3.

gravitational effects and losses due to gears and bearings. By using trajectories where
only one axis is moved and selecting parts of the measurement with constant velocity, the
effects of acceleration-, centrifugal- and Coriolis forces are avoided. Since gravitation only
has an influence on axes 2 and 3 of the manutec-r15 robot, a set of some measurements at
slow speeds are used to establish an overall gravitational model. It is used to compensate
the respective torques of these joints. Thus the following measurements can be assumed
to solely reflect the influence of friction in gears and bearings.

Frictional losses of robotic joints are usually modeled as a torque which is a function
of its own rotational joint speed. This nonlinear function for one joint $Q_{i,f}$ is mostly
described by the sum of a term for viscous damping and another for dry friction

$$Q_{i,f1} = a_1\dot{q}_i + a_2\text{sign}(\dot{q}_i). \tag{1}$$

Although such simple model suffices for a number of applications it needs to be refined
for more precise modeling. Fig. 1 shows measured friction torques of some axes of a robot
over their full speed range, normalized to their maximum speed and maximum torque. It
can be seen that all axes show significant degressive characteristics, not covered by the
simple model given in (1). A better description of the measured friction characteristics
can be found by using one of the following equations:

$$Q_{i,f2} = a_1 + a_2\dot{q}_i + a_3\text{sign}(\dot{q}_i) + a_4\dot{q}_i^{\frac{1}{3}}, \tag{2}$$

$$Q_{i,f3} = a_1 + a_2\dot{q}_i + a_3\text{sign}(\dot{q}_i) + a_4\arctan(a_5\dot{q}_i). \tag{3}$$

For a given measured friction characteristic an analytical description with a minimal
square model error can easily be calculated for (2) since it has only linear dependencies in
its parameters. Whereas for model (3) a nonlinear optimization procedure is applied. As
depicted in Fig. 2 both models lead to much better results than the classic approach (1).
The mean quadratic error between the measured and modeled torque is used to decide
whether model (2) or (3) gives a better description of the measured characteristics.
It has to be kept in mind that modeled parameters found in this way no longer present
mechanical models (such as dry friction or viscous damping) but merely force the sum of
all effects of the multistage gears into a mathematical description.

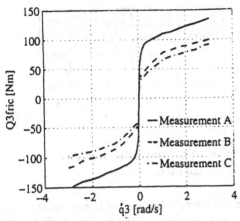

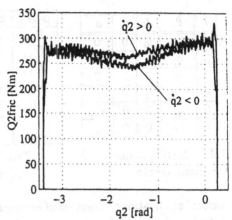

Fig. 3 Variation of friction characteristic for different operating conditions.

Fig. 4 Friction torque of axis 2 at constant velocity vs. joint angle.

In order to evaluate the quality of the identified friction models a number of measurements are made. The most dominant effect is shown in Fig. 3 . Measurement **A** is made immediately after turning on the 'cold' robot. Measurement **B** is made after one hour of continuous operation at 70% of maximum speed of the axis. Measurement **B** and **C** were made just right before and after the measurement of a relatively fast trajectory. The relatively large time constants imply that that these effects arise mostly due to temperature variation in the gears and bearings. The difference between measurements **B** and **C** is regarded as a measure for possible changings of the friction model while measuring a single trajectory such as for identifying a dynamic model.

The verification of the gravitation model that was used to isolate the friction characteristics for axes 2 and 3 is shown in Fig. 4 . It shows the absolute value of the friction torque for a trajectory where axis 2 is moved at constant speed in its full operating range. It is depicted versus the joint angle of the axis. A non-compensated gravitational effect would lead to significant differences between the line for $\dot{q}_2 > 0$ and $\dot{q}_2 < 0$.
Furthermore, this presentation shows another unmodeled effect. At the horizontal position of the axis (0 rad, π rad) higher torques can be seen than at the upright position ($\frac{\pi}{2}$ rad), which implies a friction model dependency of driving torque. Similar effects can also be found for other axes. They lead to systematic errors in the identification of the dynamic model.

Dynamic model with base parameters of the manutec-r15

For deriving the robots equation of motion, algorithm [2] is used which yields a description of the dynamics based on a set of base parameters. The coordinate frames for the links are defined by the Modified Denavit-Hartenberg (MDH) notation [3] and a parameter vector $p_{link,i}$ of 10 mechanical parameters is assigned to each link

$$p_{link,i} = [J_{i,xx}, J_{i,xy}, J_{i,xz}, J_{i,yy}, J_{i,yz}, J_{i,zz}, m_i r_{i,x}, m_i r_{i,y}, m_i r_{i,z}, m_i]^T, \qquad (4)$$

where the $J_{i,..}$ are the elements of the inertia tensor, m_i is the mass of each link and the $r_{i,.}$ are the distances of the center of mass to the coordinate frame fixed to the link. In

i	θ_i	d_i	a_i	α_i
1	$-q_1$	l_0	0	0
2	q_2	0	0	$-\frac{\pi}{2}$
3	q_3	0	l_1	0
4	$-q_4$	l_2	0	$\frac{\pi}{2}$
5	q_5	0	0	$-\frac{\pi}{2}$
6	q_6	0	0	$\frac{\pi}{2}$

$P_{link,1}$	$P_{link,2}$	$P_{link,3}$	$P_{link,4}$	$P_{link,5}$	$P_{link,6}$
$J_{1,xx}$	$J_{2,xx}$	$J_{3,xx}$	$J_{4,xx}$	$J_{5,xx}$	$J_{6,xx}$
$J_{1,yy}$	$J_{2,xz}$	$J_{3,yy}$	$J_{4,yy}$	$J_{5,yy}$	$J_{6,yy}$
$J_{1,zz}$	$J_{2,yy}$	$J_{3,zz}$	$J_{4,zz}$	$J_{5,zz}$	$J_{6,zz}$
$m_1 r_{z1}$	$J_{2,zz}$	$m_3 r_{y3}$	$m_4 r_{z4}$	$m_5 r_{y5}$	$m_6 r_{z6}$
m_1	$m_2 r_{x2}$	m_3	m_4	m_5	m_6
	$m_2 r_{z2}$				
	m_2				

Tab. 1 MDH parameters for the manutec-r15

Tab. 2 Non-zero link parameters according to (4) for the manutec-r15

practical applications a number of elements in (4) are zero because of the orientation and position of the coordinate frame as well as symmetric shapes of the bodies. In Tab. 1 the definition of the MDH parameters for the manutec-r15 are given and Tab. 2 shows the remaining elements of the $p_{link,i}$ for all links. Applying algorithm [2] to these definitions leads to a parameter vector

$$\mathbf{p}_r = \begin{bmatrix} p_1 \\ p_2 \\ p_3 \\ p_4 \\ p_5 \\ p_6 \\ p_7 \\ p_8 \\ p_9 \\ p_{10} \\ p_{11} \\ p_{12} \\ p_{13} \\ p_{14} \end{bmatrix} = \begin{bmatrix} J_{1zz} + J_{2yy} + J_{3yy} + l_1^2(m_3 + m_4 + m_5 + m_6) \\ J_{2xx} - J_{2yy} - l_1^2(m_3 + m_4 + m_5 + m_6) \\ J_{2xz} \\ J_{2zz} + l_1^2(m_3 + m4 + m5 + m_6) \\ m_2 r_{x2} + l_1(m_3 + m_4 + m_5 + m_6) \\ J_{3xx} + J_{4yy} + 2l_2 m_4 r_{z4} + l_2^2(m_4 + m_5 + m_6) - J_{3yy} \\ J_{3zz} + J_{4yy} + 2l_2 m_4 r_{z4} + l_2^2(m_4 + m_5 + m_6) \\ m_3 r_{y3} - m_4 r_{z4} - l_2(m_4 + m_5 + m_6) \\ J_{4xx} + J_{5yy} - J_{4yy} \\ J_{4zz} + J_{5yy} \\ J_{5xx} + J_{6xx} - J_{5yy} \\ J_{5zz} + J_{6xx} \\ m_5 r_{y5} - m_6 r_{z6} \\ J_{6zz} \end{bmatrix}. \qquad (5)$$

The full set of equation of motion can be written for each link i of a n-dof robot as

$$Q_i = \mathbf{A}_i(\ddot{\mathbf{q}}, \dot{\mathbf{q}}, \mathbf{q})\mathbf{p}_r, \qquad i = 1 \dots n, \qquad (6)$$

where Q_i is the driving torque, $\mathbf{q} = [q_1 \dots q_n]$ is the vector of joint angles and $\mathbf{A}_i$ is a nonlinear function of joint angles, velocities and accelerations. For the given system the symbolic computation of the $\mathbf{A}_i$ is done by MAPLE which leads to an optimized C-Code with 10 trigonometric functions, 651 multiplications, 412 summations and 332 assignments. Because of their size, a verification of the equations was done by numerical comparison, using MATLAB and its Robotics-Toolbox with an recursive Newton-Euler approach.

The links of most industrial robot can be divided into three main axes for positioning and three hand axes for orientation of the endeffector. Due to this structure the elements of the base parameter vector can be divided into three different groups, which are the following for the example given in (5). The movement of the hand axes ($q_4 \dots q_6$) is only influenced by parameters p_{10} to p_{14}. Moving the main axes $(q_1 \ldots q_3)$ and leaving the hand axes at rest only parameters p_1 to p_8 have a non-constant influence in $A_i, (i = 1 \ldots 3)$ and can be identified by the algorithm described in the next section.

Parameter p_9 is the only 'dynamic' coupling parameter which is excited when main- *and* hand axes are being moved. For the manutec-r15 the almost symmetric shape of link 4 $(J_{4xx} \approx J_{4yy})$ and a small moment of inertia J_{5yy} leads to a very small influence of this parameter to the dynamics and could not be identified by the presented approach.

Identification procedure and optimal trajectory generation

A common identification method in robotics is the multivariable linear least square (L.S.) method based on a representation of robot dynamics as in equation (6). For a certain point in time Ti, measurements of $m \leq n$ different axes $\gamma_1 \ldots \gamma_m$ are combined as

$$Q_{Ti} = [Q_{\gamma_1, Ti} \cdots Q_{\gamma_m, Ti}]^T \quad \text{and} \quad \Psi_{Ti} = [A^T_{\gamma_1, Ti} \cdots A^T_{\gamma_m, Ti}]^T. \tag{7}$$

Further combination of measurements at r different time steps leads to a vector equation

$$Q = \Psi p + e, \quad Q = [Q^T_{T1} \cdots Q^T_{Tr}]^T \quad \text{and} \quad \Psi = [\Psi^T_{T1} \cdots \Psi^T_{Tr}]^T \tag{8}$$

with measurement vector Q, observation matrix Ψ, parameter vector p and the unknown error e. Minimizing the error in the L.S. sense leads to an estimated parameter vector

$$\hat{p} = \arg \cdot \min_p(\|e\|) = (\Psi^T \Psi)^{-1} \Psi^T Q. \tag{9}$$

This basic form of L.S. estimation can be refined by use of parameter weighting or by the instrumental variable method.

The main concern in using (8) and (9) is a proper choice of 'measurements' in order to ensure the excitation of all parameters. This is achieved by minimizing the condition of the observation matrix cond(Ψ) by means of some optimization methods. The intention in this paper is to use a method with few parameters for optimization in order to reduce convergence problems and to keep computation time low. Therefore, initial points $q_{i,t0}$, final points $q_{i,t2}$ and one intermediate point $q_{i,t1}$ are defined for each axis and 7th-order polynomial trajectories are calculated to connect them. At the initial and final points velocities and acceleration are set to zero. The interpolated trajectories for each axis consist of two parts

$$q(t) = \underbrace{\sum_{i=0}^{6} a_i t^i}_{A} + (d_1 + d_2 t) \underbrace{\sum_{i=1}^{6} b_i t^i}_{B} \tag{10}$$

where the a_i and b_i are defined such that part A solves the given boundary conditions, and part B solves the homogeneous boundary conditions. By varying d_1 and d_2 a number of different trajectories can be generated (Fig. 5) .

The use of the 7th-order polynomials ensures shock- and jerkless trajectories such that the elasticities of joints are not excited by driving torques. The intermediate points force the robot to some extent to remain within the boundary of the workspace. Computing a trajectory for all six robot axes leads to an optimization problem with 12 parameters and a large number of constraints for maximum joint velocities and maximum joint accelerations. Optimization is executed using the BFGS Quasi-Newton method with a mixed

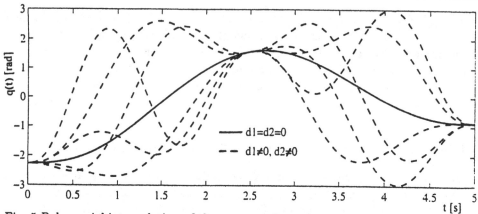

Fig. 5 Polynomial interpolation of three supporting points

quadratic and cubic line search procedure (MATLAB optimization toolbox). As can be expected, no global minimum can be found by the identification scheme, although the condition of the observation matrix could be reduced by magnitudes of up to two decades.

Identification of a model for the manutec-r15

The identification of the robot dynamics is executed in two main steps. As a preparation trajectories are generated for optimal excitation of the parameters: For the identification of the parameters p_{10} to p_{14} only axes 4 to 6 are moved, whereas, for identification of the parameters p_1 to p_8 only axes 1 to 3 are moved. If (for the general case) p_9 has also to be identified, the first trajectory described above must include motion of the main axes.

For the first step, friction characteristics for axes 4 to 6 are measured and a friction model for each of them is calculated according to (2). Then the robot is moved according to the optimal trajectory and joint angles and joint torques are measured. The measurements are filtered by means of a phase neutral 4th-order Butterworth filter and velocities and accelerations are derived. Additional scaling factors for the friction models are introduced to the parameter vector $p_{hand} = [p_{10}, \cdots p_{14}, p_{s4}, p_{s5}, p_{s6}]^T$ and the observation matrix is assembled with

$$
\Psi_{Ti} = \begin{bmatrix} A_{4,Ti} & Q_{4,f2,Ti} & 0 & 0 \\ A_{5,Ti} & 0 & Q_{5,f2,Ti} & 0 \\ A_{6,Ti} & 0 & 0 & Q_{6,f2,Ti} \end{bmatrix}. \tag{11}
$$

The use of the scaling factor takes into account, that during the time between the measurement of the friction characteristics and the identification trajectory, the friction coefficients might have changed slightly. The results for three different optimal trajectories are shown in Tab. 3 . As expected, the magnitude of the scaling parameters is close to 1 and furthermore, small variations of parameters for different trajectories already indicate a good identification result.

For the second step, again the friction characteristics of the active axes $(1 \ldots 3)$ are measured and models are calculated. The parameter vector is also expanded by scaling factors for the friction models $p_{main} = [p_1 \cdots p_8, p_{s1}, p_{s2}, p_{s3}]^T$. Before starting the identification procedure the influence of p_{10} to p_{14} is compensated in the measured torques.

P_{hand}	Traj. I	Traj. II	Traj. III
p_{10}	1.0180	1.0488	0.9560
p_{11}	0.0491	0.0043	0.1659
p_{12}	0.4353	0.4317	0.5374
p_{13}	-0.2063	-0.2148	-0.0827
p_{14}	0.1088	0.1124	0.0612
p_{s4}	1.0938	1.0999	1.0260
p_{s5}	1.0478	1.0427	1.0158
p_{s6}	1.0707	1.0847	0.9984

P_{main}	Traj. IV	Traj. V	Traj. VI
p_1	40.315	40.483	38.417
p_2	-24.493	-24.349	-23.344
p_3	-2.759	-3.462	-0.776
p_4	76.301	75.825	61.083
p_5	47.105	47.331	49.149
p_6	7.946	6.768	11.652
p_7	22.431	23.164	18.722
p_8	-14.321	-14.105	-14.477
p_{s1}	1.015	1.030	1.049
p_{s2}	0.901	0.959	1.016
p_{s3}	0.960	0.987	1.016

Tab. 3 Identified hand axes parameters for different trajectories

Tab. 4 Identified main axes parameters for different trajectories

Identification results for three different optimal trajectories are shown in Tab. 4. Again, a good agreement between the different trajectories can be seen.

To verify the quality of the identified model a number of different trajectories are used. For each trajectory a friction characteristic is measured and the angles and driving torques along the movement are recorded. In Fig. 6 the result of some of these measurements are shown. Each graph belongs to a different trajectory where all robot axes are moving slowly, whereas the depicted axis moves at high velocities and acceleration. Hereby, the amount of torque due to dynamics of the links is maximized compared to the torques due to friction. Still, large differences between (a) and (b) indicate the large influence of friction and underline the need of its proper compensation. Furthermore, comparing (b) and the torques predicted by using the identified parameters (c) shows that for the axes with lower contribution of friction (as for the main axes) the model leads to better results. The otherwise good agreement of the predicted and measured torques indicate the good quality of the identified base parameters.

Conclusion

The use of extended friction models leads to good descriptions of measured friction characteristics of robot joints. By their compensation an identification procedure can be applied, that leads to the identification of a set of base parameters for a 6-dof industrial robot. The results of the identification are evaluated by predicting the driving torques for a set of trajectories, and show the good quality of the identification. The large influence of the friction especially in the hand axes implies even further refinement of the friction modeling of robotic gears and thereby further improvements in the identification results.

References

[1] Daemi, M.; Heimann, B.: Inverse Dynamik und Identifikation von Robotermodellen, Technische Mechanik, Band 15, Heft 2, 1995, S.107-117.
[2] Gautier, M.; Khalil, W.: A Direct Determination of Minimum Inertial Parameters of Robots, Proc. IEEE Int. Conf. on Rob. and Autom., pp. 1682-1687, 1988.
[3] Khalil, W.; Kleinfinger, J.F.: A new geometric notation for open and closed loop robots, Proc. IEEE Int. Conf. on Robotics and Automation, pp. 1174-1180, 1986.

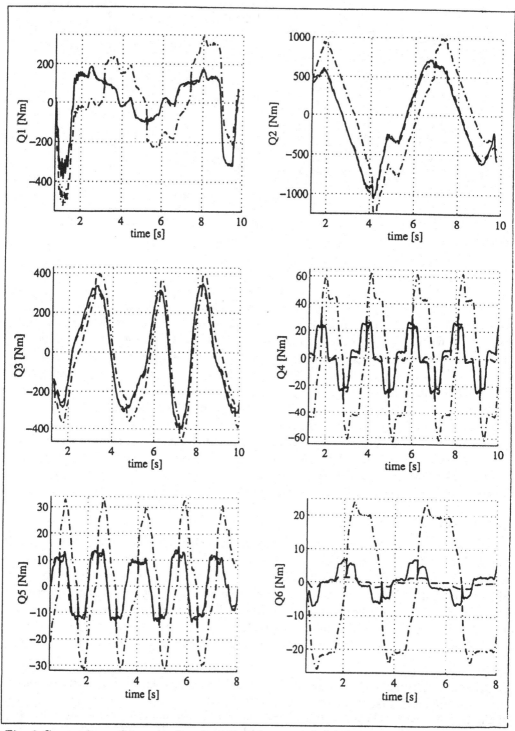

Fig. 6 Comparison of measured and predicted torques for different trajectories
—·(a) measured torques   — (b) measured torque without friction
—–(c) predicted torque without friction

Chapter III
Synthesis and Design

DESIGNING MANIPULATORS FOR BOTH KINEMATIC AND DYNAMIC ISOTROPIC PROPERTIES

R. Matone and B. Roth
Stanford University, Stanford, CA, USA

ABSTRACT

This paper is concerned with the design of manipulators for good distribution of both velocity and acceleration throughout the workspace. We derive inequalities between the condition numbers of the Jacobian matrix and matrices that represent dynamic characteristics of such robots. We show that designing a manipulator for good isotropicity of velocity (acceleration) distribution may lead to an arm with poor uniformity of acceleration (velocity) distribution. We present necessary and sufficient conditions for having both good kinematic and dynamic isotropic properties, and illustrate these concepts with results obtained for a 2R planar manipulator.

1 INTRODUCTION

The designers of robot manipulators must select the kinematic structure, link dimensions, mass distribution, actuator size and location, and drive mechanisms. These choices determine important aspects of manipulator performance. Of interest to us are two distinct types of properties, namely, the kinematic and the dynamic properties.

The main tool that researchers have been using to quantify kinematic characteristics of a manipulator is the analysis of its Jacobian matrix J [1-8], i.e., the matrix relating joint speeds to end-effector velocity. Many indices for kinematic performance have been proposed based on this matrix, for instance, the value of the determinant [1], manipulability [2], minimum singular value [3] and condition number [4]. In positions where J is singular, the end-effector's degrees-of-freedom decrease instantaneously. Alternatively, if J is isotropic, i.e., all its singular values are equal, the end-effector can move with the same velocity in all the Cartesian directions, for any normalized joint speed vector. One means to design robot arms for good kinematic characteristics is to select structural parameters that make the Jacobian matrix as isotropic as possible in the workspace [4-8]. Then, for a given joint-speed norm, the velocity distribution would be as uniform as possible.

On the other hand, dynamic performance can be characterized by the acceleration capability of the end-effector perceived at the end-effector [9] or at the actuators [10]. This ability is indicated by matrices defined in [9] and [10], we refer to these matrices as M_1 and M_2, respectively. If M_i (i=1,2) are isotropic, then by applying forces directly to the end-effector or the joint actuators one can accelerate the end-effector equally well in all the Cartesian directions. Similar to the kinematic case, one means to design manipulators for good dynamic characteristics is to select structural parameters that make the M_i matrices as isotropic as possible in the workspace. Then, the acceleration distribution would be as uniform as possible.

There has been an increasing need for manipulators that can at the same time work at high speeds, deal with heavy payloads and work with good accuracy. These manipulators must have both favorable kinematic and dynamic characteristics. It would be desirable to have the distribution of both velocity and acceleration, throughout the workspace, as isotropic as possible. This paper is concerned with the design of such manipulators. We derive inequalities between the condition numbers of the Jacobian matrix and the M_i matrices. Then we show that designing a manipulator for good isotropicity of velocity (acceleration) distribution may lead to an arm with poor uniformity of acceleration (velocity) distribution. Next we present necessary and sufficient conditions for both good kinematic and dynamic properties. Finally, we illustrate these concepts with results obtained for a 2R planar manipulator.

2 KINEMATIC ISOTROPY

In this section we recall briefly the concept of kinematic isotropy. We start by considering an n-degrees-of-freedom serial manipulator. The joint variables are given by the $n \times 1$ joint vector q, and the end-effector position is given by the $m \times 1$ ($m \leq 3$) position vector x. If this manipulator is used to position a point of its end-effector, disregarding its orientation, the mapping of joint velocities ($\dot{q}$) into end-effector velocity ($\dot{x}$) is given by $\dot{x} = J\dot{q}$, where J is the $m \times n$ Jacobian matrix.

The condition number of J as a measure of kinematic performance was introduced by Salisbury and Craig [4]. The condition number is defined as the ratio of the maximum (σ_{J_m}) to the minimum (σ_{J_1}) singular value of J, i.e., $\kappa_J \equiv \dfrac{\sigma_{J_m}}{\sigma_{J_1}}$. It measures the propagation of relative errors in the computation of $\dot{q}$ using $\dot{x} = J\dot{q}$, when there are uncertainties in both the $\dot{x}$ vector and the J matrix. The error propagation is minimum when κ_J is minimum [14]. The condition number also measures proximity of singular configurations. When the robot is approaching such a configuration, κ_J approaches infinity. By contrast, the minimum value of κ_J is one. In this situation, all the singular values of matrix J are equal. Joint coordinates for which κ_J=1 are considered the most distant from singularities, and they define the so-called isotropic configurations [4-8]. In these postures the manipulator can move with the same velocity in all directions of the operational space, when $\dot{q}$ is normalized to be a constant length vector. To see this property we use the constraint $\dot{q}^t\dot{q} = 1$ and $\dot{x} = J\dot{q}$ to obtain an ellipsoid in dimension m. The equation of this ellipsoid is $\dot{x}^t (JJ^t)^{-1} \dot{x} = 1$. In an isotropic configuration this ellipsoid becomes a sphere. Then the velocity vectors have the same norms in all the Cartesian directions.

In this paper isotropic configurations are termed kinematically isotropic configurations. This is done to contrast them with the dynamically isotropic configurations, which are introduced in the next section.

3 DYNAMIC ISOTROPY

The characterization of dynamic performance of robot manipulators is a difficult task. Indeed, the complexity of robot dynamics has led to the formulation of several performance measures [9-13]. Most of these measures are related to the acceleration capabilities of the end-effector. Asada [9] has introduced the Generalized Inertia Ellipsoid (GIE). If the GIE is isotropic, the equivalent inertia of the end-effector is the same in all the Cartesian directions. Yoshikawa [10] has defined dynamic manipulability and has proposed the Dynamic Manipulability Ellipsoid (DME). If the DME is isotropic, the actuators can accelerate the end-effector equally "easily" in all the Cartesian directions. Khatib and Burdick [11] have defined the Hyper-Parallelepiped of Acceleration (HPA) and formulated a cost function to optimize the dynamic design of robot arms. Isotropic accelerations were found by inscribing spheres in the HPAs. Graetinger and Krogh [12] have defined the acceleration radius and computed it as an optimization problem. Desa and Kim [13] have dealt with non-linearities in an analytical fashion. They have derived expressions for isotropic acceleration and maximum acceleration for a 2R planar mechanical arm.

Note that isotropic is a common term in the foregoing articles. While the motivation to use this term in robot dynamics is clear, i.e., to make reference to properties that are uniform in all directions, it has been interpreted to mean two different physical concepts. The first one designates configurations in which equal-norm accelerations of the end-effector can be produced in every Cartesian direction for a given input effort (for instance, a given length of the joint torque vector or end-effector force). The second one refers to the amount of end-effector acceleration that is achievable in every Cartesian direction, for given bounds on the joint torque vector. Properties like uniformity of the GIE [9] and DME [10] belong to the first category, while isotropic acceleration [11-13], uniformity of the HPA [11] and acceleration radius [12] belong to the second one.

In this work, dynamic isotropy refers to the first category and it is always associated with the uniformity of the singular values of a certain matrix. Two matrices are analyzed, namely, the ones used to define the GIE and the DME. Configurations in which the considered matrix has a condition number of one are called dynamically isotropic (with regard to the property defined by that matrix.) In what follows we discuss the meaning of isotropicity of each matrix. In the next section we introduce and discuss conditions for having both good kinematic and dynamic isotropic properties.

We begin with the operational space equations of motion, i.e.,

$$f = M_1(x)\ddot{x} + v(x, \dot{x}) + p(x). \tag{1}$$

Here $\ddot{x}$ is the end-effector acceleration, $M_1(x)$ is the kinetic energy matrix, $v(x, \dot{x})$ is the centrifugal and coriolis force vector, $p(x)$ is the gravity force vector, and f is the operational force vector. The relation between $M_1(x)$ and the joint space kinetic energy matrix, $A(q)$, is given by

$$M_1 = [J(q)A^{-1}(q)J^t(q)]^{-1}. \tag{2}$$

In non-singular configurations, both matrices $A(q)$ and $M_1(x)$ are symmetric and positive. Matrix $M_1(x)$ is the basis of the Generalized Inertia Ellipsoid. For a manipulator in the horizontal plane and at rest, eq. (1) gives $f = M_1(x)\ddot{x}$. Using the constraint $f^tf=1$, we obtain $\ddot{x}^t M_1^t M_1 \ddot{x} = 1$, which is the equation of an ellipsoid of dimension m, that has axes with semi-lengths $\sigma_{M_{11}}, \sigma_{M_{12}}, ..., \sigma_{M_{1m}}$, where $\sigma_{M_{11}} \leq ... \leq \sigma_{M_{1m}}$ are the singular values of matrix M_1.

Hence, for any unit-norm force vector f, the tips of all possible end-effector acceleration vectors of a manipulator at rest, in the horizontal plane, lie on an ellipsoid, the Generalized Inertia Ellipsoid. If the condition number of matrix M_1 is unitary ($\kappa_{M_1} =1$), then the ellipsoid becomes a sphere and a human operator applying forces directly to the end-effector can accelerate the end-effector equally "easily" in all the Cartesian directions. Here we call configurations in which $\kappa_{M_1} =1$, dynamically isotropic according to matrix M_1.

The Dynamic Manipulability Ellipsoid is a similar tool in the sense that it also indicates the "easiness" to accelerate the end-effector. However this "easiness" is given in relation to the actuators instead of a human operator. To see this property we invoke the relation between joint torques τ and end-effector force f, i.e., $\tau= J^t f$, and substitute it into eq. (1). Then, we obtain

$$\tau = M_2(q)\ddot{x} + c(q,\dot{q}) + g(q), \tag{3}$$

where $g(q)=J^t(q)p(x)$, $c(q,\dot{q})=J^t(q)v(x, \dot{x})$, and

$$M_2(q) \equiv J^t(q)M_1. \tag{4}$$

For non-redundant manipulators ($n=m$), we can write $M_2(q)= [J(q)A^{-1}(q)J^t]^{-1} = A(q)J^{-1}(q)$.

For a manipulator in the horizontal plane and at rest, eq. (3) gives $\tau = M_2(q)\ddot{x}$. Using the constraint $\tau^t\tau=1$, we obtain $\ddot{x}^t M_2^t M_2 \ddot{x} = 1$, which is the equation of an ellipsoid of dimension m that has axes with semi-lengths $\sigma_{M_{21}}, \sigma_{M_{22}}, ..., \sigma_{M_{2m}}$, where $\sigma_{M_{21}} \leq \sigma_{M_{22}} \leq ... \leq \sigma_{M_{2m}}$ are the singular values of matrix M_2.

Hence, for any unit-norm joint torque vector, τ, the tips of all possible end-effector acceleration vectors, of a manipulator at rest in the horizontal plane, lie on an ellipsoid, the Dynamic Manipulability Ellipsoid. If the condition number of matrix M_2 is unitary ($\kappa_{M_2} =1$), then the ellipsoid becomes a sphere and the actuators can accelerate the end-effector equally "easily" in all the Cartesian directions. Here we say configurations in which $\kappa_{M_2} =1$ are dynamically isotropic according to matrix M_2.

4 CONDITIONS FOR BOTH KINEMATIC AND DYNAMIC ISOTROPY

In what follows we will make use of the following theorems.
Theorem 1 (extracted from [8]): Let $A_{m,p}$ and $B_{p,n}$ be two given matrices. If the condition numbers of A, B, and AB are κ_A, κ_B, and κ_{AB}, respectively, then, $\kappa_{AB} \leq \kappa_A \cdot \kappa_B$.
Theorem 2: If the square matrix $A_{m,m}$ (or $B_{m,m}$) is isotropic, then matrix AB is isotropic if, and only if, the square matrix $B_{m,m}$ (or $A_{m,m}$) is also isotropic.
To verify Theorem 2, we multiply AB by its transpose, obtaining

$C=(AB)(AB)^t=ABB^tA^t$. Assuming that B is isotropic, BB^t is a multiple of the mxm identity matrix. Since we started with A being isotropic, AA^t is also a multiple of the mxm identity matrix, then matrix C must also be a multiple of the mxm identity matrix. Thus matrix AB is isotropic. In the other direction, if AB and A are isotropic, the same must happen with matrix B, by the same reasoning as above.

__Theorem 3__: If the square, symmetric and positive matrices $A_{m,m}$ and $C_{m,m}$ are isotropic, then matrix $B_{m,m}$ is also isotropic, where $A=BCB^t$.

To prove this theorem we recall that since matrices A and C are symmetric, we can factor them as [14] $A=Q_AA_AQ_A^t$ and $C=Q_CA_CQ_C^t$, where matrices Q_P $(P=A,C)$ are orthogonal and matrices A_p are diagonal with the eigenvalues of matrix P as elements. Since both matrices A and C are also positive, their eigenvalues are equal to their singular values. Now, using the fact that matrices A and C are isotropic, all the singular values of each matrix are equal. Thus, we can write $A=\sigma_A I_{m,m}$ and $C=\sigma_C I_{m,m}$, where σ_A and σ_C are the singular values of matrices A and C, respectively, and $I_{m,m}$ is the mxm identity matrix. Finally, we can write $A=BCB^t$ as $BB^t=\dfrac{\sigma_A}{\sigma_C}I_{m,m}$. Thus, matrix B is isotropic.

Note that Theorem 3 is valid only if both matrices A and C are symmetric and positive. Let's assume, for instance, that $A=\begin{bmatrix} 0 & 2 \\ -2 & 0 \end{bmatrix}$, $C=\begin{bmatrix} 0 & 1 \\ -1 & 0 \end{bmatrix}$, and $B=\begin{bmatrix} 0 & 2 \\ 1 & 0 \end{bmatrix}$. In this case matrix B is not isotropic, even though both matrices A and C are isotropic and $A=BCB^t$.

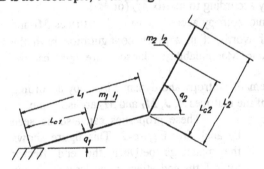

From eqs. (2) and (4), and Theorems 1, 2 and 3, it is clear that having a unitary condition number for one of the matrices J, M_1 or M_2, does not imply that the condition numbers of the remaining matrices are also unitary. In other words, a given configuration can be kinematically (dynamically) isotropic and not dynamically (kinematically) isotropic. Examples 1 and 2, below, illustrate these possibilities. Both examples refer to the 2R planar manipulator shown in Figure 1.

Figure 1: 2R Planar Manipulator

For this 2R system, matrices $J(q)$ and $A(q)$ are given by

$$J=\begin{bmatrix} -L_1\sin(q_1) & -L_2\sin(q_2) \\ L_1\cos(q_1) & L_2\cos(q_2) \end{bmatrix} \quad A=\begin{bmatrix} I_1+m_1L_{c1}^2+m_2L_1^2 & m_2L_1L_{c2}\cos(q_1-q_2) \\ m_2L_1L_{c2}\cos(q_1-q_2) & I_2+m_2L_{c2}^2 \end{bmatrix}$$

where m_i, I_i, and L_{ci}, designate, respectively, the mass, inertia and location of the center of mass of link i.

__Example 1__

Here we consider the case in which $L_1=L_2=1m$, $q_1=0$, $q_2=90°$, $L_{c1}=L_{c2}=0.5m$, $m_1=m_2=10kg$, $I_1=5kgm^2$, and $I_2=2kgm^2$. Under these conditions the 2R manipulator has kinematic isotropy. Matrices J, A, M_1, and M_2 are given by

$$J=\begin{bmatrix} 0 & -1 \\ 1 & 0 \end{bmatrix}, \quad A=\begin{bmatrix} 17.5 & 0 \\ 0 & 4.5 \end{bmatrix}, \quad M_1=\begin{bmatrix} 4.5 & 0 \\ 0 & 17.5 \end{bmatrix}, \quad M_2=\begin{bmatrix} 0 & 17.5 \\ -4.5 & 0 \end{bmatrix}.$$

The condition numbers of matrices J, M_1, and M_2 are, respectively, $\kappa_J = 1$, $\kappa_{M_1} = 3.89$, and $\kappa_{M_2} = 3.89$. Although the manipulator has kinematic isotropy, it is far from having dynamic isotropy according to the M matrices.

Example 2

Now we have $L_1 = 1m$, $L_2 = 4m$, $q_1 = 0$, $q_2 = 90°$, $L_{c1} = 0.5m$, $L_{c2} = 3m$, $m_1 = 20Kg$, $m_2 = 10kg$, $I_1 = 8kgm^2$, and $I_2 = 2kgm^2$. Under these conditions, matrices J, A, M_1, and M_2 are given by

$$J = \begin{bmatrix} 0 & -4 \\ 1 & 0 \end{bmatrix}, \quad A = \begin{bmatrix} 23 & 0 \\ 0 & 92 \end{bmatrix}, \quad M_1 = \begin{bmatrix} 5.75 & 0 \\ 0 & 23 \end{bmatrix}, \quad M_2 = \begin{bmatrix} 0 & 23 \\ -23 & 0 \end{bmatrix}.$$

The condition numbers of matrices J, M_1, and M_2 are, respectively, $\kappa_J = 4$, $\kappa_{M_1} = 4$, and $\kappa_{M_2} = 1$. In this case the manipulator has dynamic isotropy according to matrix M_2, and does not have kinematic isotropy nor dynamic isotropy according to matrices M_1.

From Theorem 1 and eqs. (2) and (4), we can state the following inequalities, for a general manipulator:

$$\frac{\kappa_A}{\kappa_J^2} \le \kappa_{M_1} \le \kappa_J^2 \kappa_A, \quad \frac{\kappa_A}{\kappa_J} \le \kappa_{M_2} \le \kappa_J \kappa_A, \quad \frac{\kappa_M}{\kappa_J} \le \kappa_{M_2} \le \kappa_J \kappa_{M_1}.$$

From Theorems 2 and 3, and eqs. (2) and (4), we can state the following properties for non-redundant manipulators:

Property 1: If a non-redundant arm has kinematic isotropy and dynamic isotropy according to matrix M_1 (or M_2), it also has dynamic isotropy according to matrix M_2 (or M_1).

Property 2: If a non-redundant arm has dynamic isotropy according to both matrices M_1 and M_2, it also has kinematic isotropy. In other words, if in a given configuration both the Generalized Inertia Ellipsoid and the Dynamic Manipulability Ellipsoid are spheres, the Jacobian matrix is isotropic.

Property 3: A non-redundant arm has both kinematic isotropy and dynamic isotropy according to matrices M_1 and M_2, if and only if any two of the matrices A, J, M_1 and M_2 are isotropic.

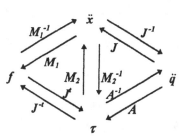

Figure 2: Mappings between f, $\ddot{x}$, τ and $\ddot{q}$

These properties can also be seen by analyzing Figure 2. This figure shows the mappings between the end-effector force f, the end-effector acceleration $\ddot{x}$, the joint torques τ and the joint accelerations $\ddot{q}$, for a non-redundant manipulator. Note that when the manipulator is at rest, the mapping of $\ddot{q}$ into $\ddot{x}$ is given by the Jacobian matrix, i.e., when $\dot{q} = 0$, we can write $\ddot{x} = J\ddot{q}$.

For instance, let's assume that matrices J and M_1 are isotropic. Since J is isotropic, a sphere in the $\ddot{q}$ space is transformed into a sphere in the $\ddot{x}$ space. Since M_1 is also isotropic, this second sphere is converted into a third one, now in the f space. This third sphere is transformed into a fourth one, in the τ space, since J^t is isotropic. Thus the first sphere, in the $\ddot{q}$ space, can be directly converted into the last one, in the τ space. This implies that matrix A is isotropic. Matrix M_2 is also isotropic since the second sphere, in the $\ddot{x}$ space, can be directly converted into the fourth one, in the τ space.

The same reasoning can be applied to verify the three foregoing properties, excepting part of Property 3. By analyzing Figure 2, we cannot see that if matrices A and M_1 are isotropic, matrices J and M_2 must also be isotropic. This also follows because both matrices A and M_1 are symmetric and positive, besides being isotropic, as per Theorem 3. However, these characteristics of matrices A and M_1 were not used to construct Figure 2.

Example 3, below, shows a situation in which we have kinematic and dynamic isotropicity according to both M matrices.

Example 3

Here we have $L_1=L_2=1m$, $q_1=0$, $q_2=90°$, $L_{c1}=0.5m$, $L_{c2}=0.8m$, $m_1=4kg$, $m_2=10kg$, $I_1=3.4kgm^2$, and $I_2=8kgm^2$. Under these conditions, matrices J, A, M_1, and M_2 are given by

$$J = \begin{bmatrix} 0 & -1 \\ 1 & 0 \end{bmatrix}, \quad A = \begin{bmatrix} 14.4 & 0 \\ 0 & 14.4 \end{bmatrix}, \quad M_1 = \begin{bmatrix} 14.4 & 0 \\ 0 & 14.4 \end{bmatrix}, \quad M_2 = \begin{bmatrix} 0 & 14.4 \\ -14.4 & 0 \end{bmatrix}$$

The condition numbers of matrices J, M_1, and M_2 are all equal to 1.

Figures 3 and 4 illustrate the ellipses obtained for each of the matrices J, M_1, and M_2 for the manipulators of Examples 1 and 3. Notice the improvement in uniformity in the M matrices' ellipses in Example 3, as compared to Example 1. The scales used to draw the ellipses of Figures 3 and 4 are: a) 3:1; b) 60:1; c) 60:1.

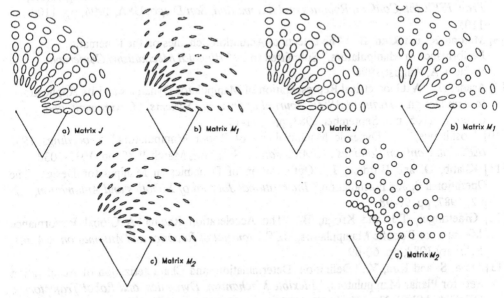

Figure 3: Manipulator of Example 1   Figure 4: Manipulator of Example 3

5 CONCLUSIONS

Inequalities between the condition numbers of the Jacobian matrix and matrices that represent dynamic characteristics of manipulators have been derived. Necessary and sufficient conditions for having both kinematic and dynamic isotropic properties for non-redundant manipulators have been introduced. We have shown that designing a manipulator for good isotropicity of velocity (acceleration) distribution may lead to an arm with poor uniformity of acceleration (velocity) distribution. These concepts have been illustrated with results obtained for a 2R planar mechanical arm.

REFERENCES

[1] Paul, R.P. and Stevenson, C.N., "Kinematics of Robot Wrists," *The International Journal of Robotics Research*, v.2, n.1, 1983, pp. 31-38.

[2] Yoshikawa, T., "Manipulability of Robotic Mechanisms," *International Journal of Robotic Research*, v.4, n.2, 1985, pp. 3-9.

[3] Klein, C.A. and Blaho, B.E., "Dexterity Measures for the Design and Control of Kinematically Redundant Manipulators," *The International Journal of Robotics Research*, v.6, n.2, 1987, pp. 72-83.

[4] Salisbury, J.K. and Craig, J.J., "Articulated Hands: Force Control and Kinematic Issues," *The International Journal of Robotics Research*, v.1, n.1, 1982, pp. 4-17.

[5] Angeles, J., Ranjbaran, F. and Patel, R.V., "On the Design of the Kinematic Structure of Seven-Axes Redundant Manipulators for Maximum Conditioning," *Proc. IEEE International Conf. on Robotics and Automation*, Nice, France, May 1992, pp. 494-499.

[6] Angeles J., "The Design of Isotropic Manipulator Architetures in the Presence of Redundancies," *The International J. of Robotics Research*, v.11, n.3, 1992, pp. 196-201.

[7] Kircanski, M.V., "Robotic Isotropy and Optimal Robot Design of Planar Manipulators," *Proc. IEEE Int. Conf. on Robotics and Automation*, San Diego, USA, 1994, pp. 1100-1105.

[8] Matone, R. and Roth, B., "The Effects of Actuation Schemes on the Kinematic Performance of Manipulators," Submitted to *24th Biennial Mechanisms Conference*, Irvine, CA, August 1996.

[9] Asada, H., "A Geometrical Representation of Manipulator Dynamics and Its Application to Arm Design," *Trans. of ASME, Journal of Dynamic Systems, Measurement, and Control*, v.105, n.3, September 1983, pp. 131-135.

[10] Yoshikawa, T., "Dynamic Manipulability of Robot Manipulators," *Procedings 1985 IEEE Int. Conf. on Robotics and Automation*, St. Louis, March 1985, pp. 1033-1038.

[11] Khatib, O. and Burdick, J., "Optimization of Dynamics in Manipulator Design: The Operational Space Formulation," *International Journal of Robotics and Automation*, v.2, n.2, 1987, pp. 90-98.

[12] Graettinger, T.J. and Krogh, B., "The Acceleration Radius: A Global Performance Measure for Robotic Manipulators," *IEEE Journal of Robotics and Automation*, v.4, n.1, February 1988, pp. 60-69.

[13] Desa, S. and Kim, Y., "Definition, Determination, and Characterization of Acceleration Sets for Planar Manipulators," *Flexible Mechanism, Dynamics, and Robot Trajectories*, DE v.24, ASME, New York, NY, 1990, pp. 207-215.

[14] Golub, G.H. and Van Loan F., *Matrix Computations*, John Hopkins U. Press, 1989.

GROUP THEORETICAL SYNTHESIS OF BINARY MANIPULATORS

G.S. Chirikjian

Johns Hopkins University, Baltimore, MD, USA

Abstract

This paper addresses a paradigm based on binary (two-state) actuation which may lead to lower cost and higher reliability for robotic manipulators. Binary manipulators constructed from pneumatic cylinders are both light weight and inexpensive, requiring minimal feedback hardware and trivial computer interfaces. However, for the benefits of binary actuation to be realized, methods developed in the pure mathematics literature over that past thirty years must be used to make the design and inverse kinematics of these manipulators tractable. As is demonstrated, the Fourier transform of functions on the Euclidean motion group is a powerful tool which can be used to this end.

1 Introduction

Mechanisms and robotic manipulators with discrete actuators can be used reliably without feedback, and thus have the potential to be much cheaper than manipulators with standard continuous actuation. In fact, the "binary" manipulator constructed from 24 pneumatic cylinders shown in Figure 1 was built together with a simple computer interface for approximately $ 1500 in parts and materials - much cheaper than if continuous actuation had been used. Each actuator is driven to its stops, and thus has two states (one bit). Each leg has two bits.

The bottleneck in the use of binary manipulators has been the planning of relatively smooth motion, and overcoming the problems associated with the combinatorially explosive number of configurations that result from numerous actuators. In fact, this is a major reason why open-loop discrete actuation is not more widely used in robotics.

In order for the benefits of binary actuation to be realized, methods developed in the pure mathematics literature over that past thirty years must be used to make the design and inverse kinematics of these manipulators tractable. This paper addresses one aspect of the problem of how to "synthesize" a binary manipulator so that the finite number of frames reachable by its end-effector are distributed in a desirable way.

If a binary manipulator is composed of a serial cascade of n actuated modules each with K states, the total number of configurations attainable is K^n. If for instance each

Figure 1: Hardware Implementation of a 24-bit Variable-Geometry-Truss

module is a Stewart platform with six legs where each leg consists of two binary actuators (resulting in 4 discrete lengths), then $K = 4^6 = 4096$, and the number K^n grows very quickly, e.g. when $n = 3$, then $K^n = 6.8 \times 10^{10}$. In Figure 1, $K = 4096$ and $n = 2$.

The problem of synthesizing a binary manipulator's workspace consisting of a large but finite number of frames is made tractable if it can be "localized" to each of the n modules in the structure. This is formulated most naturally by specifying the desired number of reachable frames per unit task space volume (i.e., the *density* of frames) for the whole manipulator instead of storing the information for each of the K^n frames individually. The next step is to determine the corresponding density of frames for each manipulator module that results in the overall workspace density of the binary manipulator.

Both the density of frames for a single module and density of the whole manipulator are positive real-valued functions on the Euclidean Motion Group, SE(N), which describes all frames and how they can be composed [5, 7]. $N = 2$ and $N = 3$ correspond to planar and spatial motion respectively. The problem of determining the density of the workspace of the whole manipulator is stated mathematically as the convolution product of the density functions of individual modules, where the convolution product is defined relative to the group operation. The inverse problem of determining what the density for each module must be for a given workspace density is solved using harmonic analysis on the group $SE(N)$. This is basically a generalization of classical Fourier analysis. A generalization of the convolution theorem allows us to solve for individual module density functions, from which point the synthesis problem can be solved (e.g., as in [1]).

In the following sections we review mathematical concepts dealing with convolution of functions and harmonic analysis on the Euclidean group. This provides the definitions and tools needed to use the concept of convolution applicable to the current problem.

2 Workspace Generation by Convolution

For discretely actuated manipulators the *density* of reachable frames in $SE(N)$ determines how accurately a random position and orientation can be reached. This density information is also extremely important in planning the motions of discretely actuated manipulator arms [3]. Density is calculated directly by dividing a compact subset of $SE(N)$ containing the workspace into finite but small volume elements. The number of frames reachable by the end of the manipulator which lie in each volume element is stored. Dividing this number by the volume element size gives the average density in each element. An efficient method for the calculation of this density histogram is given in [2]. A smooth density function can be used to approximate the shape of this density histogram as in [4]. Note that the density function is always real-valued and positive.

It is an important aspect of the manipulator design problem to specify the density of reachable frames throughout the workspace. That is, areas which must be reached with great accuracy should have high density, and those areas of the workspace which are less important need less density. For relatively few actuators, synthesis is achieved by enumerating reachable frames and using an iterative procedure as discussed in [1]. However, to compute this workspace density function using brute force and iterating is computationally intractable for large n, e.g., it requires K^n evaluations of the kinematic equations relating actuator state to the resulting end frame for a manipulator with n actuated modules each with K states.

If instead we imagine that the manipulator is divided into two connected parts, then a density function $k(g)$ can be associated with those frames reachable by the end of the lower half of the manipulator, and a density function $f(g)$ can be associated with the end of the upper half of the manipulator. The variable g represents an arbitrary element of $SE(N)$. Rigid body motions are composed as $g_1 \circ g_2$, or by the multiplication of the corresponding homogeneous transforms as $H_1 H_2$. The function $k(g)$ is defined relative to the base frame, and $f(g)$ treats the frame at the end of the lower segment as the base frame. That is, $k(g) = f(g)$ when the manipulator is cut into two equal parts and there are an even number of identical modules. However, k and f will not be the same function in more general scenarios. By adjusting kinematic parameters such that actuator strokes

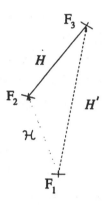

Figure 2: Concatenation of homogeneous transformations

are limited or extended, the set of reachable frames (and thus the density) is altered. This is achieved mechanically by simply inserting or removing rigid stoppers that specify the physical actuator length corresponding to the discrete states.

While it may not be possible to calculate K^n frames to compute the density function of the workspace, it is often feasible to compute $K^{n/2}$ frames for each of the two segments. For example, in Figure 1 the density can be generated using brute force since $(4096)^2 \approx 1.67 \times 10^7$. If however four modules are concatenated instead of two, then $(4096)^4 \approx 2.81 \times 10^{14}$ frames cannot be enumerated using brute force. Instead the density of the whole workspace can be generated by the convolution of the density functions corresponding to the two halves of the manipulator:

$$\int_{SE(N)} k(g_1) f(g_1^{-1} \circ g_2) d\mu(g_1) = h(g_2). \qquad (1)$$

The geometry of why this is so follows below. Suppose there are two frames of reference - the base frame and one attached to the middle of the manipulator. Quantities described in the the base frame, $\mathcal{F}_1$, are denoted with a prime, ', while quantities described in the one in the middle of the manipulator, $\mathcal{F}_2$, are denoted without a prime. Let the homogeneous transform matrix H describe the position and orientation of a third frame, $\mathcal{F}_3$, with respect to $\mathcal{F}_2$, and let $\mathcal{H}$ describe the position and orientation of $\mathcal{F}_2$ with respect to $\mathcal{F}_1$. Then the position and orientation of $\mathcal{F}_3$ with respect to $\mathcal{F}_1$ is

$$H' = \mathcal{H}H,$$

as illustrated in Figure 2.

If a density function $f(H)$ is defined in the moving frame fixed to the middle of the manipulator, then the same physical quantity can be represented in the fixed base frame as $f'(H')$ where

$$f'(H') = f(H) \quad \text{or} \quad f'(H') = f(\mathcal{H}^{-1}H').$$

Thus, the natural way to think of the workspace density of the whole manipulator is as the convolution of the functions $k(\cdot)$ and $f(\cdot)$ on the Euclidean group. That is, the density of the distal half of the manipulator $(f'(H) = f(\mathcal{H}^{-1}H))$ is weighted by the density of the proximal half $(k(\mathcal{H}))$ and integrated over all values of $\mathcal{H}$:

$$(k * f)(H) = \int_{SE(N)} k(\mathcal{H}) f(\mathcal{H}^{-1}H) d\mu(\mathcal{H}).$$

This is exactly the same as Equation 1, except that homogeneous transform notation is used and so the group operation o is simply matrix multiplication. This calculation does not depend on the number of actuated modules in the manipulator, and allows us to compute the workspace density much more efficiently than brute force. We can further subdivide the manipulator into smaller parts and perform multiple convolutions for even faster workspace density function generation.

Now, there are two inverse problems of interest that naturally arise:

(1) Suppose that the proximal half of the manipulator has been designed ($k(g)$ in Equation 1 is specified), and the problem is to design the distal half of the manipulator (find $f(g)$) so that the density function for the workspace of the whole manipulator comes as close as possible to a desired density function $h(g)$. One must then solve the inverse problem

$$(k * f)(g) = h(g). \tag{2}$$

(2) Suppose that we are free to design each module, but that the structure is a concatenation of n identical modules. In this case, the problem is to solve for $f(g)$ such that

$$\underbrace{(f * \ldots * f)}_{n \; times}(g) = h(g). \tag{3}$$

The remainder of this paper is devoted to the task of elegantly solving the two above stated inverse problems which are very important for binary manipulator design.

3   The Fourier Transform on $SE(N)$

It is very natural to try to transform the convolution equations (2) and (3) into algebraic equations in a different domain in much the same way that classical Fourier analysis is used in other engineering problems.

The Fourier transform of a function $f(g)$ for $g \in SE(N)$, and the inverse transform are defined as [6]:

$$\mathcal{F}(f) = \hat{f}(a) = \int_{SE(N)} f(g) U(g^{-1}, a) d\mu(g)$$

and

$$f(g) = \mathcal{F}^{-1}(\hat{f}) = \int_0^\infty \mathrm{trace}(\hat{f}(a) U(g, a)) a \, da.$$

The linear operator $U(g, a)$ for $a \in \mathbb{R}^+$ is called a *representation* of the group $SE(N)$, and has the property that

$$U(g_1 \circ g_2, a) = U(g_1, a)U(g_2, a) \tag{4}$$

for all $g_1, g_2 \in SE(N)$. This plays an analogous role as the exponential function for the Fourier transform on the real line.

This generalized Fourier transform is a powerful tool for binary manipulator workspace generation because

$$\mathcal{F}(f_1 * f_2) = \mathcal{F}(f_2)\mathcal{F}(f_1),$$

where the convolution is defined as in Equation 1. This convolution theorem results directly from the homomorphism property of $U(g, a)$ given in Equation 4.

For the case of $SE(2)$, $U(g, a)$ is expressed as an infinite dimensional matrix with elements

$$u_{mn}(g(r, \phi, \theta), a) = i^{n-m} e^{-i[n\theta + (m-n)\phi]} J_{n-m}(ar) \quad m, n \in \mathbf{Z} \tag{5}$$

where the group elements are expressed in polar coordinates:

$$g = g(r, \phi, \theta) = \begin{pmatrix} \cos\theta & -\sin\theta & r\cos\phi \\ \sin\theta & \cos\theta & r\sin\phi \\ 0 & 0 & 1 \end{pmatrix}$$

and $J_\nu(x)$ is the νth order Bessel function [6].

4 Applications to Workspace Synthesis

Consider the planar case when we are given a desired workspace density function

$$h(g) = H_1(r)(C_h + 2A_h \cos\alpha) + 2B_h H_2(r) \cos\phi \tag{6}$$

where $A_h, B_h, C_h \geq 0$ and $C_h^2 \geq 4(A_h^2 + B_h^2)$ are given real numbers which ensure that $h(g) > 0$ for all $g = g(r, \phi, \theta) \in SE(2)$. A density function of this form allows us the flexibility to design workspaces where the density is highest when the end-effector orientation angle is closest to zero, and tapers to a lower value for high orientation angle. It also allows us to consider workspaces which vary with the angle ϕ in the plane.

Due to orthogonality, the only nonzero part of the Fourier transform matrix is the block

$$\hat{h}^{3\times3}(a) = \int_{SE(2)} h(g) U^{3\times3}(g^{-1}, a) d\mu(g) = \begin{pmatrix} \hat{h}_{-1,-1} & \hat{h}_{-1,0} & \hat{h}_{-1,1} \\ \hat{h}_{0,-1} & \hat{h}_{0,0} & \hat{h}_{0,1} \\ \hat{h}_{1,-1} & \hat{h}_{1,0} & \hat{h}_{1,1} \end{pmatrix} = \begin{pmatrix} \gamma_1(a) & 0 & 0 \\ \beta_1(a) & \kappa_1(a) & \beta_1(a) \\ 0 & 0 & \gamma_1(a) \end{pmatrix} \tag{7}$$

where $\gamma_1(a) = A_h \int_0^\infty H_1(r) J_0(ar) r \, dr \equiv A_h \epsilon_1(a)$, $\kappa_1(a) = C_h \int_0^\infty H_1(r) J_0(ar) r \, dr \equiv C_h \epsilon_1(a)$, and $\beta_1(a) = -i B_h \int_0^\infty H_2(r) J_1(ar) r \, dr \equiv -i B_h \epsilon_2(a)$.

4.1 Solving $(f * \ldots * f)(g) = h(g)$

In this subsection we seek to find $f(g)$ such that $(f * \ldots * f)(g) = h(g)$ for given $h(g)$.

This is solved by taking the Fourier transform of both sides, and observing that fractional powers of the matrix in Equation 7 are of the form

$$\hat{f}^{3\times3}(a) = \left(\hat{h}^{3\times3}(a)\right)^{\frac{1}{2^m}} = \begin{pmatrix} \gamma_1^{1/2^m}(a) & 0 & 0 \\ \dfrac{\beta_1(a)}{\epsilon_1^{(1-\frac{1}{2^m})}(a)\prod_{j=1}^m(A^{1/2^j}+C^{1/2^j})} & \kappa_1^{1/2^m}(a) & \dfrac{\beta_1(a)}{\epsilon_1^{(1-\frac{1}{2^m})}(a)\prod_{j=1}^m(A^{1/2^j}+C^{1/2^j})} \\ 0 & 0 & \gamma_1^{1/2^m}(a) \end{pmatrix}.$$

Thus, when there are $n = 2^m$ identical modules, each module must have a density

$$f(g) = \int_0^\infty \mathrm{tr}\left(\left[\hat{h}^{3\times3}(a)\right]^{\frac{1}{2^m}} U_a(g)\right) a\, da =$$

$$(2A_h^{\frac{1}{2^m}}\cos\alpha + C_h^{\frac{1}{2^m}})\int_0^\infty J_0(ar)\epsilon_1^{\frac{1}{2^m}}(a)a\,da + \frac{2B_h\cos\phi}{\prod_{j=1}^m(A_h^{1/2^j}+C_h^{1/2^j})}\int_0^\infty J_1(ar)\frac{\epsilon_2(a)}{\epsilon_1^{(1-\frac{1}{2^m})}(a)}a\,da.$$

By determining the module density function $f(g)$ for given workspace density $h(g)$, the relatively few states of a single module can be altered using techniques presented in [1], and the combinatorially explosive design problem of treating the manipulator as a single unit is avoided.

4.2 Solving $(k * f)(g) = h(g)$

Taking the Fourier transform of both sides of Equation 2, the original problem becomes

$$\hat{f}(a)\hat{k}(a) = \hat{h}(a) \quad \text{or} \quad \hat{f}(a) = \hat{h}(a)[\hat{k}(a)]^{-1}.$$

The desired function is then recovered using the inverse transform as:

$$f(g) = \mathcal{F}^{-1}\left(\hat{h}(a)[\hat{k}(a)]^{-1}\right).$$

To make the discussion concrete consider the following functions:

$$h(g) = H_1(r)(C_h + 2A_h\cos\theta) + 2B_hH_2(r)\cos\phi$$
$$k(g) = K_1(r)(C_k + 2A_k\cos\theta) + 2B_kK_2(r)\cos\phi, \tag{8}$$

The Fourier transform matrices corresponding to these functions have the same form as in Equation 7. Defining γ_2, κ_2, and β_2 for $\hat{k}^{3\times3}$ in the same way that γ_1, κ_1, and β_1 were defined for $\hat{h}^{3\times3}$, the inverse of $\hat{k}^{3\times3}$ is easily calculated and matrix multiplication yields:

$$\hat{f}^{3\times3} = \hat{h}^{3\times3}\left[\hat{k}^{3\times3}\right]^{-1} = \begin{pmatrix} \gamma_1/\gamma_2 & 0 & 0 \\ \beta_1/\gamma_2 - (\beta_2\kappa_1)/(\gamma_2\kappa_2) & \kappa_1/\kappa_2 & \beta_1/\gamma_2 - (\beta_2\kappa_1)/(\gamma_2\kappa_2) \\ 0 & 0 & \gamma_1/\gamma_2 \end{pmatrix}.$$

From the definition of the matrix elements of the unitary representation of $SE(2)$ given in Equation 5, we get

$$\text{tr}(\hat{f}^{3\times3}U_a^{3\times3}) = 2\cos\theta J_0(ar)(\gamma_1/\gamma_2)+J_0(ar)(\kappa_1/\kappa_2)+2i\cos\phi J_1(ar)(\beta_1/\gamma_2-(\beta_2\kappa_1)/(\gamma_2\kappa_2))$$

and so

$$f(g) = 2F_1(r)\cos\theta + F_2(r) + 2F_3(r)\cos\phi,$$

where $F_1(r) = \int_0^\infty (\gamma_1(a)/\gamma_2(a))J_0(ar)a\,da$, $F_2(r) = \int_0^\infty (\kappa_1(a)/\kappa_2(a))J_0(ar)a\,da$, and $F_3(r) = i\int_0^\infty (\beta_1(a)/\gamma_2(a) - (\beta_2(a)\kappa_1(a))/(\gamma_2(a)\kappa_2(a)))J_1(ar)a\,da$. Since $\beta_1(a)$ and $\beta_2(a)$ are pure imaginary, $F_3(r)$ is real (as are $F_1(r)$ and $F_2(r)$).

5   Conclusion

In this work, it is shown how the workspace density of discretely actuated manipulators are prescribed using the concepts of the convolution product and Fourier transform of scalar functions on the Euclidean Motion Group. While motivational examples were presented which yield closed form solutions, much work remains in developing general numerical codes.

Acknowledgements

This work was supported by a Presidential Faculty Fellow Award from the National Science Foundation. Thanks go to Mr. Joshua Houck for constructing the manipulator shown in Figure 1, and to Ms. Imme Ebert-Uphoff for her input in Section 2.

References

[1] Chirikjian, G.S., "Kinematic Synthesis of Mechanisms and Robotic Manipulators with Binary Actuators," ASME Journal of Mechanical Design, Vol. 117, No. 4., pp 573-580, Dec. 1995.

[2] Ebert-Uphoff, I., Chirikjian, G.S., "Efficient Workspace Generation for Binary Manipulators with Many Actuators," Journal of Robotic Systems, June, 1995, pp 383-400.

[3] Ebert-Uphoff, I., Chirikjian, G.S., "Inverse Kinematics of Discretely Actuated Hyper-Redundant Manipulators Using Workspace Densities," Proc. IEEE Int. Conf. on Robotics and Automation, April 1996.

[4] Ebert-Uphoff, I., Chirikjian, G.S., "Discretely Actuated Manipulator Workspace Generation by Closed-Form Convolution," Proc. ASME Mechanisms Conference, August 1996.

[5] Murray, R. M., Li, Z., Sastry, S.S., A Mathematical Introduction to Robotic Manipulation, CRC Press, Ann Arbor MI, 1994.

[6] Orihara, A., "Bessel Functions and the Euclidean Motion Group," Tohoku Mathematical Journal, Vol. 13, 1961, pp. 66-71.

[7] Park, F.C., Brockett, R.W., "Kinematic Dexterity of Robotic Mechanisms," The International Journal of Robotics Research, Vol. 13, No. 1, February 1994, pp 1-15.

DISTRIBUTED KINEMATIC DESIGN
FROM TASK SPECIFICATION

S. Regnier, F.B. Ouezdou and P. Bidaud
University of Paris 6, Vélizy, France

Abstract : *This paper addresses the problem of designing a robotic mechanism or manipulator such that its end effector frame comes closest to reaching a set of desired goal frames. We formulate this non-linear problem as a distributed optimization problem, in which the kinematic parameters are computed to minimize the distance between the end-effector frame and each goal frame. A local objective function is defined with a Frobenius norm of the difference between current position and orientation and the goal frame. This norm must be expressed in a local frame in order to get the analytical and **generic** form of the dimensional or joint parameter which allows the minimization of the local objective function (i.e. the joint does one's best). The main contribution of this paper is a distributed method for the kinematic design from task specification of all kinds of manipulators. The proposed methodology is illustrated with the synthesis of planar mechanisms and a 6-degrees-or-freedom manipulator.*

1 Introduction

Analytical method for kinematic design have been the object of lot of studies in conventional mechanism design and more recently, in robotics. Closed form precision point synthesis techniques guarantee that all the synthesized linkage reach exactly a given set of points. Most of the exact methods for linkage synthesis have their roots in Burmeister's work [6], which involved graphical analysis of finite displacement of rigid-body planar linkages. Analytical methods have a limited applicability. Their major drawbacks are that they provide a solution only for an exactly constrained problem. For underconstrained problems, a family of solutions can be obtained and the remaining problem is then the choice of an optimal solution. For overconstrained problem, no analytical solution exists. So, only a limited number of points can be prescribed [3] or only few classes of manipulators can be considered [4]. Although the methods differ widely in their goal and approach, researchers share the common feature of eventually reducing the synthesis problem to an optimization problem. For mechanisms, many methods have been proposed allowing to describe a desired curve [11] [5]. Although many published studies are available in the areas of mechanism design, references to robot manipulator are limited. However, an explicit expression for the gradient function with respect to the kinematic parameters [8] or a global optimization with simulated annealing [7] are used to solve the problem. But, no single optimization method able to solve synthesis of mechanisms **and** manipulators exists [2].

Previous researches have shown the attractive of multi-agent systems or distributed methods to solve inverse kinematics of manipulators with or without closed loops [10] [9]. The aim of this paper is to generalize this distributed method of resolution to the kinematic design of any kinds of manipulators. We formulate this non-linear problem as a distributed optimization problem, in which the kinematic parameters are computed to minimize the local distance between the end-effector frame and each goal frame. Due to the assumption that at each step, the parameters of only one link (i.e. agent) can be modified, a new formulation of the initial problem is proposed for each link. An iterative procedure leads step by step to compute the parameters of each link and to obtain a mechanism which is a solution of the problem. Constraints can also be prescribed on the kinematic parameters (i.e. the joint values must remain within the specified limits, the manipulator should not pass through any of the specified obstacles).

The first part of the paper is devoted to the problem statement. After, we focus on the kinematic design of manipulators from task specification. The third part concerns the same approach applied to the synthesis of mechanisms. The synthesis of four-bar mechanisms and a 6-degrees-of-freedom manipulator are examples of the implementation.

2   Problem Statement

To facilitate the development of our distributed approach, we define kinematic, dimensional or joint, spaces.

2.1   Definition of the kinematic spaces

We use classical Denavit-Hartenberg notations [1] to describe the structure of the manipulator. The 4×4 transformation matrix relating the jth coordinate system to the $(j-1)^{th}$ coordinate system is noted $T_{j-1}^{j} = \left(\begin{array}{c|c} R_{j-1}^{j} & U_{j-1}^{j} \\ \hline 0 & 1 \end{array} \right)$.

Three parameters called **dimensional parameters** are used to describe each link and the last one is the **joint parameter**. We define the DH-configuration space as a $3n$ dimensional space with the $3n$ DH-parameters (3 dimensional parameters for each of the n links). A manipulator with n joints can be represented by a vector P_{dh}:

$$P_{dh} = [\alpha_1, a_1, \bar{q}_1, \cdots, \alpha_{id}, a_{id}, \bar{q}_{id}, \cdots, \alpha_n, a_n, \bar{q}_n]^T \qquad (1)$$

$\bar{q}_{id}$ is either θ_{id} or d_{id}. A general form of $\bar{q}_{id}$ is given by $\bar{q}_{id} = (1 - \sigma_{id})\theta_{id} + \sigma_{id}d_{id}$ with $\sigma_{id} = 1$ for revolute joint and $\sigma_{id} = 0$ for prismatic.

The joint parameters define the posture of the manipulator. The vector q of n joint variables can be viewed as a point in the n-dimensional joint space:

$$q = [q_1, \cdots, q_{id}, \cdots, q_n]^T \tag{2}$$

with $q_{id} = \sigma_{id}\theta_{id} + (1 - \sigma_{id})d_{id}$.

A point p in the 6 dimensional task space corresponds to a physical point and direction in the reference frame of the manipulator:

$$p = [X, Y, Z, \vartheta, \varphi, \psi]^T \tag{3}$$

where $[X, Y, Z]^T$ is the cartesian position vector of the end effector with respect to the base frame and $R(\vartheta, \varphi, \psi)$ is the orientation matrix (for example, Euler's angles).

So, the problem of kinematic design is a mapping of kinematic task specifications into a kinematic manipulator configuration.

2.2 Distance Metrics

Assuming an inertial reference frame and length scale for physical space have been chosen, each frame can be assigned to an element of the special Euclidean group $SE(3)$. The problem of precisely "closeness" between frames then reduces to the equivalent mathematical problem of defining a distance metrics in $SE(3)$. Any number of arbitrary distance metrics can be defined [8] but certain features make the metrics more physically meaningful. Since any distance metrics combines position and orientation, one would like the metric to be scale-invariant. So, in this paper, we base the measure on the Frobenius norm of a matrix $\|M\|$ with a length scale L defined by Wampler [12] such as:

$$T_1 = \left(\begin{array}{c|c} R_1 & U_1 \\ \hline 0 & 1 \end{array}\right) \qquad T_2 = \left(\begin{array}{c|c} R_2 & U_2 \\ \hline 0 & 1 \end{array}\right)$$

$$d(T_1, T_2) = d((R_1, U_1), (R_2, U_2)) \quad = \quad \|R_1 - R_2\|^2 + \frac{1}{L^2}\|U_1 - U_2\|^2 \tag{4}$$

$$d(T_1, T_2) = \|R\|^2 + \frac{1}{L^2}\|U\|^2 \quad = \quad \sum_{ij} R_{ij}^2 + \frac{1}{L^2}\sum_{ij} U_{ij}^2$$

with $L = \max_q \|U_0^n\|$ which is the maximum Euclidean norm of absolute end effector position.

3 Serial Manipulators

3.1 Distributed Formulation

The problem of the kinematic design from task specification leads to calculate the joint values and the dimensional parameters such that the structure equation is satisfied:

$$T_0^1 T_1^2 \cdots T_{n-1}^n = T^h \tag{5}$$

where T^h is the task matrix. If the manipulator has to reach k different tasks, the k following systems of equations has to be satisfied:

$$\overset{j}{T}\overset{1}{_0}\,\overset{j}{T}\overset{2}{_1}\cdots\overset{j}{T}\overset{n}{_{n-1}} = \overset{j}{T}\overset{h}{}\quad \forall j = 1 \cdots k \tag{6}$$

So, the problem leads to the definition of the intersection of k kinematic spaces which can be stated as solving problem of $12*k$ non-linear equations with $(3+k)*n$ unknowns.

If only one link (its dimensional or joint parameters) is modified at each step, the previous equation can be projected in the local frame $\mathcal{R}_{id-1}$ of the joint id:

$$\overset{j}{T}\overset{id}{_{id-1}}\,\overset{i}{T}\overset{n}{_{id}} = \overset{j}{T}\overset{h}{_{id-1}}\quad \forall j \in [1 \cdots k] \tag{7}$$

With this distributed formulation, the values of the joint or dimensional parameters which decrease the distance between local position and goal position could be computed with an algebraic form.

3.2 Determination of the dimensional and joint parameters

Using the previous distance metrics, a local objective function is defined:

$$\| \overset{j}{M}_{id} \|^2 = \| \overset{j}{T}\overset{id}{_{id-1}}\,\overset{j}{T}\overset{n}{_{id}} - \overset{j}{T}\overset{h}{_{id-1}} \|^2 \quad \forall j \in [1 \cdots k] \tag{8}$$

In order to solve the problem, the k different Frobenius norm (positive) have to be null so their sum have to be null. We obtain:

$$S^k_{id} = \sum_{j=1}^{k} \left\| \overset{j}{T}\overset{id}{_{id-1}}\,\overset{j}{T}\overset{n}{_{id}} - \overset{j}{T}\overset{h}{_{id-1}} \right\|^2 \tag{9}$$

This general objective function can be written (from the expansion of $\| \overset{j}{M}_{id} \|^2$) with an expanding form of α_{id} and a_{id} as:

$$S^k_{id} = \sum_{j=1}^{k} \left(\left(DEN^j_\alpha * sin(\alpha_{id}) - NUM^j_\alpha * cos(\alpha_{id}) \right) + \left(a^j_{id} - a^j_m \right)^2 + \Delta^j_\alpha \right) \tag{10}$$

The parameters DEN^j_α, NUM^j_α, Δ^j_α et a^j_m are given in the annex. The study of S^k_{id} allows to find dimensional parameters minimizing this norm:

$$\frac{\partial S^k_{id}}{\partial \alpha_{id}} = \sum_{j=1}^{k} \left(-DEN^j_\alpha * cos(\alpha_{id}) + NUM^j_\alpha * sin(\alpha_{id}) \right) \Longrightarrow \alpha^m_{id} = \arctan \frac{\sum_{j=1}^{k} NUM^j_\alpha}{\sum_{j=1}^{k} DEN^j_\alpha}$$

$$\frac{\partial S^k_{id}}{\partial a_{id}} = a_{id} - a_m \Longrightarrow a_m = \frac{1}{k}\sum_{j=1}^{k} a^j_m \tag{11}$$

We can see that a_m is given by the average value between each task solution. As it was said, the parameter $\bar{q}_{id}$ is a dimensional parameter. So, by using an expansion with $\bar{q}_{id}$, we can find:

$$S^k_{id} = 2\sum_{j=1}^{k} \left((1 - \sigma_{id})(DEN^j_\theta * sin(\theta_{id}) + NUM^j_\theta * cos(\theta_{id})) + \sigma_{id}(d_{id} - d^m_j)^2 + \Delta^j_\theta \right) \tag{12}$$

$$\bar{q}_{id}^m = (1 - \sigma_{id})\theta_{id}^m + \sigma_{id}d_{id}^m \tag{13}$$

$$\theta_{id}^m = \arctan \frac{\sum_{j=1}^k NUM_\theta^j}{\sum_{j=1}^k DEN_\theta^j} \qquad d_{id}^m = \frac{1}{k}\sum_{j=1}^k d_{id}^j \tag{14}$$

with NUM_θ^j, DEN_θ^j, d_m^j and Δ_θ given in the annex.

If we consider the joint parameters, they are different from one task to another. We can notice that $||M_{id}||^2$ can be written:

$$||M_{id}||^2 = 2\sigma_{id}(DEN_\theta^j * sin(\theta_{id}) + NUM_\theta^j * cos(\theta_{id})) + (1 - \sigma_{id})(d_{id} - d_m^j)^2 + \Delta_\theta^j \qquad \forall j \in [1 \cdots k]$$

So, the joint values which minimize $||M_{id}^j||^2$ for all $j = 1 \cdots k$ are as following:

$$\theta_{id}^j = \arctan \frac{NUM_\theta^j}{DEN_\theta^j} \qquad d_{id}^j = d_m^j \qquad \forall j \in [1 \cdots k] \tag{15}$$

Each joint values q_{id}^j is computed for each task ($j = 1..k$). An iterative procedure for each link of the manipulator leads to compute the parameters and to obtain a mechanism which is solution of the initial problem.

3.3 Results

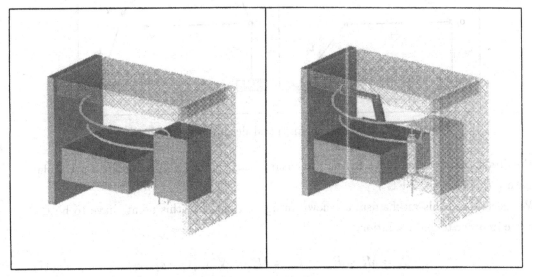

Figure 1: Prescribed path and final manipulator

In this section, it will be shown how the distributed methodology can be used to solve a kinematic design of serial manipulators. The manipulator has to reach 200 points/orientations situated between 6 obstacles (on left figure). A 6-DOF manipulator has 18 parameters so the problem has $200 * 6 + 18$ unknowns. The solution presented here (on right figure)was completed in 67 minutes of CPU times on a SPARC 5. The initial guess is a manipulator where all dimensional values are null.

4 Mechanisms

This distributed method could be applied·to manipulators with closed loops (for example, mechanisms). A closed loop is considered as two serial manipulators and the structure equation can be written for a joint id:

$$R^{id}_{id-1} U^n_{id} = U^h_{id-1} \tag{16}$$

with id the chosen joint, n the end-effector of the sub-manipulator and h the sub-task. This new problem is the same problem that the previous because we can write for each sub-equation the same distributed method. To explain this idea, we take a simple example with a planar mechanism. Dimensional synthesis of this four-bar is the determination of

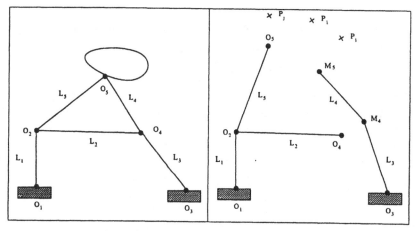

Figure 2: Four-bar mechanism and decomposed mechanism

the dimensional parameters ($L_1..L_5$) to ensure that a point (O_5) generates a particular path (P_j) (on fig. 2 left)

We decompose this mechanism as shown in fig. 2 right. So, this points have to be the same in order to get a solution:

$$O^j_5 = M^j_5 = P^j \qquad O^j_4 = M^j_4 \qquad \forall j \in [1 \cdots k] \tag{17}$$

For the joint 1 of the sub-manipulator 1, the structure equation of the closed loops can written as:

$$R^2_1 U^5_2 = U^4_1 \tag{18}$$

It's clear that this equation looks like the previous formulation established for the synthesis of serial manipulator. So, the same distributed method can be used for the synthesis of mechanisms.

The following examples refer to the synthesis of two four-bar mechanisms. In the first case, the path points to be generated are on three straight lines and after we can see the solution. In the second case, the mechanism has to reach an ellipse. The solutions determined were completed in 5 and 10 minutes of CPU times on a SPARC 5. The initial guess is two manipulators where all dimensional values are equal to 1.0.

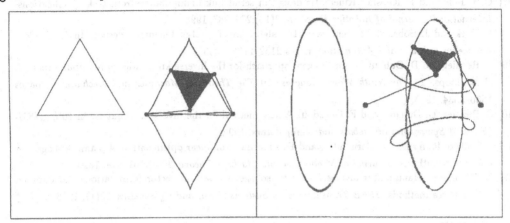

Figure 3: Prescribed path and final mechanism with its couple curve

5   Conclusion

We have presented a general formulation of the problem of kinematic design from task specification. An objective function is defined in terms of a local distance metrics parameterized by length scale between the current state and the goal state. Due to the assumption that only one link can be modified at each step, an algebraic and generic formulation of the local objective function is computed. This distributed method is employed to find 6-degrees-of-freedom manipulator and planar mechanism.

The goal of this work is to find and implement a new optimization method for non-linear problem. This distributed solving method framework involves problem decomposition, distributed formulation and multi-objectives approach. So, the problem of kinematic design can be augmented with performance objectives, such as maximizing the robot's workspace or manipulability.

References

[1] J. Denavit and R. Hartenberg. A kinematic notation for lower pair mechanisms based on matrices. *Journal Of Applied Mechanics*, 22, 1955.

[2] A. Erdman. *Modern Kinematics - Developments in The Last Forty Years*. Wiley Series in Design Engineering, 1991.

[3] S. Kota and S. Chiou. Use of orthogonal arrays in mechanism synthesis. *Journal of Mechanism and Machine Theory*, 28(6):777–794, 1993.

[4] Q. Liao and C. Liang. Synthesizing spatial 7r mechanism with 16-assembly configurations. *Journal of Mechanism and Machine Theory*, 5(28):715–720, 1993.

[5] Z. Liu and J. Angeles. A least-square method for the optimization of spherical four-bar linkages for rigid body guidance. In *The Eight World Congress on The Theory of Machines and Mechanisms*, pages 75–79, 1991.

[6] K. Luck. Computer-aided mechanism synthesis based on burmeister theory. *Journal of Mechanism and Machine Theory*, 29(6):877–886, 1994.

[7] C. Paredis and P. Khosla. Kinematic design of serial link manipulators from task specifications. *International Journal of Robotics Research*, 3(12):274–287, 1993.

[8] F. Park and J. Bobrow. Efficient geometric algorithms for robot kinematic design. In *IEEE Conference on Robotics and Automation*, pages 2132–2136, 1995.

[9] S. Régnier and D. Duhaut. A multi-agent approach for the inverse kinematics of manipulators with closed loops. In *The Ninth World Congress on The Theory of Machines and Mechanisms*, pages 1630–1634, 1995.

[10] S. Régnier, D. Duhaut, and F. Ouezdou. A new method for the inverse kinematics. In $10^t h$ *CISM-IFToMM Symposium on Robots and Manipulators*, 1994.

[11] J. Vallejo, R. Aviles, A. Hernandez, and E. Amezua. Nonlinear optimization of planar linkages for kinematic syntheses. *Journal of Mechanism and Machine Theory*, 30(4):501–518, 1995.

[12] C. Wampler. Manipulator inverse kinematic solutions based on vector formulations and damped least-squares methods. *IEEE Transactions on Systems, Man, and Cybernetics*, 16(1), 1986.

6   Annex

$$
\Delta^j_\alpha = \sin(\theta)\left(-2, T^n_{1,3}T^h_{2,3} - 2T^n_{1,2}T^h_{2,2} - 2T^n_{1,1}T^h_{2,1} - 2T^n_{1,4}T^h_{2,4}\right)
$$
$$
+ \cos(\theta)\left(-2T^n_{1,4}T^h_{1,4} - 2T^n_{1,1}T^h_{1,1} - 2T^n_{1,2}T^h_{1,2} - 2T^n_{1,3}T^h_{1,3}\right)
$$
$$
+ \sum_{i,j}^4 T^n_{i,j} + \sum_{i,j}^4 T^h_{i,j} + a^2 + d^2 - 2dT^h_{3,4} \tag{19}
$$

$$
DEN^j_\alpha = -2T^n_{2,4}T^h_{3,4} + 2\cos(\theta)T^n_{3,2}T^h_{2,2} + 2\cos(\theta)T^n_{3,1}T^h_{2,1} - 2\sin(\theta)T^n_{3,4}T^h_{1,4}
$$
$$
+ 2\cos(\theta)T^n_{3,3}T^h_{2,3} - 2\sin(\theta)T^n_{3,3}T^h_{1,3} - 2T^n_{2,1}T^h_{3,1} - 2\sin(\theta)T^n_{3,2}T^h_{1,2}
$$
$$
- 2T^n_{2,2}T^h_{3,2} - 2\sin(\theta)T^n_{3,1}T^h_{1,1} - 2T^n_{2,3}T^h_{3,3} + 2\cos(\theta)T^n_{3,4}T^h_{2,4} + 2T^n_{2,4}d
$$

$$
NUM^j_\alpha = -2T^n_{3,1}T^h_{3,1} - 2T^n_{3,4}T^h_{3,4} + 2\sin(\theta)T^n_{2,2}T^h_{1,2} - 2T^n_{3,2}T^h_{3,2} - 2T^n_{3,3}T^h_{3,3}
$$
$$
+ 2T^n_{3,4}d + 2\sin(\theta)T^n_{2,1}T^h_{1,1} - 2\cos(\theta)T^n_{2,2}T^h_{2,2} - 2\cos(\theta)T^n_{2,4}T^h_{2,4}
$$
$$
+ 2\sin(\theta)T^n_{2,4}T^h_{1,4} - 2\cos(\theta)T^n_{2,3}T^h_{2,3} + 2\sin(\theta)T^n_{2,3}T^h_{1,3} - 2\cos(\theta)T^n_{2,1}T^h_{2,1}
$$

$$
a_m = \left(2T^n_{1,4} - 2\cos(\theta)T^h_{1,4} - 2\sin(\theta)T^h_{2,4}\right)a \tag{20}
$$

$$
NUM_\theta = -2T^{sup}_{1,1}T^{dr}_{1,1} - 2T^{sup}_{1,4}T^{dr}_{1,4} - 2T^{sup}_{1,3}T^{dr}_{1,3} - 2T^{sup}_{1,2}T^{dr}_{1,2}
$$
$$
- 2aT^{dr}_{1,4} + 2\sin(\alpha)T^{sup}_{3,3}T^{dr}_{2,3} - 2\cos(\alpha)T^{sup}_{2,4}T^{dr}_{2,4} + 2\sin(\alpha)T^{sup}_{3,2}T^{dr}_{2,2}
$$
$$
+ 2\sin(\alpha)T^{sup}_{3,4}T^{dr}_{2,4} + 2\sin(\alpha)T^{sup}_{3,1}T^{dr}_{2,1} - 2\cos(\alpha)T^{sup}_{2,2}T^{dr}_{2,2}
$$
$$
- 2\cos(\alpha)T^{sup}_{2,1}T^{dr}_{2,1} - 2\cos(\alpha)T^{sup}_{2,3}T^{dr}_{2,3} \tag{21}
$$

$$
DEN_\theta = -2aT^{dr}_{2,4} - 2\sin(\alpha)T^{sup}_{3,4}T^{dr}_{1,4} + 2\cos(\alpha)T^{sup}_{2,4}T^{dr}_{1,4} - 2T^{sup}_{1,3}T^{dr}_{2,3}
$$
$$
- 2T^{sup}_{1,4}T^{dr}_{2,4} - 2T^{sup}_{1,1}T^{dr}_{2,1} + 2\cos(\alpha)T^{sup}_{2,3}T^{dr}_{1,3} - 2\sin(\alpha)T^{sup}_{3,3}T^{dr}_{1,3}
$$
$$
- 2T^{sup}_{1,2}T^{dr}_{2,2} - 2\sin(\alpha)T^{sup}_{3,1}T^{dr}_{1,1} + 2\cos(\alpha)T^{sup}_{2,2}T^{dr}_{1,2}
$$
$$
+ 2\cos(\alpha)T^{sup}_{2,1}T^{dr}_{1,1} - 2\sin(\alpha)T^{sup}_{3,2}T^{dr}_{1,2} \tag{22}
$$

$$
\Delta_\theta = T^{sup}_{2,1} + T^{sup}_{3,1} + T^{sup}_{2,2} + T^{sup}_{3,2} + T^{sup}_{2,3} + T^{sup}_{3,3} + T^{sup}_{2,4}
$$
$$
+ T^{sup}_{1,3} + a^2 + T^{sup}_{1,4} + T^{sup}_{1,2} + T^{sup}_{1,1} + 2T^{sup}_{1,4}a \tag{23}
$$

$$
d_m = -\sin(\alpha)T^{sup}_{2,4} - \cos(\alpha)T^{sup}_{3,4} + T^{dr}_{3,4} \tag{24}
$$

OPTIMAL DESIGN OF A SHAPE MEMORY ALLOY ACTUATOR FOR MICROGRIPPERS

N. Troisfontaine and Ph. Bidaud
Laboratoire de Robotique de Paris, Vélizy, France

Abstract : *In this paper, a modular microactuator based on shape memory alloy (SMA) actuation to drive the joints of a microgripper is described. Its design is a combination of two SMA actuators installed on the links of a special mechanism : one of them drives the rotation of the joint and the other one produces the necessary force to manipulate objects. An accurate model of the thermo-mechanical behavior of SMA fibers is proposed for the design of SMA actuators. The main performances of this microactuator are presented and discussed.*

1   Introduction

Today, the design of dextrous grippers for microrobotic applications constitutes a big challenge. They are key components in tele-surgery systems as well as in assembly systems of micromachines and their parts. In these applications, grippers have to satisfy accessibility constraints and be able to grasp materials of various shapes with unknown mechanical behaviors (for instance soft tissues in surgery applications). This motivates the use of dextrous articulated grippers. Appropriate actuators are essential in this context. More than geometrical constraints, they must fit force/motion transmission capabilities required by contacting tasks in terms of amplitude and precision in the control.

Microactuators such as electromagnetic, electrostatic, piezoelectric or shape memory alloy (SMA) ones, do not directly match with the requirements of these robotic systems. They produce either small displacement with high force or small force in large displacement. Moreover, while considering dimensional constraints, the design of microactuators could not be thought up just as a juxtaposition of several actuators [1].

The present article discribes a newly developed modular microactuator. Its design incorporates SMA fibers coupled with a 2 d.o.f. transmission linkage mechanism. In the following sections, after a brief presentation of the design features used, a thermo-mechanical model of SMA fibers which takes into account working conditions, is proposed and experimentally verified. This model is usefull for the design and also to achieve a precise motion control and an accurate force transmission. Additionnely, we discuss the micro-transmission design.

2   Design features.

2.1   Principle of the SMA actuation

SMA fibers have been chosen as active elements for these actuators by considering their attractive properties such as their high actuation pressure (until 250 MPa for a fractional stroke of 8%), small volume and simplicity of the actuation process (commonly obtained by Joule effect generated by an electrical power source).

However, SMA actuators suffer from major drawbacks such as :

- non linear thermo-mechanical behavior with hysteresis,
- sensitivity to ambient temperature,
- fatigue which reduces the number of actuation cycles,
- a bias force is necessary to recover its low temperature configuration,
- slow response time, limited by the cooling rate of the wire which depends directly on its diameter.

2.2   Principle of the actuator.

Actuators for dextrous grippers must produce a torque large enough to compensate the joint friction, gravity forces and to apply grasp forces when necessary. The modular microactuator, we have designed for this kind of use (figure 1) combines in series two decoupling and appropriate actuators. One drives the joint rotation and the other one produces the torque required for the grasp. Its compact and light weight structure can be integrated in a phalanx of the gripper (12 mm length and 5 mm diameter). For that, we use a mechanism with 2 d.o.f fractioned mobility which kinematics is optimized for both functions : motion and force transmissions.

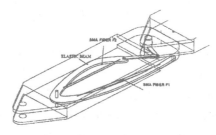

Figure 1 : Design of the modular actuator for integration in a phalanx

SMA wires are mounted on two elastic beams of the mechanism as shown on the figure 1. These elastic beams give the necessary bias force to recover the SMA fiber length at low temperature. The amplitude of the contraction of a SMA wire f_1 is amplified by the mechanism to obtain at least a rotation of 90 degrees (the transmitted force is large enough to balance friction and gravity effects). The grasping force is produced by the fiber f_2 for a given position of the joint.

3 SMA actuator

3.1 Shape memory alloy effect

The martensite transformation is produced by cooling the material or by an external stress from the austenite phase (high temperature). The stress induces a non elastic deformation by a reorientation of the different variants of martensite. This deformation disappears by heating. This effect is called "one way shape memory effect".

In the austenite phase (when the temperature $T > A_f$), the SMA fiber behavior is purely elastic while in the martensite phase (when $T < M_f$) it is a pseudo-plastic one (see figure 2 where the reference length used is the one in austenite phase due to its stability).

Between these two temperatures, both phases are present simultaneously and their ratio in volume depends on temperature T and on external stress σ.

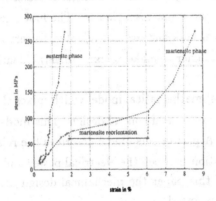

Figure 2 : Experimental identification of a NiTi wire (diameter $100\mu m$) in austenite and martensite phase.

3.2 The SMA actuator

Since, with no external stress, the length of the wire is the same in both phases, it is necessary to introduce in the actuator design an active or a passive element to produce the bias force [2]. The chosen solution consists in a preloaded elastic beam which equivalent action is represented on figure 3 :

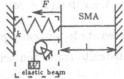

Figure 3 : Scheme of the SMA actuator

where F designates the external force, l the nominal length of the wire in the martensite phase, k is the equivalent stiffness of the elastic beam and W the load which produces the initial stress σ_W.

One of the basic difficulty in the actuator design is the definition of k, σ_W and the fiber diameter to match with the desired performances. This for, we need an accurate thermo-mechanical model of the fiber behavior which expresses the stress σ as a function of the strain (ϵ) and the supplied electrical power which must be considered as the control variable. The reason is, that the temperature T of the wire (which constitutes the natural variable) is physically difficult to measure, especially for small diameters, and is moreover, modified by the ambient temperature nearness the wire (due to convection effect). If we consider also that the electrical resistance of the fiber itself depends on the temperature and the stress, it is more convenient to use the electrical power as command variable. Different models have been proposed to describe the shape memory effect [3],[4],[5],[6],[7]. Most of them have several limitations especially for the design of control structure because they use T as variable. They consider two different scales in the observation of the phenomena : a microscopic and a macroscopic one. The first one leads to complex relations traducing basic physical laws and are not adapted to servo control system design. The second one forces the designer to make an experimental identification in the working conditions.

We propose a unidimensional mathematical model which mixes these two levels of observations for a use in the mechanical and control design. This model is limited to the "one way shape memory effect". A part of this model describes the relation $\sigma = f(\epsilon, T)$ while the other part gives the relation between the electrical power and the temperature. Since we are considering in this paper the mechanical design of the actuator, only the relation $\sigma = f(\epsilon, T)$ will be reported.

3.3   Thermo-mechanical model for the mechanical design.
3.3.1   Martensite fraction.

As said in the section 3.1., depending of the temperature T and the stress σ, the NiTi fiber can be either in a full martensite phase or in an austenite phase or both phases can be present at the same time. By using the thermodynamic statistic laws (1), the probability R that a fraction V_M of an alloy volume V is in the martensite phase, is given by [5] :

$$R(T, \sigma) = \frac{V_M}{V} = \frac{R_1}{1 + \exp K(T + c\sigma - Tm) + R_2} \tag{1}$$

with, for a transformation from martensite to austenite phase : $K = \frac{6.2}{A_f - A_s}$, $T_m = \frac{A_s + A_f}{2}$ A_s and A_f being respectively the austenite transition temperature for the begining and the end of the austenite phase.

Conversely, for a transformation starting from austenite to martensite phase $K = \frac{6.2}{M_f - M_s}$, $T'_m = \frac{M_s + M_f}{2}$, M_s and M_f being this time the martensite temperature.

$c\sigma$ represents the shift induced by external stress which comes from the Clausius-Clapeyron relation ($c = 0.145°C/MPA$ for TiNi alloy). R_1 et R_2 are introduced in (1) to insure the continuity of R when we reverse the transformation.

3.3.2 Stress versus strain in martensite and austenite phases

In this model, the length used as reference is the one in the austenite phase with no stress. In the austenite phase, since the behavior is basically elastic, the strain ϵ_A of the wire is simply expressed by :

$$\epsilon_A = \frac{\sigma}{D_A} \tag{2}$$

where D_A is the equivalent Young Modulus in the direction of σ in the austenite phase. In the martensite phase, the elastic and pseudo-plastic behavior is given by :

$$\epsilon_M = \frac{\sigma}{D_M} + \alpha(\sigma - \sigma_1) \tag{3}$$

where : $\alpha = 0$ if $\sigma \leq \sigma_1$ or $\sigma \geq \sigma_2$, else $\alpha = \frac{\sigma_2 - \sigma_1}{\epsilon_{twin}}$ if $\sigma_1 \leq \sigma \leq \sigma_2$.

ϵ_{twin} is the strain due to the full martensite transformation, D_M the Young modulus in the direction of σ in the martensite phase. σ_1 is the value of stress to initiate the reorientation of the martensite variants and σ_2 the value of stress when all the variants of martensite are reoriented.

3.3.3 Stress versus strain depending on temperature

When both phases are present at the same time, the average strain $\langle \epsilon \rangle$ of the wire is :

$$\langle \epsilon \rangle = (1 - R)\langle \epsilon_A \rangle + R\langle \epsilon_M \rangle \tag{4}$$

where $\langle \epsilon_M \rangle$ and $\langle \epsilon_A \rangle$ are respectively the average strain in the martensite and austenite phase under the stress σ. Figure 4 shows a simulation of this model in the experimental conditions used for figure 2 (the characteristic $\sigma = f(\epsilon)$ for $T > A_f^*$ represents the behavior in the austenite phase showed on figure 2, while $\sigma = f(\epsilon)$ for $T < M_f$ the behavior in the martensite phase).

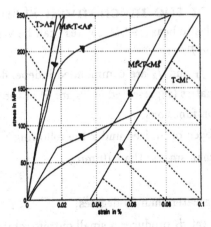

Figure 4 : Stress versus contraction for different temperatures
(* represents the transformation temperature for $\sigma = 250 MPa$).

It is clear that this kind of informations is usefull for the determination of the working conditions of SMA actuators (i.e. the electrical power range).

3.4 Application to the design of SMA actuator

The design of the SMA actuator consists to determine the main parameters k, σ_W and the diameter of the wire, to adapt its performances to the requirements.

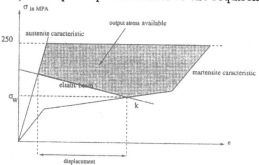

Figure 5 : Fonctionnal space for SMA actuator

As shown on figure 5, knowing the required displacement and the output force, optimization of the different parameters of SMA actuator can be deduced. To avoid the material fatigue, the amount of stress applied to the wire has to be limited at 250 MPa.

It is important to notice that, the value of k has to be chosen as low as possible to avoid loss of energy during the contraction of the elastic beam.

Moreover to obtain the same behavior, two wires can be used in parallel instead of one. This increases in other respects the response time of the actuator.

By integrating these parameters in the previous thermo-mechanical model, the relation between output force, strain and temperature can be obtained.

4   Optimization of the mechanical structure.

The mechanical structure "elastic beam + SMA wires" can be viewed as a flexible beam commanded by the parameter T.

Since the two pairs of wires f_1 and f_2 are commanded independently (f_1 is fixed when f_2 is commanded), the kinematics (represented on figure 6) of the transmission decouples force and motion transmission functions.

The linkage mechanism has been optimized under the following constraints :

- total length of the mechanism less than 10 mm,
- width less than 5 mm,
- output angle variation greater than 90 degrees,
- contraction of the segment d_3 produces a small output rotation in the same way as the contraction of d_l.
- contraction of both wires is limited to 6% to avoid fatigue.

In the optimization process, the force F produced, which is directly correlate to the diameter of the wire, is used as parameter and can be adapted to the required actuator performances. When F is detemined, the characterisics of the "elastic beam + SMA wires" can be chosen by the use of the thermo-mechanical model.

A maximun of force transmission and a maximun of output range motion is obtained for the following linkage dimensions : $d_1 = 1.5$, $d_2 = 2.2$, $d_3 = 6$, $d_4 = 1.3$, $l = 8$mm and $\Theta_{4initial} = \frac{\pi}{4}$. For these dimensions, the motion of the mechanism from initial to final position is shown on figure 7.

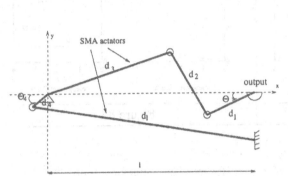

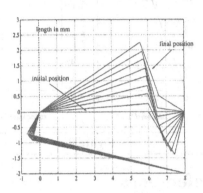

Figure 6 : Kinematics of the modular actuator

Figure 7 : Motion of the mechanism from initial to final position

This mechanism can be integrated in a body with a minimum of 4.5 mm width and 9mm length.

The performances of the motion transmission are represented on the figure 8.a and the force transmission on the figure 8.b for different values of the output position (i.e. contraction of fiber f_1), where the initial length of each fibers is taken in this case in martensite phase.

5 Conclusion

The modular design of the actuator for micro-grippers allows an adaptation to task performances (motion and force), in which each basic component can be adjusted. By adapting the characteristics of the SMA actuator, we can obtain torque performance requirements. Moreover, miniaturization of this device can be considered, limitation resulting more from technology employed for the realization than from SMA actuation.

A prototype of this mechanism and its control system is actually under realization in our laboratory.

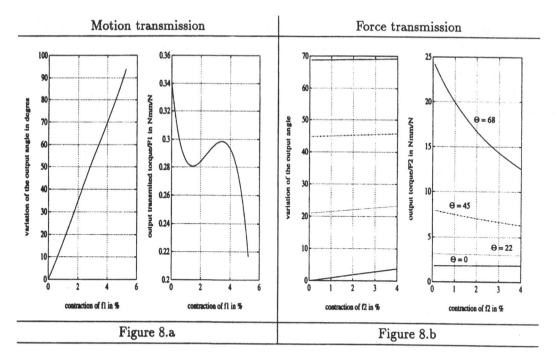

Figure 8.a   Figure 8.b

References

[1] T. Hayashi, "Special feature of micromechasnism," in *1st IFToMM Internationnal Micromechanism Symposium Tokyo*, pp. 21–24., 1993.

[2] C. Liang and C. Rogers, "Design of shape memory alloy actuators," *Journal of Mechanical Design*, vol. 114, pp. 223–230, 1992.

[3] K. Tanaka, "A thermomechanical sketch of shape memory effect : one domensional tensile behavior," *Res Mechanica*, vol. 18, pp. 251–263, 1986.

[4] C. Liang and C. Rogers, "One dinensional thermomechanical constituve relations for shape memory materials," *Journal of Intell. Marter. Syst. and Struct.*, vol. 1, pp. 207–234, 1990.

[5] K. Ikuta and H. Shimizu, "Two dimensional mathematical model of shape memory alloy and intelligent sma-cad," *IEEE*, pp. 87–92, 1993.

[6] J. Burns and R. Spies, "A numerical study of parameter sensitivities in landau-ginsburg models of phase transitions in shape memory alloys," *Journal of Intelligent Material Systems and Structuress*, vol. 5, pp. 321–332, 1994.

[7] K. Arai, S. Aramaki, and K. , Yamagisawa, "Continuous system modeling of shape memory alloy for control analysis," in *5th International symposium on micro machine and human sciences*, pp. 97–99, IEEE, 1994.

Chapter IV
Walking Machines
and Mobile Robots I

ON DIRECT-SEARCH OPTIMIZATION
OF BIPED WALKING

P. Kiriazov
Institute of Mechanics, Sofia, Bulgaria

W. Schiehlen
University of Stuttgart, Stuttgart, Germany

Abstract: Employing a set of appropriate test control functions, the optimal control synthesis task is transformed into a series of control parameter optimization problems. The direct-search optimization procedure was verified on the dynamic model of a five-link anthropomorphic walking mechanism. The structure and the shapes of the synthesized control and state functions are similar to those of the humans. The numerical results show very efficient energy-loss minimization which means that the so-called synergism with the human bipeds, can be obtained with the controlled walking machines, too.

1 Introduction

In recent years, a great deal of research work has been done in the area of legged locomotion but not so effective as in the area of robotic manipulation systems. Efforts to develop quadruped/hexapod robots are more successful in comparison with biped ones due to their better static stability. Biped mechanisms with various-type kinematic and driving schemes, Grishin *et al* [3], Formalsky [4], Furusho and Masubuchi [5], Raibert [11], Todd [12], Vukobratovich *et al* [13], have been designed and most of them are able to perform, to some extent, 2D or even 3D humanoid walking. A lot of research work is needed however to optimize the design of their mechanical, drive, sensor and control systems.

Performance requirements to dynamic walking:
◇ walking robots should be able to move at a prespecified speed;
◇ they should be able to start, perform, and stop walking with appropriate step lengths;
◇ walking should be as easy and smooth as possible, with minimum energy loss.

Control system design requirements:
1) control system is to be designed so as to fullfil all the performance requirements;
2) the control synthesis approach should be applicable to bipeds with various kinematic and driving schemes;
3) feedforward control is synthesized precisely enough, even when accurate dynamic model is not available - it means that the final adjustment of the control parameters to be possible on the robot itself.

Robot movement from a double-support (DS) state to another will be optimized following a performance-based, direct-search approach, Marinov and Kiriazov [10], Kiriazov [8]. Thus the highly complex task of robot optimal control will be reduced to a set of manageable control parameter optimization problems.

2   Problem Formulation and Solution Approach

A kinematical scheme of 7-link anthropomorphic planar biped with 6 controlled joints (hips, knees and ankles) can be taken to represent a humanoid walking in the sagittal plane, Blajer and Schiehlen [1]. The biped locomotion can be described by the angles of rotation of the links and their velocities. For the simulation purposes, we use robot dynamics models, the general structure of which is well-known

$$\ddot{q} = M(q)^{-1}(BT - C(q, \dot{q}) + g(q)) \tag{1}$$

where, q is the vector of the links' rotation angles, $M(q)$ is the inertia matrix, $C(q, \dot{q})$ stands for the velocity forces, $g(q)$ is the vector of gravitation forces, matrix B represents the actuator location, and T is the vector of actuator torques.

Considering the control synthesis problem within a one-step time-interval $[t^o; t^f]$, the boundary conditions are as follows:

initial state:
$$q(t^o) = q^o; \qquad \dot{q}(t^o) = v^o; \tag{2}$$

final state:
$$q(t^f) = q^f; \qquad \dot{q}(t^f) = v^f; \tag{3}$$

The performance indices to be optimized are the accuracy, the movement execution time $t^{st} = t^f - t^o$, and the energy-loss which quantity, for simplicity, will be taken to be proportional to the norm of the control vector: $E = cT'T$.

For such a complex optimization problem, and in the presence of modelling errors, the most reliable way in which we can find satisfactory suboptimal solutions, is the following

Performance-Based Direct-Search Approach

1. Define output variables that best describe the system performance.

2. Choose a set of appropriate test control functions.

3. Input-output pairing and solving shooting equations.

The term "appropriate" concerns the structure, the shape, and the magnitudes of the test control functions. Simple linear-spline control functions of "bang-pause-bang" or "bang-slope-bang" types can be employed as the dynamic performance of robots moving in point-to-point manner is most sensitive to their parameters. Taking such forms of the control laws is in accordance also with the Pontryagin Maximum Principle when a time/energy functional is the performance index. The theory and experience show that, one or two switching times for each control function, are enough to synthesize a satisfactory suboptimal or even optimal point-to-point motion. After a proper input-output pairing, the control synthesis problem will be transformed into a system of shooting equations with respect to several unknown control parameters.

3   Dynamic and Control Structures

For the sake of simpler explanation of the control synthesis procedure, we will assume in this paper that, during DS-phase, only the trunk moves until gets the necessary inclination. Therefore, the legged robot starts up and ends the strides with zero velocities for the state variables. Thus we can have biped locomotion without impacts, Blajer and Schiehlen [1], Daberkow, Gao, and Schiehlen [2].

During locomotion, each joint of the biped can be locked or unlocked. The latter means that the joint is powered (forward or rearward) or unpowered (free). Therefore, the structure of biped dynamics changes and, respectively, the structure of the control system has to be changed. From this point of view, four phases in performing a step can be distinguished:

- DS-phase, when the state reached at the end of the previous step is adjusted to the state needed as initial one for the next step;

- Taking-off (TO) phase, during which the leg to be transfered is pushing-off the ground. With the kinematical scheme considered, the thrust force can be produced by the actuators at the ankle or knee joints of this leg. TO-phase ends at the moment when the transfered leg leaves the ground surface;

- SS-phase, when the biped is pivoting arround the ankle joint of the supporting leg. The other leg is rotated with respect to the supporting one until it reaches the configuration required before landing at the specified foothold;

- Landing (L) phase, during which the leg which has been swinging is touching the ground and producing a force (by the same actuators as in TO-phase) thus opposing the biped motion to reduce its velocity to the required value;

In TO- and L- phases, the forces produced by the transfered leg may be so impulsive that the other leg can loose the contact with the ground. To avoid this and to minimize the movement execution time, it will be necessary to activate the hip joint of the supporting leg by a relevant torque.

If SS-motion, starting with the state final for TO-phase and ending with the state needed as initial for the L-phase, is performed accurately enough, then the control action during DS-phase will concern only the trunk attitude.

In order to avoid jerks, locking the joints is to be done with zero velocities of the corresponding joint angles. And in solving the control synthesis problems for all the phases, the continuity conditions for the global state variables $\dot{q}$ at the points of transition from any phase to the next one, must be taken into account.

It becomes clear from the above considerations that the most complicated to control is the SS-phase when almost all the joints are to be actuated and the joint motions can be strongly interacting. Moreover, the massive part of the walking mechanism is driven mainly during this time-period and the problem of time/energy optimization appears to be the most important one for this phase. Therefore, the realization of the proposed direct-search optimization approach will be explained for this particular case of SS-motion. Nevertheless, the determination of SS-motion parameters specifies some important parameters of the other phases such as their end-point states and approximate control functions.

4   Control Synthesis of SS-motion

In what follows, we assume that the mass of the feet is negligibly small and their relative motion does not considerably affect the other joint motions. Our intention is to prove the practicability and efficiency of the proposed direct-search control synthesis method on a basic biped walking mechanism consisting of a trunk and two two-link legs, Fig. 1. Here are the main points of this method in the most important case of SS-motion.

1. **Define the output variables that best describe the system performance.**

To take full advantage of the leg length, we lock the knee joint of the supporting leg in the stretched configuration. Thus the SS-motion can be described by the measured angles and angular velocities at the following joints: the ankle of the supporting leg, the knee of the trailing leg, and the two legs' hips. In our considerations, these state variables change their values during SS-phase due to three control actions: one at the knee joint of the swinging leg and two at the hip joints. In the boundary-value problem for this phase, the following final or intermediate values of the state variables will be taken as controlled outputs:

$y_1 = \theta_1(= \theta_2)$ - the angle of rotation at the ankle joint of the supporting leg, just before the L-phase;

$y_2 = \theta_2 - \theta_3$ - the angle between the thighs when (this time-moment denoted by τ_1) its velocity becomes zero and we can fix without impact this links' configuration;

$y_3 = max\theta_5$ - the angle of maximum bending at the knee of the trailing leg. Its value depends on how uneven is the terrain;

$y_4 = \theta_5 - \theta_4$ - the knee angle when (this moment denoted by τ_2) its velocity gets zero value for the second (final) time; we will lock this joint, too, assuming that the trunk motion can be damped to certain extent by the opposing forces produced at the ankle joint of this leg and at the hip joint of the supporting leg in the next L-phase;

y_5 - the maximum value of the trunk inclination; It should be within some acceptable bounds;

So, by the end of SS-motion we have two time-moments τ_1 and τ_2 at which the dynamics of the biped mechanism changes its structure. In order to synthesize the control using only one dynamic model, we will synchronize completing all the joint motions for this phase. It means that the difference between these final times, another controlled output $y_6 = \tau_2 - \tau_1$, should be zero. With this additional parameter, the number of the outputs to be controlled becomes 6. To achieve their required values, the decision variables (the control parameters) are to be of the same number.

2. Choose a set of appropriate test control functions.

From the experience on robot motion optimization problems, and observing also the structure of the control laws in the human walking, we come to the following choice: the control function at the hip joint of the supporting leg will have one switching (from acceleration to deceleration) point and each of the other two control actions (at the transfered leg) - two switching points. The magnitudes of the control functions can be determined from a set of conditions necessary to provide the robot with independent joint (decentralized) controllability , Kiriazov [7], Lunze [9].

3. Input-output pairing and solving shooting equations.

Among all the robot system parameters that are variable, we have to choose 6 which mostly influence the six controlled outputs. Some of them can be taken from the initial conditions, others will be among the parameters describing the test control functions. And it should be done in a way that existence of feasible solutions and convergence of the procedure finding them can be guaranteed.

y_1 - to be controlled by the switching time of the control torque at the hip of the supporting leg;

y_2 - by the first switching time of the control function at the hip of the transfered leg;

y_3 - can be regulated by the first switching time of the control torque at the (unlocked) knee joint;

y_4 - by the second switching time of the last control function;

y_5 - we take the initial inclination of the trunk to be the decisive parameter for this controlled output;

y_6 - by the second switching point of the control action at the transfered leg's hip;

Having the input-output pairing done, the given boundary-value problem is transformed into a system of 6 shooting equations. Under some very weak and reasonable conditions for independent parameter controllability, existence of feasible solutions and convergence of the bisection algorithm finding them can be provided, Marinov and Kiriazov [10], Kiriazov [8]. Such conditions are satisfied when the robot has good enough independent joint controllability.

So, the first level of our control optimization procedure is finding the feasible solutions of the system of shooting equations. At this stage, all control parameters that are not elements of the input-output pairs are fixed. These are the pause lengths if the test control functions are of "bang-pause-bang" type or the slope angles in the other case. At the next stage of the control synthesis procedure, all these control parameters can be varied to minimize as much as possible the time/energy functional.

5 Simulation

Computer simulation programs in MATLAB environment are developed to verify numerically the practicability and the efficiency of the proposed direct-search optimization procedure. A four-degree-of-freedom dynamic model is considered for SS- motion control synthesis. This model is obtained by a small reduction (with one degree less) from the five-degree-of-freedom dynamic model as derived in Daberkow, Gao, and Schiehlen [2], Gao [6]. The geometric and inertia data are the same, with the exception for the inertia moment of the trunk which was increased, Raibert [11], in order to provide the walking robot with the necessary independent joint controllability. The boundary value problem has been solved with accuracy of 0.02 rad for the angular controlled outputs.

At first, the given boundary value problem (transformed into a system of 6 shooting equations) is solved with zero lengths of the pause time-intervals. Thus a set of near time-optimal control functions for SS-motion are found. Then, with a required movement execution time (0.43s) (which should be greater than that obtained at the time optimization stage, 0.33s) we perform variation of the pause lengths in order to minimize the energy-loss. The structure of the obtained control functions (as shown in Fig. 2) as well as the form of the corresponding joint motions, Figs. 3 and 4, are similar to those of the human bipeds.

Comparing the results regarding the time and energy optimization, we find a quite remarkable energy minimization: with only 30 percent larger movement execution time, a reduction of the energy-loss of more than 10 times is obtained.

Thus, as in the human biped locomotion control, we can find such control laws that the counteraction between the joint motions quantified by the energy-loss function is minimized. Thus the so-called synergism can be obtained with the controlled walking machines, Vukobratovich *et al* [13], too.

6 Conclusion and Further Research

A direct-search time/energy optimization method for robot point-to-point movement has been proposed. In the face of highly complex dynamics and model uncertainties, this method proves to be very promising for attaining satisfactory feedforward control of dynamic biped walking. Following the proposed performance-based approach, other efficient methods can be developed for the following problems:
- optimization of 2-D walking control in case of stairs climbing;
- optimal control of cooperating manipulators; application to legged robots;
- feedforward control synthesis in 3-D biped locomotion;
- integrated control/structure design of walking machines;
- control synthesis of dynamic walking by learning;

Acknowledgements: This research work has been carried out at the Institute B of Mechanics, University of Stuttgart. It is supported by the German Research Foundation **DFG** and the National Science Foundation of Bulgaria, Project MM-426/94.

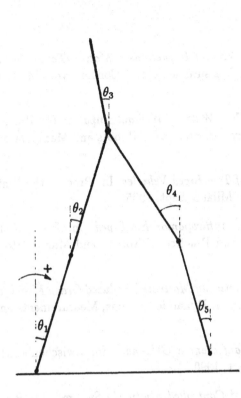

Fig. 1: Five-link biped

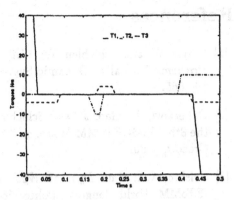

Figure 2: Control functions

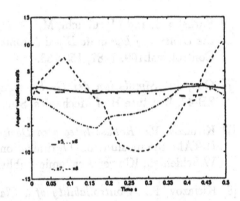

Figure 3: Velocity functions

Control functions for SS-motion:

T1 - hip torque at supporting leg
T2 - hip torque at transfered leg
T3 - knee torque at transfered leg

Angle/velocity of robot links' rotation:

x1/x5 - supporting leg's thigh & shin
x2/x6 - trunk inclination
x3/x7 - transfered leg's thigh
x4/x8 - transfered leg's shin

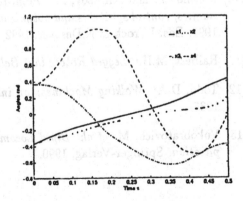

Figure 4: Angle functions

References

[1] Blajer, W. and Schiehlen, W: *Walking Without Impacts as a Motion/Force Control Problem*. Journal of Dynamic Systems, Measurements and Control, vol.114, 1992, 660-665.

[2] Daberkow, A., Gao, J., and Schiehlen, W: *Walking Without Impacts*. In: Proc. of the 8th CISM-IFToMM Symp. on Theory and Practice of Robots and Manipulators, Cracow, 1990.

[3] Grishin, A.A. *et al: Dynamic Walking of Two Biped Vehicles*. In: Proc. of the Ninth IFToMM World Congress, Politecnico di Milano, Sept. 1995.

[4] Formalski, A.M.: *Impulsive Control for Anthropomorphic Biped*. In: Proc. of the 10th CISM-IFToMM Symp. on Theory and Practice of Robots and Manipulators, Gdansk, 1994, 387-393.

[5] Furusho, J. and Masubuchi, M.: *A Theoretically Motivated Reduced Order Model for the Control of Dynamic Biped Locomotion*. J. Dynamic Systems, Measurements and Control, vol.109, 1987, 155-163.

[6] Gao, J.: *Simulation der Gehbewegung auf ebenem Gelaende*. In: Zwischenbericht ZB-47, Institute B of Mechanics, Stuttgart, 1990.

[7] Kiriazov, P.: *Robust Integrated Design of Controlled Multibody Systems*. In: Proc. IUTAM Symposium on Optimization of Mechanical Systems, Eds. D.Bestle and W.Schiehlen, Kluwer Academic Publishers, 1996, 155-162.

[8] Kiriazov, P.: *Controllability of a Class of Dynamic Systems*. ZAMM, vol. 75 SI, 1995, 85-86.

[9] Lunze, J.: *Feedback Control of Large Scale Systems*. Prentice Hall, U.K., 1992

[10] Marinov P. and Kiriazov, P.: *Point-to-Point Motion of Robotic Manipulators: dynamics, control synthesis and optimization*. In: IFAC Symposia Series, Robot Control 1991, Eds. I.Troch & K.Desoyer, 1992, 149-152

[11] Raibert, M.H.: *Legged Robots that Balance*. MIT Press Cambridge, MA, 1986.

[12] Todd, D.J.: *Walking Machines: an introduction to legged robots*. Kogan Page Ltd, 1985.

[13] Vukobratovich, M. *et al: Biped Locomotion: Dynamics, Stability, Control and Application*. Springer-Verlag, 1990.

DEVELOPMENT OF A BIPED WALKING ROBOT ADAPTING TO THE REAL WORLD

REALIZATION OF DYNAMIC BIPED WALKING ADAPTING TO THE HUMANS' LIVING FLOOR

J. Yamaguchi and A. Takanishi
Waseda University, Tokyo, Japan

Summary: In this paper, the authors introduce an anthropomorphic dynamic biped walking robot adapting to the humans' living floor. The robot has two remarkable systems: (1) a special foot system to obtain the position relative to a landing surface and the gradient of the surface during its dynamic walking; (2) an adaptive walking control system to adapt to the path surfaces with unknown shapes by utilizing the information of landing surface, obtained by the foot system. Two units of the foot system WAF-3 were produced, a biped walking robot WL-12RVII that had the foot system and the adaptive walking control system installed inside it was developed, and a walking experiment with WL-12RVII was performed. As a result, dynamic biped walking adapting to humans' living floor with unknown shape was realized. The maximum walking speed was 1.28 s/step with a 0.3 m step length, and the adaptable deviation range was from -16 to +16 mm/step in the vertical direction, and from -3 to +3 ° in the tilt angle.

1 Introduction

The control systems of conventional dynamic biped walking robots studied in the past required that the information on the accurate shapes of the paths on which the robots walked were available before implementation of dynamic walking or that of the swing leg landing control [1]-[7],[9]. However, neither the accuracy of the lower-limbs trajectory nor that of the measurement of the path surface shape demanded to maintain 3-dimensional stable dynamic walking have been referred to by any reports [1]-[8].

Although WL-12RV showed the most stable dynamic biped walking among the biped walking robots which were developed by Waseda University earlier than 1991[9], it became clear that the robot received influences of the lower-limbs trajectory deviation and the walking surface shape deviation more and more as the walking speed increased. In the pace of one step per second just like a human, the robot must satisfy the accuracy of a few millimeters in terms of the height of path surfaces and that of approximately 0.5 degree for the gradient of path surfaces, otherwise, since an adverse influence to dynamic walking caused by the foot landing position deviation occurring at the landing of the swing leg may not attenuate within the one stride so that it becomes unable to ensure the maintenance of 3-dimensional stable dynamic walking.

In contrast, in the actual environment where humans live, it is very rare that path surfaces have the characteristics of a geometric plane, with deformation by some ruggedness and undulation in almost all cases. The result shows that the robot cannot make stable dynamic walking in the environment. Therefore, the authors feel that it is time to develop a dynamic biped walking adapting to the real world.

It is considered mainly because the conventional control method of dynamic biped walking constituted an open loop control system that only indirectly controls the leg trajectory relative to the landing surface relying on internal sensors in spite of a fact that walking is a series of movements made against the landing surface[5]-[7],[9].

Therefore, the purpose of this study was to develop a foot mechanism which mounts a landing surface detection system which measures the gradient of the landing surface during dynamic walking as well as the relative position to the surface, to devise a walking control method that constitutes a closed loop control system which provides a lower-limbs trajectory control against the landing surface on the real time basis by utilizing the information of the landing surface, obtained by the above mentioned foot mechanism, and to develop a life size biped walking robot which adapt to the actual world in which the lower-limbs trajectory deviation and the surface shape deviation can not be reduced to zero.

2 Required condition of the walking surface

Required condition of the walking surface to be researched was set as follows.
(1) Height and gradient of the path surface are unknown factors.
(2) All landing spikes (see Fig. 3) in the four corners of the biped walking robot's foot must be able to be grounded.
(3) Extent of deform of the path surface and that of movement that are experienced when the machine model walks, are limited to the ranges that there is no need to make the compensating movement that considers the dynamics.

It should be noted that the gradient of lateral plane is provided to be horizontal (Accuracy $\pm 0.5°$) and known in advance. This is because the present WL-12 series has no active degree of freedom around the roll axis of the lower-limbs, regarding materials of the surface, such that does not deform at all when the machine model walks and that sags approximately 2 mm when it is standing on single leg, satisfy the conditions.

Take for instance the flooring materials which are generally used in the human living environment, it is learned that the floor which is not likely to deform, such floor covered with a carpet and other flooring materials which may deform slightly such as the tatami mattress and the like, satisfy such conditions.

3. Biped walking robot WL-12RVII

3.1 Machine model

This machine model is an improved version of a biped walking robot WL-12RVI [10], which is equipped with a trunk mechanism to compensate for three-axis moment by trunk motion and the foot mechanism WAF-2 [11] to acquire a relative position to the landing surface. The new version incorporated modifications that were required by the mounting of the newly developed foot unit WAF-3 (Waseda Anthropomorphic Foot No. 3) and was named WL-12RVII (Waseda Leg No. 12 Refined VII). The total weight of the robot is 109 kg. An assembly drawing of this machine is illustrated in **Fig. 1**. The link structure and assignment of active DOFs (degrees of freedom) are illustrated in **Fig. 2**.

3.2 Foot Mechanism WAF-3

Fig. 3 presents a rough sketch of WAF-3. In reference to the movable range (MR) of the passive DOF of human lower-limbs, a similar range was set for WAF-3. Specifically, on Cartesian coordinates as shown in **Fig. 3**, the MR on X' axis was set at ± 5.0 mm, the MR on Z' axis was set at 5.5 mm, the MR around X' axis was set at $\pm 2.3°$, and the MR around Y' axis was set at $\pm 1.7°$. The mechanisms which WAF-3 possesses are summarized below.

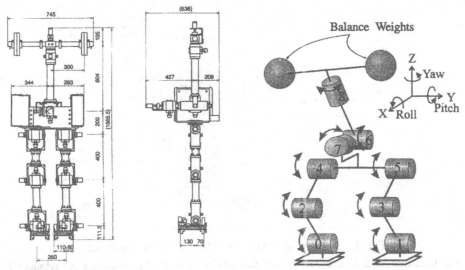

Fig. 1 Assembly drawing of WL-12RVII   **Fig. 2** Link structure of WL-12RVII

(A) Shock absorbing mechanism, (B) Support leg change stabilization mechanism, and (C) Landing path surface detection mechanism

The following describes the individual mechanisms.

(A) Shock absorbing mechanism: WAF-3 has a passive shock absorbing mechanism which uses shock absorbing material to enable "high-speed following operation to the landing surface, the buffer of the impact which occurs when the swing foot touches the ground, control of the vibration caused in the robot during dynamic biped walking", which are difficult to be handled by the feedback control with software which uses information obtained by the force sensor installed below the foot plate [8].

The shock absorbing mechanism consists of two main parts: (A-a) shock absorbing material installed between the two foot plates; (A-b) the upper stoppers. The total damping coefficient was approximately 40.6 kg•s/m in the direction of shrinkage.

(B) Stabilization mechanism of support leg change: WAF-3 has the same stabilization mechanism of support leg change as WAF-2. Combining this mechanism with shock absorbing mechanism gives a biped walking robot robustness. Refer to document [11] for details of this mechanism.

(C) Detection mechanism of landing path surface: This mechanism is capable of obtaining the following information about the landing surface.

(C-a) The information on the grounding of the lower-foot plate using microswitches

(C-b) The relative position to the landing surface (lower-foot plate) using linear potentiometers

(C-c) The gradient of the landing surface (around Y' axis) using a servo-type inclination sensor

(C-d) The angular velocity of the upper-foot plate (around Y' axis) using a rate-gyro sensor

4 Walking control method

The biped walking control method on which we report in this paper uses a three-layer structure for the stabilization of dynamic biped walking to obtain high stability as the locomotion mechanism of the real world.

Basic stabilization control of biped walking uses the biped walking control method that we have already proposed in document [9].

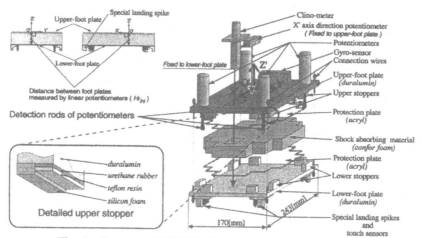

Fig. 3   Structure of WAF-3, O'-X'Y'Z': detection coordinate.

Next, stabilization devices located above the method are shock absorbing mechanism and support leg change stabilization mechanism. These mechanisms remove the influences on walking by "measurement errors of the path-surface shape and the lower-limbs trajectory deviation caused by the bend of the structural members of the machine model, etc.", which is not considered in the above-mentioned walking control method. These double stabilizing systems allow the robot to continue dynamic biped walking, even if we program-control the robot considering slightly transformable path surface like the horizontal wooden floor covered with a carpet as horizontal and smooth, when the lower-limbs trajectory deviation of the robot is within several millimeters [11].

Finally, the stabilization control located the highest is adaptive walking control system, which is proposed in this paper. This control system is closed loop control system that acquires position information of the robot in the external world from the landing path surface detection mechanism and changes the lower-limbs trajectory based on the information.

The following is an outline of the adaptive walking control system: First, a robot walks using a walking pattern (hereafter, the authors refer to this pattern as the standard walking pattern) for a flat floor. Second, the robot obtains information about landing surface using WAF-3. Using this information, the robot at the same time changes and controls the lower-limbs motion to adapt to the unknown landing surface on a real time basis. Besides, the robot's walking is suitable to the path surface with deviation in the inclination in addition to the height, without worrying about the emanation of lower-limbs motion, by combining the control for returning to the standard walking pattern every one step. By repeating this, the walk suitable to path surfaces with deviation is done in real time through all walking periods, also concerning the inclination of path surfaces in addition to the height of path surfaces. On that occasion, in the creation of the standard walking pattern, the control method of trunk-compensated, bipedal-locomotive is used, being proposed by the authors. As for lower-limb trajectories used for that, the trajectory of the lower side of foot parts is set with constant leveling. A flow chart of this walking control method is shown in **Fig. 4**.

A processing method of the information on the landing surface acquired and the control methods of lower-limb trajectories, which are vital factors in the system, are described below.

4.1 Processing method of information on landing surfaces

All kinds of information acquired by this mechanism for detecting a landing surface is sent to the computer for controlling in real time in walking, and arranged into the following information. **Fig. 5** shows each system of coordinates set to the walking system.

(1) The information on the grounding of the lower-foot plate: The grounding is judged with ON/OFF of the microswitches installed in the four corners of the lower side of the lower-foot plate.

(2) The information on the relative angle and the relative distance of landing surfaces and upper-foot plates: The four linear potentiometers are intended to indirectly measure the relative angle A' (around Y' axis) and the relative distance H_t to the path surface through measurement of the top plane of the lower-foot plate that performs landing following the shape of path surface.

(3) The gradient of landing surfaces: The gradient of landing surfaces is calculated (presumed and measured), by using the value output by the rate gyro fixed on the upper side of the upper-foot plate to be integrated and the value output by the clinometer fixed on the lower-foot plate through the attaching block and the like. On that occasion, the gradient of landing surfaces is calculated by using the expression (1).

$$A_n = \begin{cases} (Ac_n + A_{n-1} + A_{n-2})/3 \text{.......(stance phase)} \\ (Ag_n + A_{n-1})/2 \text{...............(landing phase)} \\ Ag_n \text{...................................(swing phase)} \end{cases} \quad (1)$$

A_n: The estimated value of the gradient of landing surfaces, in the detection of landing surfaces at "n" times.

Ac_n: The gradient acquired by clinometers, in the detection of landing surfaces at "n" times.

Ag_n: The result of which the relative angle of the upper- and the lower-foot plates (A'_n) is added to the inclination value obtained from the value integrated by rate-gyro sensors (100 Hz), in the detection of landing surfaces at "n" times.

4.2 Method of Lower-limbs Motion Modification

In this adaptive walking control method, the one-step phase is largely divided into the three phases such as the swing phase with no restriction, the support phase with full restriction, and the landing phase of intermediate, and the swing phase and the support phase are divided into the double support phase and the former and the latter single support phase additionally, according to the degree of which the feet of robots are restricted with path surfaces, to alter and control a lower-limbs motion. The walking action in one step walking using the walking control method is shown in **Fig. 6**. Walking motions at each phase

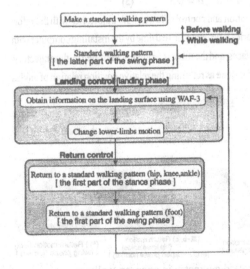

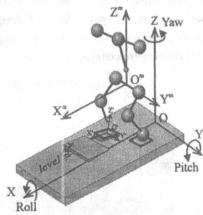

O''-$X''Y''Z''$: moving coordinate
O-XYZ : fixed coordinate
O'-$X'Y'Z'$: detection coordinate

Fig. 4 Outline of the adaptive walking control method   **Fig. 5** Coordinate systems for adaptive walking

are as follows.

I) Latter Swing Phase: The robot walks with the standard walking pattern, until the preset swing-leg landing time in the standard walking pattern, or until swing leg lands after the time.

II) Landing Phase: In this study, we define the landing phase as the period between the time when the swing leg touches the path surface or the preset landing time of the foot plate has passed on the standard walking pattern and the beginning of a double support phase on the standard walking pattern. During this phase foot lowering operation is performed using expression (2) so that the actual distance between the foot plates follow the changes in the theoretical distance between the foot plates when walking on the horizontal, smooth path on the standard walking pattern. **Fig. 7** shows the appearance of the correction of the lower-limb trajectories.

$$Z_{o'(n)} = Z_{o'(n-1)} + (H_{i(n)} \cos\alpha_n - H_{r(n-1)} \cos\alpha_{n-1})$$
$$X_{o'(n)} = X_{o'(n-1)} + (H_{i(n)} \sin\alpha_n - H_{r(n-1)} \sin\alpha_{n-1}) \tag{2}$$

$Z_{o'(n)}, X_{o'(n)}$: The Z^m and the X^m coordinates of the origins of the detection coordinates, in the detection of landing surfaces at "n" times.

$H_{i(n)}$: The simulated distance between foot plates when walking on horizontal and flat surfaces with the standard walking patterns, in the detection of landing surfaces at "n" times.

$H_{r(n)}$: The actual distance between foot plates, in the detection of landing surfaces at "n" times.

α_n: The target angle of the upper-foot plate, in the detection of landing surfaces at "n" times.

Here, explanatory to say about the method for deciding the target angle α of the upper-foot plate, actually there are deviations in also actual models caused by the deviation of structural materials, the response delay of actuators, and the slight fluctuation of the throughout model in walking, and so on, in addition to the deviation in the shape of path surfaces. Therefore, when setting a gradient of the landing surface as a target value, the robot can not land smoothly. When setting an angle between foot plates A' as a target value for smooth landing, the deviation to the deviation of path surfaces becomes larger in using as supporting legs, though it can land smoothly. The target angle is decided with the expression (3) given below, consequently. The weighting factors "a" and "b" were decided as 0.7 and 0.3 respectively, from the result of the walking experiment with the machine model WL-12RVII.

$$\alpha_n = a(A'_n + \alpha_{n-1}) + bA_n, a = 0.7, b = 0.3. \tag{3}$$

III) Support phase: **(III-a)** At the double support phase, the alteration and control of lower-limb trajectories with the information on landing surfaces are stopped and the lower-limb trajectories is generated in real time by the installing computer with the setting of total altering amounts as the deviation of landing surfaces, and the robot is program-controlled by this trajectory. **(III-b)** At the former single support phase, such returning control is done as returning the relative position relation of ankles

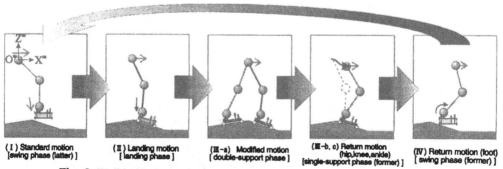

(I) Standard motion [swing phase (latter)]　(II) Landing motion [landing phase]　(III-a) Modified motion [double-support phase]　(III-b, c) Return motion (hip,knee,ankle) [single-support phase (former)]　(IV) Return motion (foot) [swing phase (former)]

Fig. 6 Walking actions adapting to the ground geometry in one step-walking

and waists accepted the trajectory alteration to the relative position relation in the standard walking pattern. The alteration of trajectories in the returning control is done by the method including an interpolation with the fifth polynomial, to serialize the position, velocity, and acceleration toward the X and the Y axes, to lower the deviation to the preset ZMP. **(III-c)** At the latter single support phase, the trajectory complies with the standard walking pattern, except for altering an angle of ankle joints (foot plates) in accordance with the inclination of path surfaces.

IV) Former Swing Phase: The angle of ankle joints (foot plate) of which the return to the standard walking pattern that does not complete is returned to the standard walking pattern, by the interpolation with the fifth polynomial.

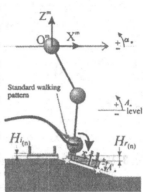

Fig. 7 Control of landing motion

5 Walking experiments

The authors conducted walking experiments using the proposed walking control method and the biped walking robot WL-12RVII with WAF-3. The following a summary of the experimental results.

1. Adaptive dynamic biped walking adapting to humans' living floor with unknown shape was realized. The walking speed was 1.28 s/step with a 0.3 m step length, and the adaptable deviation range was from -16 to +16 mm/step in the direction of Z axis, and from -3 to +3 ° in the tilt angle. This result is almost corresponding to the walking simulation result.

2. On a tatami mat characteristic in Japanese living environment, the surface of which sank about 2 mm when the machine model was placed on it with a single leg supporting, stable dynamic biped walking was realized as well as on the wooden floor that hardly transformed even if the machine model walked over it.

3. Obtaining information about the gradient of the walking surface throughout all phases of dynamic walking was successfully accomplished.

4. The adaptive dynamic biped walking success probability was approximately 100% (no falls after about 50 trials).

As one example of the walking experiment result on the path with the severest conditions, **Fig. 8** and **Fig. 9** show the measurement values of the gradient of the landing path surface and the distance between upper-foot plate and lower-foot plate when WL-12RVII with WAF-3 adapted to an unknown trapezoid placed on the horizontal, smooth floor while walking eight steps including the beginning and the end of the walk. Considering human's living environment, for the walking surface, the wooden floor covered with a short-piled wool carpet is used. This walking surface sinks about 1.5 mm in the vertical direction to the surface when the machine model is supported with a single leg.

6 Conclusions

The purpose of this study is to aim at stable and practicable dynamic walking under human's living environment by a life-size robot and to achieve stable biped walking by a life-size robot that adapts in real time to the path with deviation in its shape.

The authors first developed a foot mechanism that can measure the relative position and gradient during dynamic walking. Next, we proposed an adaptive walking control method with closed loop control system, which adapts in real time to the path with deviation in its shape by using multiple pieces of information acquired from the foot mechanism developed. In addition, performing the walking experiment using the biped walking robot developed, we achieved dynamic walking that adapts in real time to the environment with deviation in the shape of the path surface.

In conclusion, the effectiveness of the walking control method proposed in this paper and the developed walking system have been experimentally supported.

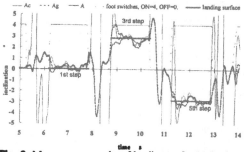
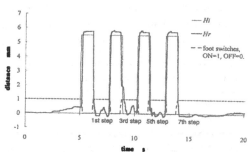

Fig. 8 Measurement results of landing surface's absolute inclination during dynamic walking adapting to an unknown trapezoidal surface. Inclination +3° (3rd step), -3° (5th step), length: 0.3 m/step, step time: 1.28 s/step, right leg.

Fig. 9 Distance between upper-foot plate and lower-foot plate during dynamic walking adapting to an unknown trapezoidal surface. Inclination +3° (3rd step), -3° (5th step), length: 0.3 m/step, step time: 1.28 s/step, right leg.

Acknowledgments

This study has been conducted as a part of the project: Humanoid at HUREL (HUmanoid REsearch Laboratory), Advanced Research Center for Science and Engineering, Waseda University. The authors would like to thank ATR, NAMCO Ltd., and YASKAWA ELECTRIC Corp., for their cooperation in this study. The authors would also like to thank ANALOG DEVICES Inc., Japan Aviation Electronics Industry, Ltd., Kuroda Precision Industries, Ltd., MOOG JAPAN Ltd., Toray Industries, Inc., and YUKEN KOGYO Co., Ltd. for supporting us in developing the hardware for a biped walking robot in the process of this study. On June 19, 1994, The authors were deprived of a joint researcher Ichiro Kato who was Professor at Waseda University. The authors would like to thank him for his advice to this study, and pray his soul may rest in peace.

References

[1] A. Sano, J. Furusho, "Realization of natural dynamic walking using the angular momentum information", *Proceedings of the 1990 IEEE International Conference on Robotics and Automation*, pp. 1476-1481, 1990.

[2] A. Sano, J. Furusho, "Control of torque distribution for the BLR-G2 biped robot", *Proceedings of the 1991 IEEE International Conference on Robotics and Automation*, pp. 729-734, 1991.

[3] S. Kajita, T. Yamaura, A. Kobayashi, "Dynamic walking control of a biped robot along a potential energy conserving orbit", *IEEE Transaction on Robotics and Automation*, Vol. 8, No. 4, pp. 431-438, 1992.

[4] S. Kajita, K. Tani, "Study of dynamic biped locomotion on rugged terrain -theory and basic experiment-", *Proceedings of the 1991 International Conference on Advanced Robotics*, pp. 741-746, 1991.

[5] A. Takanishi Y. Egusa, M. Tochizawa, T. Takeya, I. Kato, "The realization of dynamic walking by the biped walking robot WL-10RD", *Proceedings of the 1985 International Conference on Advanced Robotics*, pp. 459-466, 1985.

[6] A. Takanishi Y. Egusa, M. Tochizawa, T. Takeya, I. Kato, "Realization of dynamic biped walking stabilized with trunk motion", *RoManSy 7: Proceedings of 7th CISM-IFToMM Symposium on Theory and Practice of Robots and Manipulators, Hermes, Paris*, pp. 68-79, 1988.

[7] A. Takanishi H. Lim, O. Tsuda, I. Kato, "Realization of dynamic biped walking stabilized by trunk motion on a sagittally uneven surface", *Proceedings of the 1990 IEEE International Workshop on Intelligent Robots and Systems*, pp. 323-330, 1990.

[8] Y. F. Zheng, J. Shen, "Gait synthesis for the SD-2 biped robot to climb sloping surface", *IEEE Transaction on Robotics and Automation*, Vol. 6, No. 1, pp. 92-102, 1990.

[9] J. Yamaguchi, A. Takanishi, and I. Kato, "Development of a Biped Walking Robot Compensating for Three-Axis Moment by Trunk Motion", *Proceedings of the 1993 IEEE/RSJ International Conference on Intelligent Robots and Systems*, pp. 561-566, 1993.

[10] J. Yamaguchi, A. Takanishi, and I. Kato, "Development of a Biped Walking Robot Adapting to a Horizontally Uneven Surface", *Proceedings of the 1994 IEEE/RSJ/GI International Conference on Intelligent Robots and Systems*, pp. 1156-1163, 1994.

[11] J. Yamaguchi, A. Takanishi, and I. Kato: "Experimental Development of Foot Mechanism with Shock Absorbing Material for Acquisition of Landing Surface Position Information and Stabilization of Dynamic Biped Walking", *Proceedings of the 1995 IEEE International Conference on Robotics and Automation*, pp. 2892-2899, 1995.

SIMULATION OF AN ANTHROPOMORPHIC RUNNING ROBOT
USING THE MULTIBODY CODE MECHANICA MOTION

H. De Man, D. Lefeber, J. Vermeulen and B. Verrelst
Vrije Universiteit Brussel, Brussel, Belgium

1. Introduction

The main scope of this paper is twofold: an algorithm to control the motion of an anthropomorphic running robot is presented and a description is given of the commercial multibody code MECHANICA Motion, used for the simulations. Attention is given to the simulation code to show that it is possible to simulate rather unconventional mechanisms, such as legged robots, using general purpose multibody codes, thus avoiding the time-consuming task of writing a software code for the specific model at hand.

In the next section a description is given of the way in which a mechanical system can be built, analysed and optimized using the multibody code. In section 3 the main features of running robots are reviewed and a brief overview of theoretical and practical studies in this field is given. The mechanical model and mathematical representation of the anthropomorphic robot under study are given in section 4, as well as the control algorithms enabling the robot to run. Simulation results are presented and remarks are made in section 5. Finally, in section 6, conclusions are drawn and comments on further research are made.

2. MECHANICA Motion

MECHANICA Motion is a multibody code with a graphical interface. Parts or links of a system are created by specifying and grouping a number of previously defined points and then connected with other parts or with the world by choosing a specific type of joint, belonging to a library. Forces, also belonging to a library, are generated by applying them to a specific point or on a specific translational or rotational joint axis. Special loads which do not belong to the library can be programmed by the user in seperate routines (in FORTRAN or in C) and linked to the model. A joint axis can be forced to move in a predefined way by creating a driver on that axis.

Parameters can be added to the model, as well as measures and design variables. A parameter can be any constant relevant to the model. Changing the value of a parameter can alter a load, the position of a point, a measure, etc., without the simulation code having to regenerate the equations of motion. A measure can be any physical quantity the user wants to track during an analysis. Some measures belong to a library, any other measure can be defined by the user, using parameters and other measures. Design variables depend on a parameter and are used in sensitivity design studies.

The proper design of a specific system is greatly facilitated by the design studies MECHANICA Motion can perform, being a global sensitivity study, a local sensitivity study and an optimization study. A global sensitivity study calculates the changes in the model's measures by varying a parameter, and the corresponding design variables, over a specified range. Executing this study the user gets an insight in the dependency of the model to parameter changes. A local sensitivity study calculates the sensitivity of the measures to slight changes in one or more parameters, making it possible to determine which parameters have the most effect on the model's performance. Finally, in an optimization study the program calculates the value or values of one or more parameters needed to best achieve a specified design goal.

3. Running robots

Legged robots can be classified in two different categories: those that walk and those that run. The main difference between walking and running originates from the fact that during walking at least one of the legs supporting the robot's body is in contact with the ground, whereas during running this condition must not be satisfied. A common definition of running is that it is a locomotion pattern characterized by ballistic flight phases during which all feet are lifted off the ground.

Compared to a walking machine, a running machine can attain higher forward velocities and can take steps with a greater length and a greater height. The control problem however is more complex. A walking robot can be controlled in a kinematic way by making sure that the projection of its center of gravity lies in the polygon formed by the contour of its supporting foot or feet, neglecting the inertial forces. A running robot can only be balanced in a dynamic way, by taking into account and manipulating in the right way the inertial forces acting on it.

The robots developed at the MIT by Raibert and his team are probably the best-known running robots. The basic algorithm used to control their motion consists of three decoupled parts: energy stored in a pneumatic spring in the legs is modulated to manipulate the hopping height, forward speed is controlled by positioning the legs during the flight phase and body attitude is regulated during the stance phase. Based on this principle a one-legged hopping robot, a bipedal running robot and a quadrupedal running robot were constructed [1,2,3].

Recently a number of researchers showed renewed interest in Raibert's hopping robot, focusing on its stability from a mathematical and system theory point of view [4,5,6,7,8]. A new algorithm, controlling forward velocity, hopping height and foot placement was developed by De Man et al. [9]. An electrically actuated one-legged hopping robot has been built by Papantoniou [10] and other designs with electrical actuators are studied by Mehrandez et al. [11] and by Rad et al. [12]. A rope-hopping robot has been constructed by Gokan et al. [13].

4. Anthropomorphic running robot

4.1. Model

In figure 1 the model of the anthropomorphic robot is given, running on a horizontal plane in the gravitational field. Although the model is 3-dimensional, its motion is restricted in such a way that all its points move in parallel vertical planes.

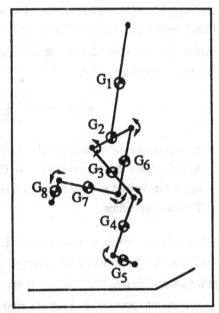

The robot consists of a trunk, a pelvis and two identical legs. Each leg consists of an upper leg, a lower leg and a foot. All the segments are connected through pin joints, except for the trunk and the pelvis, which are rigidly connected. The length of the i'th segment is l_i, its mass is m_i and I_i is the moment of inertia about the axis through the center of gravity G_i perpendicular to the sagittal plane.

figure 1: anthropomorphic running robot

The actuation of the robot consists of a passive and an active part. The passive part is formed by torsional springs, placed at the hips and the knees. The active part is built up by six rotational actuators, exerting a PD-controlled torque at the two hips, at the two knees and at the two ankles respectively. The gains of all the PD-controllers are first roughly estimated and then tuned by performing optimization studies that minimize the torques. Torque limiters keep the torques between fixed thresholds.

The ground is modelled as a spring-damper combination in the vertical direction and as a Coulomb friction force in the horizontal direction.

4.2. Control algorithms

Running is considered as alternately hopping, in a symmetrical way, on the right and the left leg. The leg that will be the next supporting leg gets a marker from a boolean operator the moment its knee passes, in the direction of motion, the knee of the actual supporting leg. At the same moment the actual supporting leg loses this marker. To guarantee the alternating and symmetrical motion between the two legs, the actuator in the hip of the unmarked leg tracks the angle with respect to the vertical of the upper leg of the marked leg, hereby reversing the angle's sign. To avoid the foot of the unmarked leg hitting the ground when passing the marked leg, the actuator in the knee of the unmarked leg tracks the angle of the knee of the marked leg, diminishing it by a fixed angle. The actuator in the ankle of the unmarked leg keeps the angle of the ankle fixed.

The average forward velocity of the robot is controlled by a slightly modified version of the algorithm developed by Raibert [1]. The horizontal position x_{np} of the neutral point where the ankle (coinciding with the heel) of the next supporting leg has to touch the ground to achieve the desired velocity, is given by:

$$x_{np} = x_G + A \frac{\dot{x}_G T_s}{2} + B (\dot{x}_G - \dot{x}_{des})$$ (1)

with: x_G = the actual horizontal position of the global center of gravity G
$\dot{x}_G$ = the actual horizontal velocity of G
$\dot{x}_{des}$ = the desired average horizontal velocity of G
T_s = the stancetime

A and B are two constants whose numerical values are determined by performing an optimization study, minimizing the average difference between the desired and the actual average forward velocity. The constant A takes the asymmetry of the trajectory of the global center of gravity into account. In Raibert's algorithm this asymmetry is ignored.

Simultaneously with the actuator in the knee, which brings the leg to a nearly stretched configuration at the moment of touch-down, and the actuator in the ankle, which keeps the foot horizontal, the actuator in the hip generates the following torque to reach the neutral point:

$$T_1 = P_1 (x_{np} - x_a) + D_1 (\dot{x}_h - \dot{x}_a)$$ (2)

with: x_a = the horizontal position of the ankle
$\dot{x}_a$ = the horizontal velocity of the ankle
$\dot{x}_h$ = the horizontal velocity of the hip
P_1 = the proportional gain
D_1 = the derivative gain

The orientation of the trunk during stance is controlled by the actuator in the hip of the supporting leg. This actuator uprights the trunk to a nearly vertical position, letting it slightly incline in the direction of motion and letting it have a backward angular velocity with respect to the direction of motion. For specific combinations of this positive inclination and this backward angular velocity, the angular momentum about the global center of gravity of the robot at take-off is close to zero, thus limiting the global rotation of the robot during the next flight phase. The generated torque is given by:

$$T_2 = P_2 \{ C - (x_{G_t} - x_h) \} + D_2 \{ D - (\dot{x}_{G_t} - \dot{x}_h) \} \tag{3}$$

with: x_{G_t} = the horizontal position of the center of gravity of the trunk G_t

$\dot{x}_{G_t}$ = the horizontal velocity of G_t

x_h = the horizontal position of the hip

P_2 = the proportional gain

D_2 = the derivative gain

The numerical values for the constants C and D are determined by performing an optimization study, minimizing the angular momentum about the global center of gravity at take-off.

Since the torque is exerted between the trunk and the upper leg, the knee tends to overstretch. To counteract this effect, the actuator in the knee keeps the knee angle fixed during the first half of the stance phase.

To maintain running, torsional springs are placed at the hips and the knees, changing kinetic energy of the robot into potential energy the first half of the stance phase and releasing this energy the second half of the stance phase. Since the robot loses a part of its energy every time impact with the ground occurs, extra energy has to be added. This is done by releasing a second, prestressed torsional spring in the knee of the supporting leg during the second half of the stance phase and by exerting a constant torque in its ankle, pressing the foot against the ground. The running height, defined as the difference between the maximum height of the global center of gravity during flight and its height at take-off, is controlled by the amount of energy that is added. To attain a specific running height, the numerical values for the spring constant of the prestressed torsional spring, for its resting angle and for the torque in the ankle, are first estimated by trial and error and then refined by performing an optimization study, minimizing the difference between the actual and the desired running height.

5. Results and remarks

To test the algorithms described above, running with a desired average forward velocity of $1m/s$ and a running height of $0.07m$ has been simulated. To initiate running, the robot is dropped with its body oriented vertically, its right leg oriented downward, its left leg oriented backward and all initial velocities equal to zero. The parameters of the model of the simulated robot are given in table 1.

trunk: $l_1 = 0.4$; $m_1 = 10$; $I_1 = 0.44$	$[l] = m$
pelvis: $l_2 = 0.25$; $m_2 = 0.5$; $I_2 = 3.4\text{e-}5$	$[m] = kg$
upper legs: $l_3 = l_6 = 0.2$; $m_3 = m_6 = 1$; $I_3 = I_6 = 3.5\text{e-}3$	$[I] = kgm^2$
lower legs: $l_4 = l_7 = 0.2$; $m_4 = m_7 = 1$; $I_4 = I_7 = 3.5\text{e-}3$	
feet: $l_5 = l_8 = 0.075$; $m_5 = m_8 = 0.5$; $I_5 = I_8 = 2.5\text{e-}4$	

table 1: model parameters

Figure 2 shows the forward velocity $\dot{x}_G$ of the global center of gravity versus time. Steady-state running, with the average velocity being equal to the desired value, is attained after a small number of steps. The vertical position y_G of the global center of gravity of the robot is given in figure 3. The running height is given by the distance between the two horizontal lines and is equal to the desired value. The angle θ_1 of the trunk with respect to the vertical and the angular momentum M_G about the global center of gravity are given in figures 4 and 5 respectively. These two figures show that the control strategy for the posture of the trunk during stance and the limitation of the global rotation of the robot during flight works effectively, since the orientation of the trunk at the end of the stance phase is nearly vertical and since the angular momentum is nearly zero during flight. Figure 6 shows the angles θ_3 and θ_6 with respect to the vertical, of the upper right leg and the upper left leg respectively. It can be seen that the legs move in a nearly perfect symmetrical alternating way.

Figures 7 to 9 show the torques $T_{1,2}$, $T_{2,3}$ and $T_{3,4}$ generated by the PD-controlled actuators in the hip of the right leg, in its knee and in its ankle. These results are given for one steady-state stride period, starting and ending at the moment of take-off on the right leg. The torques in the left and the right leg are the same, but are shifted in time over half a stride period. The peaks in the middle and at the end of the stride period are due to impact.

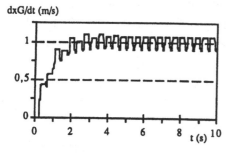

figure 2: forward velocity of G

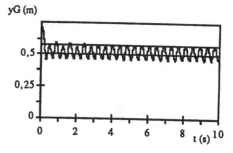

figure 3: vertical position of G

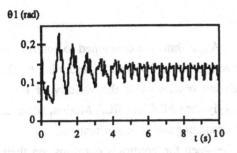

figure 4: orientation of the trunk

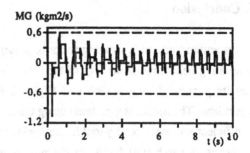

figure 5: angular momentum about G

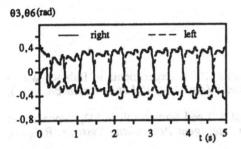

figure 6: orientation of the upper legs

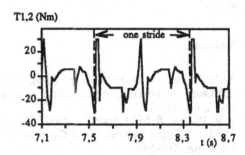

figure 7: torque in the right hip

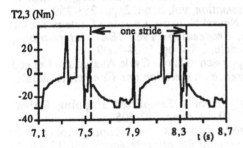

figure 8: torque in the right knee

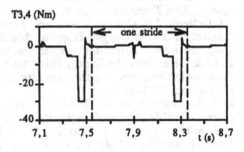

figure 9: torque in the right ankle

The figures show that the robot acts in the desired way. A few remarks can be made however. Firstly, the different algorithms of the global control strategy are not decoupled. If for example a higher or lower running speed is demanded, another angular velocity of the trunk at the end of stance will be needed to limit the robot's global rotation during flight. Secondly, the active and the passive part of the robot's actuation do not always cooperate. The active part sometimes has to perform extra work to overcome work done by the passive part. And thirdly, the control algorithms as they are formulated now, for example using the velocity of the global center of gravity to determine the position of the neutral point, may demand to much hardware and too much computer power to be used in practice. Further research has to be done on all these aspects.

6. Conclusion

An anthropomorphic running robot is presented. Algorithms are developed to control the alternating behaviour of the legs of the robot, its average forward velocity, its running height and its posture during stance and flight. Simulation results show the validity of these algorithms. The model is simulated using the multibody code MECHANICA Motion, which is an efficient and user friendly tool for designing and simulating complex mechanisms.

Future research will focus on the interaction between the control algorithms, on their practical feasability and on the interaction between the passive and the active part of the robot's actuation.

References

1 M.H. Raibert, H.B. Brown (Jr.) & M. Chepponis, 1984, Experiments in Balance with a 3D One-Legged Hopping Machine, *International Journal of Robotics Research*, vol. 3, no. 2, pp. 75-92.

2 R.R. Playter & M.H. Raibert, 1992, Control of a Biped Somersault in 3D, *Proceedings 1992 IEEE/RSJ International Conference on Intelligent Robots and Systems*, Raleigh, USA, pp. 582-589.

3 M.H. Raibert, 1990, Trotting, Pacing and Bounding by a Quadruped Robot, *Journal of Biomechanics*, vol. 23, no. 1, pp. 79-98.

4 M. Sznaier & M.J. Damborg, 1989, An Adaptive Controller for a One-Legged Mobile Robot, *IEEE Transactions on Robotics and Automation*, vol. 5, no. 2, pp. 253-259.

5 J.J. Helferty, J.B. Collins & M. Kam, 1989, A Neural Network Learning Strategy for the Control of a One-Legged Hopping Machine, *Proceedings 1989 IEEE International Conference on Robotics and Automation*, Scottsdale, USA, pp. 1604-1609.

6 Z. Li & J. He, 1990, An Energy Perturbation Approach to Limit Cycle Analysis in Legged Locomotion Systems, *Proceedings 29th Conference on Decision and Control*, Honolulu, Hawaii, pp. 1989-1994.

7 D.E. Koditschek & M. Bühler, 1991, Analysis of a Simplified Hopping Robot, *International Journal of Robotics Research*, vol. 10, no. 6, pp. 587-605.

8 R.T. M'Closkey & J.W. Burdick, 1993, Periodic Motions of a Hopping Robot with Vertical and Forward Motion, *International Journal of Robotics Research*, vol. 12, no. 3, pp. 197-218.

9 H. De Man, D. Lefeber, F. Daerden & E. Faignet, 1996, Simulation of a New Control Algorithm for a One-Legged Hopping Robot (Using the Multibody Code MECHANICA Motion), *Proceedings International Workshop on Advanced Robotics and Intelligent Machines*, April 2-3, Manchester, U.K.

10 V.K. Papantoniou, 1991, Theoretical and Experimental Approach on Actively Balanced Legged Robots, Ph.D dissertation, Université Libre de Bruxelles, Brussels, Belgium.

11 M. Mehrandezh, B.W. Surgenor & S.R.H. Dean, 1995, Jumping Height Control of an Electrically Actuated, One-Legged Hopping Robot: Modelling and Simulation, *Proceedings 34th IEEE Conference on Decision and Control*, New Orleans, USA, pp. 1016-1020.

12 H. Rad, P. Gregorio & M. Buehler, 1993, Design, Modeling and Control of a Hopping Robot, *Proceedings 1993 IEEE/RSJ International Conference on Intelligent Robots and Systems*, Yokohama, Japan, pp. 1778-1785.

13 M. Gokan, K. Yamafuji & H. Yoshinada, 1994, Postural Stabilization and Motion Control of the Rope-Hopping Robot, *JSME International Journal Series C*, vol. 37, no. 4, pp. 739-747.

GAIT CONTROL METHOD OF ROBOT
AND BIOLOGICAL SYSTEMS BY USE OF HOJO-BRAIN

Y. Sankai
University of Tsukuba, Tsukuba, Japan

Abstract: The purpose of this research is to propose and develop a new control method in the robotic field and the bio-medical field, which is configured by the robotic/biological simulator, analytical motion mode, sensory feedback and the artificial CPG which is constructed by recurrent neural network (RNN) and genetic algorithm (GA). We call such controller "HOJO-Brain", which means supplementary brain for motion control. We apply this method in the robotic field and the bio-medical field. In the robotic field, this HOJO-Brain is applied to a 5-DOF legged locomotion robot. In the bio-medical field, this HOJO-Brain is applied to animals as the FES (Functional ElectroStimulation) controller. This FES control system with HOJO-Brain would have a possibility to realize more effective and emergent motion control for severely physically handicapped persons such as the quadriplegia. By the computer simulations and the simple actual experiments using animals, we could confirm the fine adaptivity and emergence for the motion control. This HOJO-Brain strategy might be regarded as essential for the total motion control.

Introduction

Patients who damaged spinal cords or physically handicapped person who has motility disturbance by cerebral infantile palsy are restrained to the wheel chair or the bed and inconvenient life is forced on them. In the robotic field, it is difficult to identify the parameters of nonlinear or time-variant systems. Moreover, it is also difficult to design the controller for the robot whose motion is unknown. The purpose of this research is to propose and develop a new control method in the robotic field and the bio-medical field, which is configured by the robotic/biological simulator, analytical motion mode, sensory feedback and the artificial CPG[1,2] which is constructed by recurrent neural network (RNN) and genetic algorithm (GA). We call such controller "HOJO-Brain", which means supplementary brain for motion control. In this research, it is the one to develop the control strategy by the supplementary brain "HOJO-Brain" artificially added besides the motion control system which the living body originally possesses for the physically handicapped person who has the trouble in the signal transmission from the motion control system, based on the conception of newly building in an adaptive supplementary brain "HOJO-Brain". We apply this method in the robotic fields and the bio-medical field. In the robotic field, this HOJO-Brain is applied for a 5-DOF legged locomotion robot. In the bio-medical field, this HOJO-Brain is applied to animals as the FES (Functional ElectroStimulation) controller[3]. The special mark of this research shown in Fig.1 is *cybernic controller* which consists of analytical motion modes and *self-organized artificial CPG/PSs = Recurrent Neural Network(RNN) + Genetic Algorithm(GA)*.

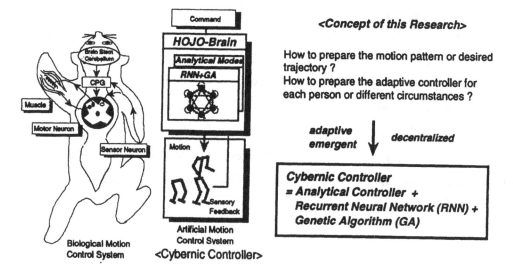

Fig.1 Basic concept of this research

Method and Results

The configuration of HOJO-Brain is shown in Fig.2. The artificial central pattern generator (CPG/PGs) and the analytical controller are main parts of this system. Sensory signal such as displacement, velocity, acceleration, tactile information are fed back to the CPG/PGs and the controller to perform self-organization, parameter tuning in some motion modes, state feedback. In this level, we can choose the status in the CPG/PGs, i.e., we can use the plain CPG (no learning) or pre-learned CPG by use of the simulator described as virtual human or biological system or robot system. The purpose of this virtual biosimulator is to realize pre-learning and to reduce the actual repetition. It is very similar to the image training.

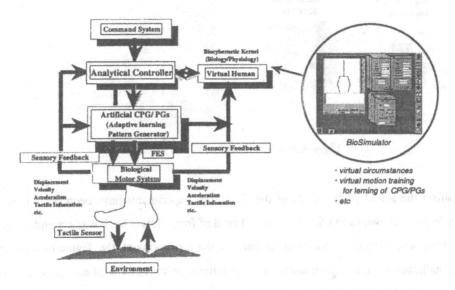

Fig.2 Configuration of "HOJO-Brain"=CPG/PGs and analytical controller

The neuron model adopted here is recurrent neural network (RNN). Furthermore, each neuron has a time constant and mutual linkages. Therefore, this network can learn the dynamic motion pattern. The state equation of *i-th* neuron is shown in Fig.3. The synaptic weights and time constant are modified by using genetic algorithm (GA). In this research, RNN weight matrix as the gene is not integer but real number. Fig.4 is GA calculation and the algorithm in this research.

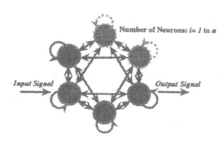

State Equation of i th Neuron:

$$\tau_i \frac{dx_i}{dt} = -x_i(t) + \sum_{j=1}^{n} w_{ij} y_j(t) + u_i(t)$$

$$y_i(t) = f(x_i(t)) \qquad f(x) = \frac{1}{1 + e^{-x}}$$

$x_i(t)$: *state of i th neuron,* $u_i(t)$: *input signal to i th neuron,*
$y_i(t)$: *output from j th neuron to i th neuron*
τ_i: *time constant of i th neuron,* w_{ij}: *synaptic weight coef. of i th neuron*
$f(x)$: *sigmoid function, n: number of neurons*

Fig.3 Architecture of neural network and state equation

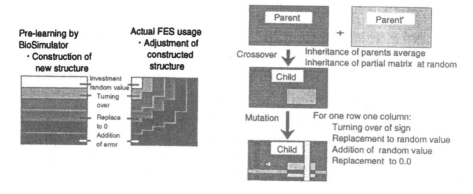

Fig.4 Genetic Algorithm in this research

To confirm the actual performance of the CPG/PGs, experiments have been performed for twenty frogs as shown in Fig.5. Every spinal cord of frog's lower back was cut and every leg below knee was connected to FES controller. In these experiment, the degree of freedom is one. Conditions in these experiments are set as follows; RNN[number of neuron unit=8, reset inner states of CPG=0], GA[number of pieces=64, mutation rate in pre-learning= 0.15(cf. lerrorl>10% then 0.3), mutation rate in after-learning=0.05(cf. lerrorl>10% then 0.3), performance index of fitness=only regulation(desired value=50 degree)]. Results are shown in Fig.6. In this case, pre-learning of the artificial CPG by use of the simulator was performed previously before the actual experiment. In the simulation, we confirm the advantage of this method for parameter perturbation (e.g. 30%). Of course, parameters of living frogs are unknown and different. After experiments, relatively fine results are obtained only 15--19 generations as shown in Fig.6.   In the left figure in FIg.6, we can find a overshoot, but amplitudes of input signals are relatively small.

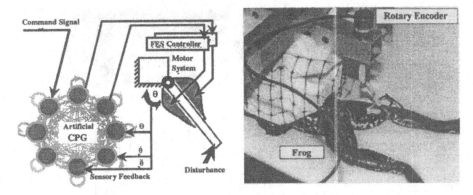

Fig.5 Experiment

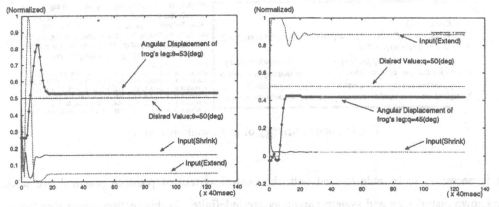

Fig.6 Results of adaptation after 15--19 genaerations

In the right figure(Fig.6), there is no overshoot but amplitudes of input signals are relatively large. In this experiment, performance index is only regulation value. If we need other indices, we can prepare some indices for judgment of fitness, e.g., "regulation, overshoot, input amplitude".

Next, we try to apply this controller to the biped locomotion robot in simulations. In this simulation shown in Fig .7, the model of biped robot has 3-DOF and the knee is restrained (0 degree to 180 degree: free). The control frames, i.e., strategy and modes are designed by designers, but parameters and motion timing are searched by the HOJO-Brain controller by the self-organized mechanism. The control frame is shown in Fig.7. After virtual trial and error, stable gait(walking) control is obtained. Searched parameters are also shown in Fig.7.

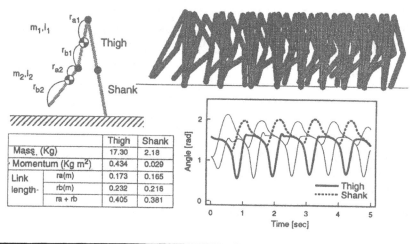

		Thigh	Shank
Mass (Kg)		17.30	2.18
Momentum (Kg m²)		0.434	0.029
Link length	ra(m)	0.173	0.165
	rb(m)	0.232	0.216
	ra + rb	0.405	0.381

Strategy:

$$\tau_h = K_1(v_{hr} - v_h)$$

$$\tau_k = K_2(h_{tr} - h_t) + K_3(\theta_{kr} - \theta_k) + D_1(\dot{\theta}_{kr} - \dot{\theta}_k) + T_1 + K_4(v_{hr} - v_h)$$

Fitness: $J = A_1 x_h + A_2 y_h + A_3 n$

τ_h: *hip joint torque* θ_h: *hip joint angle* $\dot{\theta}_h$: *hip joint angular velocity*
τ_k: *knee joint torque* θ_k: *knee joint angle* $\dot{\theta}_k$: *knee joint angular velocity*
v_h: *hip joint velocity* (x – *axis direction*) x_h: *hip joint position* (x – *axis*)
h_t: *hight of toe* n: *number of steps* y_h: *hip joint position* (y – *axis*)

<Mode1> the first half of right leg: free, left leg: support
<Mode2> the latter half of right leg: free, left leg: support
<Mode3> both of legs: support, right leg: forward
<Mode4> the first half of left leg: free, right leg: support
<Mode5> the latter half of left leg: free, right leg: support
<Mode6> both of legs: support, left leg: forward
$(K_1, K_2, K_3, K_4, T_1, D_1, v_{hr}, h_{tr})$
< Mode1 > = (7.1, – 49.9, 509, – –, – –, 49.5, 0.87, 0.11)
< Mode2 > = (6.2, – 49.2, 495, 12.5, – 1.9, 51.1, 0.76, 0.1)
< Mode3 > = (– 6.5, 50.1, 503, 7.9, 0.78, 50.5, 0.88, 0.11)
< Mode4 > = (– 7.0, 49.2, 500, – –, – –, 48.7, 0.75, 0.1)
< Mode5 > = (– 6.3, 50.9, 504, – 10.2, 0.54, 50.2, 1.2, 0.11)
< Mode6 > = (6.1, – 50.7, 502, – 8.4, – 1.2, 49.9, 0.82, 0.11)

Fig.7 Control strategy of gait : walking control

It is necessary to be able to deal with the control system of possible practical use when unknown disturbance and system parameter are indefinite. In this section, some simulations are performed to confirm the performance of the HOJO-Brain controller. The aim of this simulation is to obtain effective attitude/posture control methods such as not only straddling or holding but also walking when the unknown disturbance is added to biped robot or human. 5-DOF model is adopted here. Control frame is shown in Fig.8.

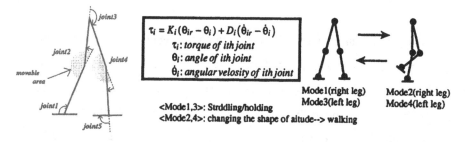

$$\tau_i = K_i(\theta_{ir} - \theta_i) + D_i(\dot{\theta}_{ir} - \dot{\theta}_i)$$

τ_i: *torque of ith joint*
θ_i: *angle of ith joint*
$\dot{\theta}_i$: *angular velosity of ith joint*

Mode1(right leg)   Mode2(right leg)
Mode3(left leg)   Mode4(left leg)

<Mode1,3>: Strddling/holding
<Mode2,4>: changing the shape of aitude--> walking

Fig.8 Control strategy of gait : posture/attitude and walking control

Parameters(K,D) and motion timing are searched by the similar way mentioned above. Mode1,3 are the straddling or holding states and Mode2,4 are the states of changing the shape of attitude. In the simulation, joint(1,2,4,5) have torque saturation assumed 7[N] and joint3 has torque saturation assumed 14[N]. In Fig.9, state of the mode is switching according to the disturbance and biped system stand still after disturbance. In case of continuous disturbance shown in Fig.10, the biped system is walking according to the disturbance. This disturbance is similar to power assist. In case of cyclic disturbance shown in Fig.11, gait control mode is switched according to the state of the attitude.

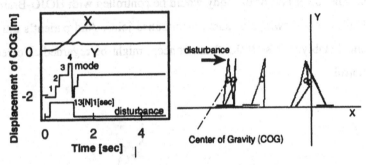

Fig.9 Simulation: disturbance 13[N]1[sec]

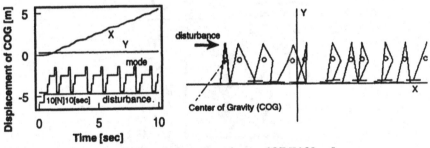

Fig.10 Simulation: disturbance 10[N]10[sec]

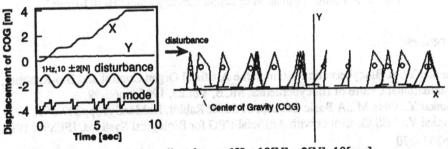

Fig.11 Simulation: disturbance 1Hz, 10[N]+-2[N], 10[sec]

Discussion and Conclusions

We could confirm the excellent adaptivity and emergence for the motion control by using this supplementary brain "HOJO-Brain", i.e., cybernic controller which consists of CPG.PGs and analytical controller. The combination of analytical controller and CPG/PGs would be suitable structure. Especially, it must be practical good use to apply this HOJO-Brain for the physically handicapped person who has the trouble in the lumbar vertebra (Fig.12). In this example, the upper half of the body and the lower half of the body will have a separate control system. The lower half of the body would be controlled with HOJO-Brain even when the upper half of the body swings or some disturbance joins. And patient's attitude/posture will be maintained stably. This HOJO-Brain concept might be regarded as essential for the total motion control.

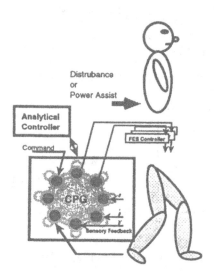

Fig.12 A future application of supplementary brain "HOJO-Brain"

References

[1] Sankai Y.: A Biocybernetic System in the Artificial Organ Control and Bio-Robotics, International Centre of Biocybernetics, MCB, Vol.23, 1995, pp66-99.
[2] Sankai Y., Ohta M..:A Basic Concept of Super Rabbit, RoManSy9, pp349-359, 1993.
[3] Sankai Y.: FES Controller with Artificial CPG for Biological Systems, IFES95, 1995, pp267–270

GAIT RHYTHM GENERATORS
OF A TWO LEGGED WALKING MACHINE

T. Zielinska

Warsaw University of Technology, Warsaw, Poland

Abstract: The gait of currently designed two-legged walking machines differs from the humans', although the kinematic structures of these machines' legs frequently imitate human limbs. The paper presents the method of generating the trajectories of hip and knee joint angles resulting in a gait pattern similar to that of a human. For that purpose the solutions of coupled van der Pol oscillator equations are utlised. In the view of many researches these equations can be treated as a good model of the Central Pattern Generator generating functional (also locomotional) rhythms in living creatures. The oscillator equations are solved by numerical integration. The method of changing the type of gait by changing adequate parameter values in oscillator equations is presented (change of velocity and trajectory of leg-ends). The obtained results enable an enhancement of two-legged walking control systems by including gait pattern generators which will assume a similar role to that of biological generators.

Current state of research

The kinematic structures of biped walking machines usually imitate the structure of human skeleton. The gait of machines, as well as human gait, is dynamically stable [9, 10]. The scientists elaborating methods of controlling the motion of biped walking machines use different models and follow different ways of solving dynamics problems (an interesting overview is presented by Shuuji Kajita et.al. [7]). Generally speaking, the solution to the dynamics problem suplies the information about the force and torque transients, which should assure the proper posture (i.e. vertical) and the forward motion of the machine. The desirable changes of leg coordinate trajectories result from:

- a *priori* assumption of the leg-end trajectory shape, or other geometrical considerations (e.g. determination of joint angle changes based on an assumed motion of the device's centre of gravity [2]),

- computation of these trajectories on the basis of the solution of the walking machine dynamics equations [4] supplied with additional criteria (minimal energy consumption, motion comfort criteria),
- application of supervisory control laws (planning the foot mark locations by inference rules),

or

- result from the assumption that the motion of leg links resembles the oscillations of an inverted pendulum [9]),

so the method of machine motion (gait pattern) differs from the human gait pattern. The above mentioned situation induced the author to investigate biped machine gait generators. The author decided to look closely at biological patterns and to utilise the idea of Central Pattern Generator (CPG).

The Central Pattern Generator consists of groups of neurons (usually in the spinal cord) which collectively realise sequences of cyclic excitations [3, 6]. These neurons can cause excitation without feedback from the musco-skeletal system and without control signals generated by the brain [5, 8]. This statement has been verified by experiments with animals having their cortical steam cut. Pattern generation can be periodically initiated, terminated or modulated by external control inputs.

Out of the group of the most widespread oscillator models, van der Pol oscillators [1] are investigated as the human locomotion rhythm generators (CPG).
The presented paper deals with four van der Pol oscillators, namely the analysis of influence of parameter changes on reproducing the two-legged gait. The functions resolving the coupled equations of these oscillators describe the changes of angles in the hip and knee joints.

The influence of the change in oscillator parameter values was investigated. The generated gait pattern was compared with human gait (earlier such comparisons have not been done). Analysis of gait rhythm generators resulted in the formulation of conclusions which will be presented in the paper.
The changes in angles during motion are described by equations of four coupled van der Pol oscillators:

$$\ddot{x}_1 - \mu_1 \cdot (p_1^2 - x_a^2) \cdot \dot{x}_1 + g_1^2 \cdot x_a = q_1$$
$$\ddot{x}_2 - \mu_2 \cdot (p_2^2 - x_b^2) \cdot \dot{x}_2 + g_2^2 \cdot x_b = q_2$$
$$\ddot{x}_3 - \mu_3 \cdot (p_3^2 - x_c^2) \cdot \dot{x}_3 + g_3^2 \cdot x_c = q_3$$
$$\ddot{x}_4 - \mu_4 \cdot (p_4^2 - x_d^2) \cdot \dot{x}_4 + g_4^2 \cdot x_d = q_4$$

where
$$x_a = x_1 - \lambda_{21} \cdot x_2 - \lambda_{31} \cdot x_3$$
$$x_b = x_2 - \lambda_{12} \cdot x_1 - \lambda_{42} \cdot x_4$$
$$x_c = x_3 - \lambda_{13} \cdot x_1 - \lambda_{43} \cdot x_4$$
$$x_d = x_4 - \lambda_{24} \cdot x_2 - \lambda_{34} \cdot x_3$$

These equations have 24 parameters. The values of parameters have been selected in such a way that transients of independent variables x_1, x_2, x_3, x_4 model the adequately defined angles in lower limbs.

The equation of each oscillator was partitioned into two equations of the first order. The so obtained system of eight equations was solved by second order Runge-Kutta method.

In the investigations of oscillators it was assumed that the variables x_1, x_2, x_3, x_4 represent the values of angles expressed in degrees. The variables were scaled. The method of scaling is presented in [11]. The choice between degrees and radians is unimportant from the point of view of oscillators functioning as gait generators (pulses fired by the groups of neurons are indirectly related to angles). When a stable numerical solution have been obtained, commences the search for such values of oscillator parameters that a proper image of gait was generated. The parameters were selected heuristically, because in the case of four coupled oscillators it is not possible to univocaly determine the relationship between the values of parameters and the independent variable transients. The validity of parameters is checked by observing the image of limb motion model (Fig.1). The values of generated angles were compared with the values measured on a walking man. The values of satisfied oscillator parameters are given in **Table 1** and the gait image obtained by numerical solution of oscillator equations for such a set of parameters is shown in **Fig. 2**.

direction and sense of gait

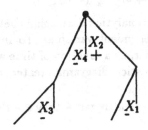

Figure 1: Denotation of angles obtained from the gait generator

TABLE 1

Oscillator parameter values			
$\mu_1 = 2$	$\mu_2 = 2$	$\mu_3 = 2$	$\mu_4 = 2$
$p_1^2 = 2$	$p_2^2 = 1$	$p_3^2 = 2$	$p_4^2 = 1$
$g_1^2 = 24$	$g_2^2 = 19$	$g_3^2 = 24$	$g_4^2 = 19$
$q_1 = 12$	$q_2 = -20$	$q_3 = 12$	$q_4 = -20$
$\lambda_{21} = -0.2$	$\lambda_{12} = -0.2$	$\lambda_{24} = 0.2$	$\lambda_{42} = 0.2$
$\lambda_{31} = 0.2$	$\lambda_{13} = 0.2$	$\lambda_{34} = -0.2$	$\lambda_{43} = -0.2$

By observing the limb positions it was ascertained that always at least one leg has contact with the ground, and this is the necessary condition of correct walking.

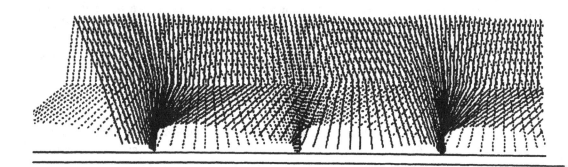

Figure 2: The gait image obtained by numerical solution of oscillator equations

The proportions between the limb dimensions, the distance traversed, gait period (125 integration steps) were taken into account. This and the proportions of an adult man's body (the length of legs is approximately 48% of the body height) and the initial condition were the basis for estimation of the step length and motion velocity. Step length (the distance between consecutive foot marks of the left and right leg) was $0.325m$ (body height of $1.8m$ was assumed) and the velocity was $0.67 \frac{m}{s}$ ($2.4 \frac{km}{hour}$).

As it was mentioned, it is impossible to determine analytic relationships between oscillator parameters and the obtained transients of independent variables. To investigate these relationships a set of phase diagrams and plots of x_i as a function of time were elaborated [11]. The observed regularities were verified by stick diagrams created by varying the values of selected parameters.

The oscillator coupling parameters determine the type of gait – they are the coordinators of motion of both limbs.

Due to that, after the initial selection of coupling parameter values, only the other parameters should be varied in such a way that the required gait image is obtained.

For the remaining parameters a regularity between the changes of these parameters and the changes of phase plot shapes can be spotted.

The alteration of parameters p_i^2, g_i^2, μ_i influences the shape of trajectory X_i, and introduces a phase shift, so it is extremely helpful in gait generation.

The parameters p_i^2, g_i^2, q_i, μ_i, first of all, determine the shape of the trajectory of a single leg-end. To a lesser extent they influence the type of gait (jumping, pathological gait). During gait rhythm generation these parameters can be altered to finally correct the leg trajectory (as it was done to increase the gait velocity).

In real gait periodical vertical displacements of hip joints can be observed. The hip joint is at its lowest just before the leg makes contact with the ground. In the support phase the hip joint is located a bit higher. **Fig.3** presents the real gait image taking into account the waddle of the hips. It also contains the the plot of the distance of the hip center point from the ground.

In the next stage of research the vertical displacements of hip joints were also introduced

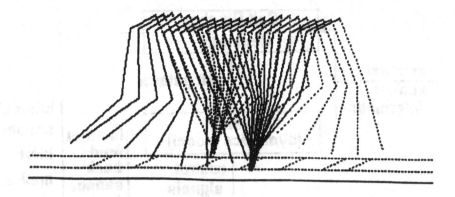

Figure 3: The real gait image taking into account the waddle of the hips

into the generated gait (**Fig.4**).

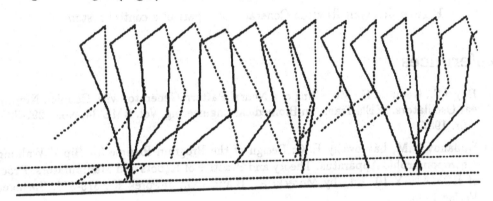

Figure 4: The generated gait image taking into account the waddle of the hips

Conclusions

Coupled oscillators can be used in the walking machine control systems as the generators of patterns similar to real gait. The generated leg position patterns prior to execution should be adequately corrected taking into account the stability conditions, and the analysis of information obtained through sensors and other factors.**Fig.5** presents a general structure of a biped machine gait control system.

This work is funded by Polish Scientific Research Committee Grant no. 3 3001 9203, by Polish-American Maria Sklodowska-Curie FUND II (MEN/NSF-94-159) and Institute of Aircraft Engineering and Applied Mechanics Grant.

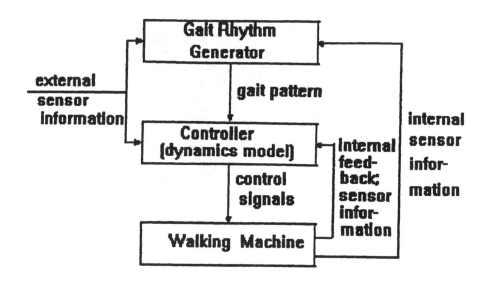

Figure 5: Gait Rhythm Generator as a part of a control system

References

[1] Bay J.S., Hemami H.: Modelling of a Neural Pattern Generator with Coupled Nonlinear Oscillators. IEEE Trans. on Biomedical Engineering, Vol. BME-34, No.4, 297-306, Apr. 1987.

[2] Budanov V.M, Lavrowsky E.K.: Design of the Program Regime for Biped Walking Antropomorphic Apparatus. Theory and Practice of Robots and Manipulators. Proc. of ROMANSY 10. Ed,. by Morecki A., Bianchi G., Jaworek K., 379-386. Springer Verlag 1995.

[3] Cohen A.H., Rossignol S., Griller S: Neural Control of Rhythmic Movements in Vertebrates. John Wiley and Sons, New York, Chichester, Brisbane, Toronto, Singapore 1988.

[4] Formalsky A.M.: Impulsive Control for Antropomorphic Biped. Theory and Practice of Robots and Manipulators. Proc. of ROMANSY 10. Ed,. by Morecki A., Bianchi G., Jaworek K., 387-394. Springer Verlag 1995.

[5] Grillner S.: Control of Locomotion in Bipeds, Tetrapods and Fish. Hanbook og Physiology. Ed. by Brookhat J.M., Mountcastle V.B., 1179-1236, American Physiological Society, 1981

[6] Hertz J., Krogh A., Palmer R.G.: Introduction to the Theory of Neural Computation (in Polish). WNT, Warsaw 1993.

[7] Shuuji Kajita, Tomio Yamaura, Akira Kobayashi: Dynamic Walking Control of a Biped Robot Along a Potential Energy Conserving Orbit. IEEE Trans. on Robotics and Automation, vol.8, no.4, 431-438, 1992.

[8] Kandel E.R., Schwartz J.H. Jessel T.M.: Principles of Neural Science. Elsevier, New York 1991.

[9] Raibert M.H., Brown H.B., Murthy S.S.: 3-D Balance Using 2-D Algorithms? The Int. Journal of Robotics Research, no.1, 215-224, MIT Press 1984.

[10] Vukobratovic M.: Legged Locomotion and Anthropomorphic Mechanism. Michailo Pupin Institute. Beograd 1975.

[11] Zielińska T.: Coupled Oscillators Utilised as Gait Rhythm Generators of Two Legged Walking Machine. Biological Cybernetics. Springer Verlag, vol.4, no.3, pp.263-273, 1996

Chapter V
Walking Machines
and Mobile Robots II

DEVELOPMENT OF A MECHANICAL SIMULATION OF HUMAN WALKING

K. Kedzior, A. Morecki, M. Wojtyra, T. Zagrajek and T. Zielinska

Warsaw University of Technology, Warsaw, Poland

A. Goswami

INRIA, Grenoble, France

M. Waldron and K. Waldron

Ohio State University, Columbus, OH, USA

ABSTRACT

The primary objective of the proposed project is the development of an enhanced understanding of human locomotion by means of the construction and testing of a electromechanical biped designed expressly to model the human locomotion system as closely as possible.

1. Introduction

The development of a teleoperated computer coordinated biped of human scale has many potential applications in its own right. There are many situations where a teleoperated vehicle would be attractive but conventional mobility systems cannot be used because the environment has been built for human locomotion. For example, a remotely controlled system for rescue work in fires, high radiation environments, or in chemically poisoned environments such as my occur in mining accidents, is attractive. However, in order to be effective the machine must be able to negotiate doorways, catwalks and stairs and ladders designed for human beings. Routine maintenance within the containment of nuclear reactors is another application with similar constraints. The proposed system would provide a great deal of information applicable to the design of such systems, but the main objective would be study of human locomotion [1, 6].

2. Computer Model of a Human Limb Musculo-Skeletal Model

A computer model of a human limb musculo-skeletal system has been developed. The mathematical model is based on the finite element method and is suitable for ergonomic applications and gait analysis. The mathematical model was coded into a computer package in order to make calculations possible [2].

The model consists of rigid bodies, flexible elements and actuators. Bones are modeled as rigid bodies. Joints are modeled as flexible elements. Every flexible element consists of six springs; three for translations and three for rotations. If a movement in the biological joint is small or impossible, then the stiffness of the spring for this direction of this movement in the model is made large. For directions of movement, which are allowed by the biological joint, the stiffness of the spring in the model is small. Muscles are modeled as actuators which act along geometrically defined lines of action. The lines of action are obtained by connecting the point of origin of the muscle with the point at which the muscle inserts onto a bone. Some of the muscles are modeled by two lines: the first being the line from the muscle origin to a point, at which the muscle starts to wrap around a bone. The second line is from the point at which the muscle unwraps from the bone to the point of insertion. The force exerted by a muscle is a function of three parameters: its excitation, its instantaneous length, and the velocity with which it is contracting. The muscle force is modeled as a function of muscle excitation, expressed as a number between 0 and 1, and time. The problem of muscle recruitment, or apportionment of force among several available muscles, is solved by an optimization procedure. Several non-linear merit criteria were tested, such as the soft saturation criterion, and energy criterion. The model consists of 32 muscles for each lower extremity and 64 muscles for each upper extremity (including half of the shoulder girdle) [2], Figure 1

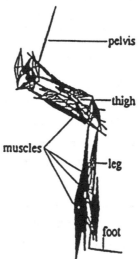

Figure 1. The model of the right lower extremity

Many tests were made in order to validate the model parameters. The parameters were obtained from published measurements of human body mass and geometric properties. It was shown that using a model of the human lower extremity it is possible to simulate free motion of the leg (flexion and extension), squats, and gait without raising the feet. Simulation of typical human gaits is planned for the future. The model is a unique model which fully describes the dynamics of phenomena occurring during the working of the human musculo-skeletal system.

3. Design of an Electromechanical Model

Design of an electromechanical system ideally necessitates several simultaneous decisions. In order to design a robot manipulator, for example, we need to specify the kinematic parameters: the number of degrees of freedom, the number of joints, the link lengths, the joint displacement ranges, etc.; the dynamic parameters: the masses and the moments of inertia of the members; the actuation system configuration: electric motors, hydraulic or pneumatic actuation; and the sensor system: joint position encoders, velocity, and acceleration sensing etc. Moreover, given the functional objectives of the robot: that is, what exactly we want it to do; its performance: that is, how well it does it depends both on the efficiency of its controller, which is a function both of the raw computing power and the cleverness of the algorithms, and of the kinematic and the dynamic parameters selected earlier. In reality, therefore, the design becomes a highly iterative process, later decisions affecting and modifying the earlier ones. Since very few of the system components may be selected unilaterally without regard to the other components, a tricky question is: "Where to start?"

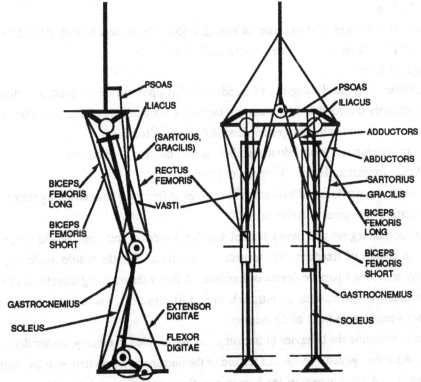

Figure 2. General schematic view of anthropomorphic biped robot

One philosophy, that followed here, is to start designing those components on which we have the least amount of initial information. Recalling that the choice of one component depends on the selection of the others, selecting the least known components first is the most efficient procedure in terms of minimizing the changes that have to be made throughout the design. This is thus a reasonable approach to adopt.

For an anthropomorphic biped robot such as the one we set out to design here, many of the above-mentioned decisions have, fortunately, already been implicitly made. The very inclusion of the word "anthropomorphic" in the description of the robot gives us some clear direction as to what the geometric dimensions, the kinematic structure, and the inertial parameters of the machine are going to be [3], Figure 2.

We will shortly discuss typical dimensions and mass properties of the leg. We will attempt to minimize the number of parameters to be included using as a guideline the geometric and inertial properties of a 2 m tall person. The lengths and weights of different leg links of a typical human being with height H and weight W are as follows [6]:

Thigh: 0.24 H, 0.10 W, center of mass at 0.567 T up from knee joint. Thigh length: T = 0.489 m.

Shank: 0.25 H, 0.045 W, center of mass at 0.567 S up from the ankle joint. Shank length: S = 0.448 m.

Foot: 0.04 H (from ankle joint to center of mass), 0.15 M (from heel to toe), 0.0015 W, center of mass at 0.571F back from the toe tip. Foot length: F = 0.29 m.

Pelvis breadth: 0.28 m.

The natural leg has 30 degrees of freedom and is operated by 66 muscle actions. It is necessary to simplify this model. Let us assume the following numbers of degrees of freedom:
- torso: rotation in the suggested plane (6 degrees of freedom)
- hip: extension/flexion, adduction/abduction (2 degrees of freedom)
- knee: extension/flexion (6 degrees of freedom)
- ankle: plantarflexion/dorsiflexion (6 degrees of freedom) passive spring restraining toe rotation (6 degrees of freedom).

Thus for each leg we will have a total of 4 active degrees of freedom. In addition, the torso has one active degree of freedom with respect to the pelvis. For the whole model we have 9 degrees of freedom plus 1 passive degree of freedom. A list of the main leg muscles is as follows: Hip (11 muscles), thigh rotation (1 muscle), knee (10 muscles), ankle (4 muscles), foot (2 muscles), toe (4 muscles): a total of 32 muscles.

In order to mimic the behavior of human joints we decided to employ tendon driven joints in our robot. Also the agonist - antagonist feature of the human muscle system will be emulated as far as is possible. A third aspect of the human muscle system that we intend to include in our design in the presence of multi-joint muscles, particularly two-joint muscles. The advantage of two-joint muscles is clearly to facilitate the transfer of energy from one joint of the system to another. Assuming up to a two function level of interconnection of the muscles, a full set of tendons will have the ability to move every joint, in both the positive and the negative directions, independently as well as having the ability to control all possible combinations of movements between a joint and its adjacent joints. We need two tendons on the agonist and antagonist sides of a joint to fully control it. The problem arises of how many actuation tendons should be included in the machine. Because we will have 9 active degrees of freedom: 1 at the torso, 2 × (2 at the hip, 1 at the knee, 1 at the ankle). Actuation of these degrees of freedom in the robot requires 18 tendons. So far we have only accounted for the one-joint muscles. There are three two-joint muscles in the model. Two of them involve both the hip and the knee. For two legs we need four more tendons. Also each leg has one two-joint muscle connecting the knee and the ankle. Two more tendons are therefore needed for the two legs. In summary, we will need 24 active tendons for the anthropomorphic biped robot. We have not included in our model a two-joint tendon performing knee extension and ankle dorsiflexion.

If we take a close look at the human anatomy we find that most muscles actuate limb motion in several directions. For example, a shortening of the sartorius muscle will cause both hip flexion and medial/lateral thigh rotation. Each joint is acted upon by several muscles which act in synergy. Obviously, the most drastic functional simplification would be to ignore how the joints are actuated in the biological system, except that one should be able to actuate them in any direction typically by using actuators, such as electric motors, acting directly on the joints. In this case there is a synergy between the electric motors since they do not act against each other. Our approach lies somewhere in between. We will have several tendons acting on a single joint but not as many as we see in a human leg. Based on this short discussion the following list of actively driven tendons was assumed

1. Pelvis forward tilt
2. Pelvis backward tilt
3. Hip extension: *Gluteus Maximus*
4. Hip flexion: *Iliopsoas*
5. Hip abduction: *Gluteus Medius, Gluteus Minimus*
6. Hip adduction: *Adductor Longus, Adductor Brevis, Adductor Magnus*
7. Knee extension: *Vastus Lateralis, Vastus Medialis, Vastus Intermedius*
8. Knee flexion: *Biceps Femoris (short head), Semitendinosus, Semimembranosus*
9. Ankle plantarflexion: *Soleus, Tibialis Posterior*
10. Ankle dorsiflexion: *Tibialis Anterior*
11. Hip extension/knee flexion: *Rectus Femoris*

12.   Hip flexion/knee extension: *Biceps Femoris (long head)*

13.   Knee flexion/ankle plantarflexion: *Gastrocnemius*

Total: $11 \times 2 + 2 = 24$ driven tendons plus unactuated Spring/Tendons:

1.   Hip medial/lateral rotation

2.   Toe extension/flexion: Extension (*Extensor Hallucis Longus, Extensor Digitorum Longus*), and flexion (*Flexor Hallucis Longus, Flexor Digitorum Longus*)

We will now discuss the problem of motor selection. An important design goal is to configure the actuation system so that digital control techniques can be used to faithfully mimic the actuation characteristics of the muscles, even though the tendon geometry, and certainly the actuator mechanics, may be somewhat different. This requires a system adapted to precise control with relatively high bandwidth, and easily interfaced to a digital system.

We may consider the following actuation options:

A.   ELECTRIC MOTORS

1. Brushless DC: Direct drive, tendon drive or compliant tendon.

2. Commutated DC: Direct drive, tendon drive or compliant tendon.

3. AC Induction motor: Direct drive, tendon drive or compliant tendon.

4. Stepping motor: Direct drive, tendon drive or compliant tendon.

The advantages of electric actuation are flexibility in application together with a clean system with a compact power supply and moderate maintenance. Use of an electric motor in combination with a ball-screw produces a stiff and efficient linear actuator.

The disadvantages are poor force to weight ratio making it necessary to run at a high speed to generate the required power. For this reason speed reducers are usually necessary. The side effects of use of speed reducers include increased complexity and weight, compliance and backlash.

B.   HYDRAULIC ACTUATION

1. Linear actuator with direct drive, tendon drive or compliant tendon.

2. Rotary actuator with direct drive, tendon drive or compliant tendon.

Advantages include the highest force/weight ratio available. There is no need for speed reducers. hydraulic actuation is suitable for heavy-duty applications.

Disadvantages include the potential for fluid leakage, difficult maintenance, and usually a bulky power supply.

C.   PNEUMATIC ACTUATION

1. Linear actuator with direct drive, or tendon drive.

2. Air motor with direct drive, tendon drive or compliant tendon

Advantages include inexpensive components and a convenient power supply requiring only a compressor or compressed air supply.

Disadvantages of pneumatic cylinders include very low force/weight ratio and high inherent compliance in the system. The characteristics of air motors are much like electric motors in requiring high operating speed to produce effective power, and hence needing speed reducers.

One choice that is popular among the robotics community is the use of brushless DC motors. The main advantages of the brushless motors over commutated motors are lower rotor inertia, lower friction, higher output torque for equal motor volume, and reliability and long life due to the absence of brush wear. The principle disadvantage is the complex electronics necessary for commutation using rotor position feedback to control currents in the coils. Another disadvantage is lower torque for equal motor volume. Also brushless motors exhibit reluctance cogging [4].

The next problem to be discussed is that of how to arrange so many actuators on the legs? The objective is to find whether an actuation system capable of delivering sufficient power in order to produce human like motion to a biped robot with human-like geometric and dynamic properties can be configured. While considering this question it should be kept in mind that the actuators may need to carry their own weights, and that the inertial properties of the biped will be closely related to the placement of the actuators.

After selecting the types of joints and the type of actuators we face the problem of the actual physical placement of the actuators. This is guided by our objective that the inertial properties of the robot should resemble, as much as possible, those of human leg. The first principle is perhaps to position an actuator to be as close as possible to the joint (joints for multi-joint tendons) which it is driving. This objective may not always be realizable since the total weight of the actuators responsible for a joint may exceed the anthropomorphic weight of the adjacent limb. In this case we may need to collect the actuators in the pelvis level so that the weight is counted as part of the trunk. This strategy is, in fact, present in the biological system since the important foot actuators are all on the shank, and many of the major hip actuators are in the trunk. This preserves the passive mechanical properties (such as the period of oscillation) of the legs taken separately. In case the total structural weight of a limb and the associated actuators becomes less than the anthropomorphic weight of the limb, extra weight can be added to the limb to compensate for the difference.

It is not easy to quantify the departure of the robot dynamic model from the human model as caused by the heavier actuators. In order for us to gain some insight into human locomotion from experiments performed on an electro-mechanical simulacrum we should be able to quantify or estimate the expected departure of the dynamic behavior of the machine from that of a perfect human-like model. We intend to "pack" all the motors together at the torso, this being the position with the best choice of displacement/velocity/acceleration.

Displacement: The heaviest portion of the robot should be, in general, located as centrally as possible. Displacements away from the joints would cause large moments about them. The maximum inertial moment should happen around the front leg ankle at the beginning of the single support phase, since in this configuration the torso is farthest from the ankle. However, the ankles don't need to be the strongest of the leg joints because in this configuration the ankles don't need to support the moment. The robot moves on to the next phase due to its inertia (this resolves the apparent contradiction that the base motor of an arm manipulator is the largest of all joints but the "base" motor of a walking robot and a human being is not).

Velocity: A significant part of the mechanical energy expenditure of the robot is due to its kinetic energy. The forward velocity is not a concern in itself in the placement of the motors. The change in its magnitude, i.e., the forward acceleration is.

Acceleration: The forward and vertical acceleration of the center of mass are closely related to the amount of power spent by the machine. It turns out that in a single gait cycle the body accelerates and decelerates and this effect diminishes as we go higher up on the body with the eyes having the lowest displacements and velocity fluctuations (this has been transformed to an optimality criterion for the human locomotion).

The mechanical structure of the hip joint could be a ball rod end (similar to a spherical joint or a spherical bearing). The "collar" under the hip joint in Figure 2 is simply a means to provide the correct anthropomorphic moment arms to the muscles activating the joints. The knee joint should be a revolute, or better a linkage to simulate the migration of the joint axis, and the ankle joint might be a revolute, or perhaps a flexure. Options for the structural material include aluminum, steel, and composites: either glass epoxy or graphite epoxy. Configuration options that might be explored include a rigid torso, a compliant foot, a metatarsal joint, a movable head, and movable arms. Although the decision of placing all the motors at the trunk is attractive and a mechanically sound one, there is a strong argument for a more distributed motor placement. Let's review the relative pros and cons.

Each leg mass is about 15% of the total body mass. The HAT (head, arms, trunk) of a person is about 70% of the total body mass. Each leg of a 80 kg person will have mass about 12 kg with the thigh mass being 8 kg, the shank mass 4 kg and the foot mass 0.4 kg. For our robot the total motor mass is about 50 kg (about 22 kg per leg plus two 3 kg motors for pelvic tilt). Therefore if we put all the motors in the trunk 50 kg = 70% of body mass. Each leg should therefore have mass about (50 / 0.7) × 0.15 = 11 kg (approximately). Given our selection of material (square cross-section aluminum tubing with 75 mm side and 2 mm thickness) each leg has mass about 2 kg (0.78 kg for each of the thigh and shank, and 0.47 kg for the foot).

Now in order to make the inertia properties of the robot resemble that of a human being, we would need to put extra mass into the thigh, shank, and foot. Therefore, why not remove some of the motors from the trunk and place them in the limbs? The robot will be carrying more useful mass in that case.

However, it should be noted that there are wiring and cable routing problems to be encountered in such a modified system. The thigh contains the knee extensor, and knee flexor/ankle extensor actuators (3.3 +1.8 = 5.1 kg). The shank contains the knee flexor and ankle extensor actuators (0.3 + 2.2 = 2.5 kg). The robot should be equipped also with necessary sensors such as position sensors (potentiometers, resolvers, incremental encoders, absolute encoders), velocity sensors (tachometers), force sensors (load cells, strain gauges, capacitative sensors) as well as attitude sensors (inclinometers, directional gyros), angular rate sensors (rate gyros), acceleration sensors (accelerometers) and exteroceptive sensors to model the environment in which it is operating.

4. A Method Of Reference Trajectory Generation

Movement of an anthropomorphic biped should be similar to that of a human. The structure of its control system should include levels of reference (pattern) trajectories construction. The realization of such trajectories will result in a humanoid gait. We propose to use an Artificial Neural Network to generate reference trajectories. We recognize that the functioning of the neural network should imitate the Central Pattern Generator located in the spinal cord [7].

In the work presented here a recurrent back propagation neural network was chosen for gait pattern generation. The learning algorithms for such a network use least square error optimization methods. Errors detected in the outputs of the network are used to adjust the neural network parameters, specifically the weights and activation functions.

The Central Pattern Generator of a living organism is developed while the organism is growing. The performance of the generator depends on the dimensions and mass parameters of the organism. It is assumed that information about the difference between the generated signal and the realized movement is utilized for learning in the networks generating these patterns in nature. The memory in these neural networks is believed to be built via network feedback.

A three layer recurrent back propagation neural network was used in our work. The input layer contains 5 neurons, one for each knee and hip angle, and the fifth one for the time information. Note that only motion in the sagittal plane has been modeled at the present time. The hidden layer contains 5 neurons, and the output layer contains 4 neurons, one for each knee and hip joint angle at the next time step. The input neurons are straight feed-forward neurons.

The neurons of the hidden and output layers have sigmoidal activation functions. The input training data that was used was the image of a real gait. The outputs from the output layer were connected back to the input layer as is shown in Figure 3.

The inputs to the back propagation network were discrete values of 2 knee angles, 2 hip joint angles (left and right leg) at time j and the time value j (Figure 3). Whereas the outputs are the subsequent knee and hip angles at the next time step in the gait. These values are fed back to the input.

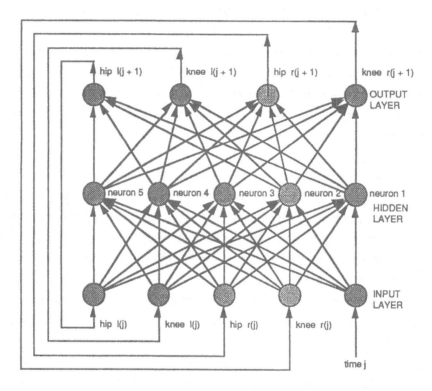

Figure 3. The structure used for the neural network. The notation "knee r/l(j)" denotes the knee joint angle of the right/left leg at time interval j. Similarly "hip r/l(j)" denotes hip joint angle of the right/left leg at time step j.

The network was considered to be trained when the average error declined to less than 0.0001 and the maximum error did not exceed 0.04. (All the input-output values were normalized to values between 0 and 1).

Once trained, the neural network generated a gait pattern by receiving the initial 5 input values from the biped's environment. The network based on its stored (during training) patterns generated angle values for the next time step. These values, together with the discrete value of the next time step, were treated as the inputs for the following time step. In Figure 4 the training pattern (human), and gait image generated by the system is displayed.

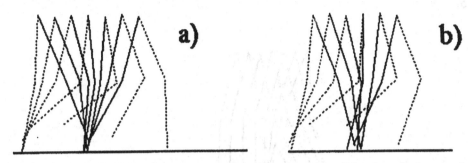

Figure 4. Gait images: a) natural (human) gait, b) gait generated by the neural network

The differences in the angle values for the pattern and the generated gait are not greater than 5°. Such errors we believe to be acceptable because they are common in normal (non pathological) gait. In our imitation of the Central Gait Pattern Generator we are not interested in the detailed imitation of personal, human gait images but rather in capturing regularities in the angle changes, that is, in capturing the synergy of the pattern generator. Coordination of the angles of different joints is dependent on the changes of values that are learned by the network during training.

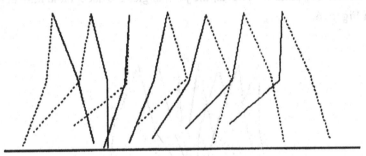

Figure 5. Decreasing θ_{1j}, θ_{2j} or increasing θ_{5j} results in a run.

It must be noted that at present we do not known how the changes of natural neuron activations actually initiate changes in vertebrate gait.

In our research we examined the relationship between changes in threshold parameter θ_j in the hidden layer and the angle changes of the gait image.

For the neurons in the hidden layer, when θ_{1j} (the threshold for neuron number 1) , and θ_{2j} (threshold for neuron number 2) are decreased, or θ_{5j} (for neuron number 5) is increased, the movement velocity is increased. The gait then transforms from a walk to a run as is shown in Figure 5.

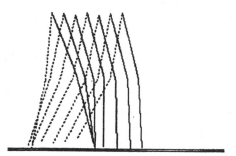

Figure 6. Increasing θ_{1j}, θ_{2j}, θ_{4j} , decreasing θ_{5j} and any change of θ_{3j} results in constant joint angles

Whereas if we increase θ_{1j}, θ_{2j}, or θ_{4j}, or decrease θ_{5j} or make any change in θ_{3j} there is no resultant gait that is generated. In this case the network produces constant values of the joint angles resulting in no possible movement of the legs. However, the resulting position of the legs is an acceptable normal leg pattern. That is, the joint angles are acceptable human leg positions. This is shown in Figure 6.

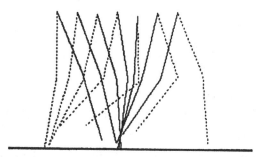

Figure 7. Decreasing θ_{4j} results in a walk with higher velocity than for original training network (as in Figure 4b).

Decreasing of θ_{4j}, on the other hand, does result in increased motion velocity but the leg movement remains that for a walk and not a run (that is one leg remains on the ground as is shown in Figure 7). However, the velocity is greater than that of the original training network (as in Figure 4b).

5. Conclusions

A computer model of a Human Limb Musculo-Skeletal Model was proposed. Design of an electromechanical biped as a simulacrum of the human system to access variables that are not accessible in *in vivo* experiments was discussed. The concept includes the use of tendons to simulate the action of the more important muscles, with digital control techniques being used to compel the mechanical actuators to mimic the mechanical behavior of the biological muscles.

The investigation shows that the synergy of a recurrent artificial neural network movement pattern generator lies in the activation functions of the hidden layer neurons. By the changes of thresholds in the hidden layer we can control changes in the gait without losing generator synergy. Observations of gait images show that changes of gait invoked by changes of thresholds look similar to the changes observed in the natural human gait.

In future this network can be expanded to include the ground reaction forces and the dynamic properties of the entire biped mechanical and control system. For such a system, the simple gait generation presented here can act as an input and the output of the complex mechanical system can provide the error values to retrain the entire machine.

Acknowledgment This study has been supported by the Polish-American Marie Sklodowska-Curie Fund, Grant Number MEN/NSF-9-159.

References

[1] A. Morecki, Biomechanical Modeling of Human Walking. Proceedings of Ninth World Congress on the Theory of Machines and Mechanisms. Milan, Italy, 1995, pp. 2400-2404.

[2] K. Kedzior, M. Wojtyra, T. Zagrajek, Dynamic Model of Human Lower Extremity. Book of Abstracts XVth Congress of the ISB, Jywaskyla, July 1995, pp. 58-59.

[3] A. Morecki, K. Waldron *et al.*, Development of a Mechanical Simulation of Human Walking. Second Polish-American Maria Sklodowska - Curie Foundation Grant Number MEN/NSF-9-159, Report November 1995.

[4] K. Waldron and A. Goswami, Design of a Tendon-driven anthropomorphic biped robot, Internal Report, Ohio State University, 1995.

[5] Mechanical Design of Manipulators and Robots (Class Notes for Course ME 752) By Kenneth J. Waldron, 1992, Department of Mechanical Engineering, Ohio State University.

[6] R.D. Cope, "A Methodology for the Understanding of Time-Varying Leg-Muscle Forces During Human Walking," doctoral dissertation, The Ohio State University, 1986.

[7] T. Zielinska, Utilization of Human and Animal Gait-Properties in the Synthesis of Walking Machine Motion. Monograph: Institute for Biocybernetics and Biomedical Engineering No. 40, Warsaw 1995, 132 Pages.

MECHANICAL DESIGN OF AN ANTHROPOMORPHIC ELECTROPNEUMATIC WALKING ROBOT

G. Figliolini and M. Ceccarelli
University of Cassino, Cassino, Italy

Abstract - Anthropomorphic walkers have been recognised as versatile walking systems although their mechanical structures may be very complex. In this paper we report a mechanical design which has been developed at the Laboratory of Robotics of the University of Cassino with the aim to construct a prototype for an anthropomorphic walking robot with low-cost and easy-running components. The proposed mechanical solution with an electropneumatic actuation is illustrated and basic design characteristics are discussed mainly looking at the walking stability.

1. Introduction

In the last years the research in the field of non industrial robotics has made many progresses in order to substitute or help the man in different dangerous works, such as in nuclear power plants and chemical industries, or for inspection underwater terrain or planetary surfaces. Many research Institutes and Universities have developed solutions for the movement of a robot by using a single leg or combinations of legs, wheels and caterpillar tracks. However the walking robots seem to be the more suitable for the applications on a very rough terrain, [1]. Many solutions of walking robots have been developed: for example, biped robots [2-5], quadruped robots [6,7], six-legged robots [8-10]. Since stability during the motion is one of the most critical issues, several walking robots have been built with four or six legs to have at least three contacts with the terrain. Thus, particular attention has been taken to the stability of hexapod or quadruped walking robots during the motion on irregular terrain or in an environment with obstacles [6-9]. Four or six legs configuration ensures a static stability, the biped walker may work only within dynamic stability conditions with one leg in contact with the ground giving some more complexity for the motion control [2-5].

This paper concerns with the design and the construction of a pneumatically activated anthropomorphic robot controlled electronically with a PLC in a digital environment. An important issue concerns with the components which are low-cost and easy-running commercial means. At the moment the first prototype can walk along a generic polygonal trajectory on a rigid flat plane. The equilibrium of the robot during the motion has been ensured by the activation of two suction cups which are installed on the underside of each foot. When the suction cups are activated, a foot is pushed onto the rigid plane, and the stability of the robot is ensured both in longitudinal and transversal direction.

2. The leg mechanism

Each leg has been built as a 1 d.o.f. mechanism by using a pantograph togheter with an articulate double parallelogram, Fig.1. The pantograph has the fundamental function of generating a suitable trajectory of the foot through a proper path of the actuation point H. The double parallelogram ensures the pure translation of the foot in a vertical plane. The contact with the ground is stabilized by means of two suction cups which are installed on bottom side of the foot. Their action is always correct since the translational motion due to the parallelogram makes possible to have proper orientation of the foot with respect to a flat plane of the ground. The sizes of the leg mechanism have been reported in Table 1 referring to Fig.1. The relative trajectory of a foot contact point can be represented as a flat egg-shaped path which is composed of a straight line segment and a circular arc. The first one corresponds to the forward motion of the walking

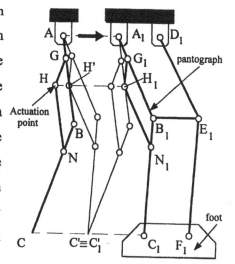

Fig. 1 - A kinematic sketch of the leg mechanism at different configurations during the walking motion.

robot body and the latter concerns with the forward motion of the free foot. Therefore, it has been thought useful to reproduce these basic path characteristics for the C foot point. A previous solution has been provided by using a four-bar linkage through a suitable synthesis of the coupler curve [10]. We have thought more convenient to exploit the pantograph scaling property by using a suitable cam profile to guide the actuation point H so that the path of the foot C point can be imposed by designing or selecting a proper cam according to a given application (for example, walking on a flat plane or a rough plane with given peaks, climbing up or descending steps). Although the kinematics of the actuation can be quite easy, some problems may arise with the dynamics of the motion along the cam profile, since the cam geometry cannot define the direction of the motion and additional devices could be needed. At the moment we have solved the problem as a very first solution by using a straight-line for H path so that we have used just one linear actuator and a straight-line trajectory have been obtained also for the foot C point. Thus, by designing the pantograph with a scale factor 4 and by giving a stroke of 50 mm for H, a linear relative step displacement of 200 mm for $CC'=C_1C'_1$ has been obtained for each foot of the walking robot.

3. A prototype and the electropneumatic actuation

A prototype of the anthropomorphic electropneumatic walking robot has been built with a mechanical design maintaining the kinematic simplicity of the leg mechanism by using an electropneumatic actuation with a commercial PLC. The kinematic structure of the legs has been built by means of commercial light section bars 30x10 mm. The knee joint of each leg has been obtained by a revolute connection along the pin of a brass bearing among the upper bar and two lower bars, Fig.2. Two lower bars have been used to ensure a centred reaction force at the knee joint. In addition each leg has been built with two parallel leg mechanisms of Fig.1 at a distance of 150 mm in order to obtain a proper transversal stiffness, too. The actuation has been provided through a linear pneumatic cylinder for each leg motion and a

Table 1 - Dimensions of a built prototype of an anthropomorphic electropneumatic walking robot, Fig.3.

	AB	BC	GH	HN	AC	$A_1 D_1$	f	h	b

rotative pneumatic cylinder for an ankle motion of each foot. A pneumatic actuation energy has been chosen to have light actuators and an easy digital control and autonomy. The autonomy can be easy ensured by equipping the walking robot of a small compressor module and a PLC control kit. The built prototype, Fig.3, is 900 mm tall and weights 280 N approximately and it can carry a payload of about 320 N. The pneumatic linear actuators are commercial double effect cylinders Festo DZH-32-50, [11], and they work at a pressure of 6 bar. The pneumatic rotative actuators are commercial rotative double effect cylinders Festo DSR-25-180, [11]. Thus, the compressor module must provide an air flux of about 150 l/min at a pressure of 6 bar.

The foot can be considered as a critical component for the walking robot since it ensures the equilibrium during the motion and performs the rotation of the walking robot in order to follow a polygonal trajectory . Its mechanical design is shown in Fig.4. Commercial means have been used for the construction of each foot, [11]: two suction cups Festo VAS 1/4" of diameter 75 mm, two ejectors Festo VAK 1/4", one pneumatic rotative double effect actuator Festo DSR-25-180, two unidirectional flux regulators Festo GR-M5. The ankle joint has been built by using an axial ball bearing between a plate fixed to the foot house and a mobile plate connected to the leg structure. Thus, when one foot is attached to the ground because of the suction cups, the corresponding ankle actuator can perform the rotation of the whole walking robot about a vertical axis so that a polygonal walking trajectory can be easy obtained.

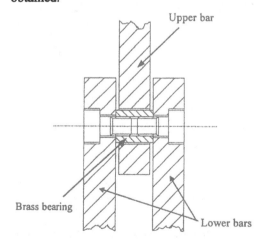

Fig. 2 - Mechanical design of a knee joint.

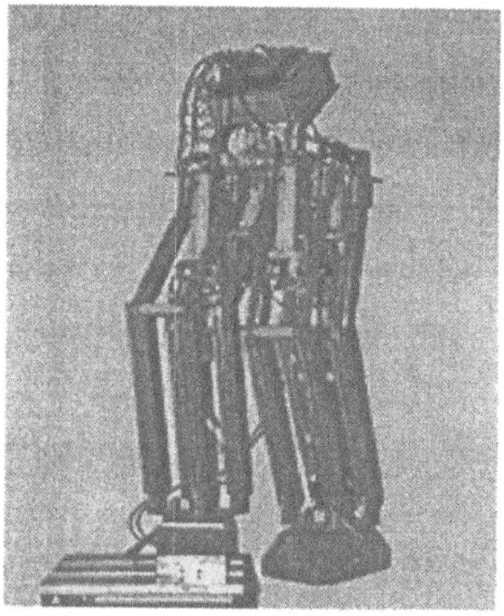

Fig. 3 - A built prototype of the walking robot in a stable equilibrium configuration on one foot with rotated ankle.

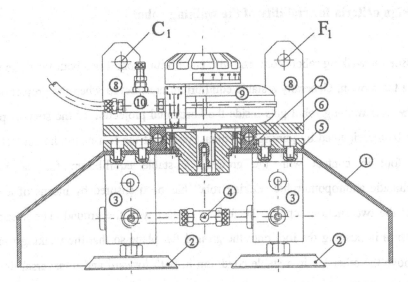

Fig.4 - Mechanical design of a foot: 1) Foot house; 2) Suction cups; 3) Ejectors; 4) Flux pipeline to ejectors; 5) Fixed plate of ankle joint; 6) Mobile plate of ankle joint; 7) Axial rolling bearing of ankle joint; 8) Connection coupling to the leg; 9) Pneumatic rotative cylinder; 10) Flow regulator.

The electropneumatic circuit for each leg is shown in Fig.5. It is based on: one linear double effect cylinder with a magnetic oval piston; one rotative double effect cylinder; two flow regulators for the linear cylinder and other two for the rotative cylinder; two bistable digital electrovalves four way-two position for the activation of the linear cylinder and the rotative cylinder; two suction cups; two ejectors equipped with a release device; and one unstable digital electrovalve three way-two position for the suction cups. A PLC Festo 101, [11], has been used to program and control the walking of the robot by properly regulate the running of the cylinders and the suction cups.

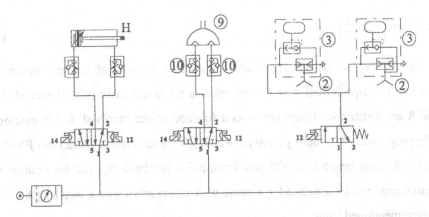

Fig.5 - The electropneumatic actuation of each leg of the walking robot.

4. On design criteria for stability of the walking robot

A basic issue for walking robots concerns with equilibrium conditions both for static position and during the motion. Generally a stable equilibrium is achieved when the projection of the mass centre of a walking robot gets inside the horizontal projection of the support polygon, which can be individuated by joining the contact points of the feet or by the contact area of the only foot in contact with the ground. A stable equilibrium for the proposed electropneumatic anthropomorphic walking robot has been ensured by means of a suitable working of the two suction cups of the foot in contact with the ground. The suction cups action consists in pushing the foot onto the ground flat plane so that the walking robot may stand or move the other leg in a stable condition in both forward and transversal directions, Fig.6. In fact each suction cup may act with a force F to counterbalance the upsetting action of the global weight P, which can be considered as the sum of the robot weight and the payload. P has assumed as acting along a vertical line crossing $A_1 D_1$ in the middle. Thus, referring to Fig.6 a) the transversal equilibrium is obtained by a foot when the total moment about the upsetting rotation axis acrossing Q is null, i.e.

$$F \cdot h - P \cdot \left(\frac{b}{2} - h \right) = 0 \qquad (1)$$

This condition is independent from the position of the other leg. For the forward motion, Fig.6 b), a stable equilibrium is obtained by foot in the worst condition with the other leg farest as possible when the total moment about the upsetting rotation axis through R or S is null, to give for both the cases, assuming $f/3 = A_1 D_1$, the expression:

$$F \cdot f - P \cdot \left(\frac{CC' - f}{2} \right) = 0 \qquad (2)$$

Eqs. (1) and (2) express design conditions for a stable running of the anthropomorphic walking robot. The equilibrium may be ensured also when the ankle joint is rotated, Fig.3. Figs.7 and 8 are illustrative design charts as a function of the involved design parameters. Indeed, the proportions of a built prototype, Fig.3, give from Fig.7 a force ratio P/F=3 and thus from Fig.8 a step length CC'=500 mm. From P/F value the design and the running work of the suction cups can be obtained for a stable walk of the robot with a step length CC' less than the abovementioned value.

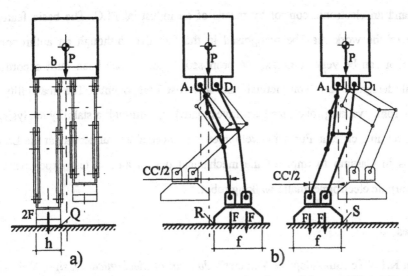

Fig.6 - A scheme of the action of the suction cups for the walking stability: a) transversal equilibrium; b) forward motion equilibrium.

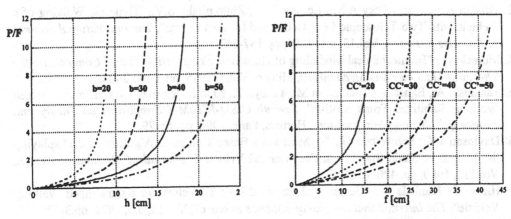

Fig.7 - Effect of the dimensions b and h on the transversal stability.

Fig.8 - Effect of the dimensions CC' and f on the longitudinal stability.

Indeed the built prototype may perform a step of 200 mm which gives a wide margin with respect to the stability limit. Eqs. (1) and (2), in the form of Figs. 7 and 8, are also useful to determine the feasible payload as a function of design parameters and running performances.

5. Conclusion

In this paper a mechanical design and a built prototype for an anthropomorphic electropneumatic walking robot have been presented with the characteristic of a pneumatic actuation and an electronic control by means of an industrial PLC. The basic features and originality of the work may be recognised in the fact that, although an anthropomorphic walking robot can be very complex, the proposed prototype has been built according to a mechanical design to use commercial and low-cost components. The feasibility of the anthropomorphic walking robot has been investigated also through a stability analysis, giving design and running criteria. Further developments are needed and underway at the Laboratory of Robotics in Cassino to improve the mechanical design and to test experimentally the anthropomorphic electropneumatic walking robot.

References

1. Waldron K.J., "Terrain Adaptive Vehicles", *Journal of Mechanical Design*, Vol.117, June 1995, pp.107-112.
2. Figliolini G., Papa L., "Biped Walking Electropneumatic Robot", *4th International Workshop on Robotics in Alpe-Andria Region*, Portschach, 1995, pp. 245-248.
3. Grishin A.A., Formal'sky A.M., Lensky A.V., Zhitomirsky S.V., "Dynamic Walking of a Vehicle with Two Telescopic Legs Controlled by Two Drives", *The International Journal of Robotics Research*, Vol.13, n.2, 1994, pp.137-147.
4. Morecki A., "Biomechanical Modelling of Human Walking", *Ninth World Congress on the Theory of Machines and Mechanisms*, Milano, Vol.3, 1995, pp. 2400-2403.
5. Takanishi A., Egusa Y., Tochizawa M., Takeya T., Kato I., "Realization of Dynamic Biped Walking Stabilized Trunk Motion", *Seventh CISM-IFToMM Symposium on Theory and Pratice of Robot and Manipulators*, Hermes, Paris, 1990, pp. 68-79.
6. Davinson J.K., Schweitzer G., "A Mechanics-Based Computer Algorithm for Displaying the Margin of Static Stability in Four-Legged Vehicles", *Journal of Mechanical Design*, Vol.112, 1990, pp. 480-487.
7. Hirose S., Kunieda O., "Generalized Standard Foot Trajectory for a Quadruped Walking Vehicle", *The International Journal of Robotics Research*, Vol.10, n.1, 1991, pp.3-12.
8. Xiding Q., Yimin G., Jide Z., "Analysis of the Dynamics of a Six-Legged Vehicle", *The International Journal of Robotics Research*, Vol.14, n.1, 1995, pp. 1-8.
9. Nagy P.V., Desa S., Whittaker W.L., "Energy-Based Stability Measures for Reliable Locomotion of Statically Stable Walkers: Theory and Application", *The International Journal of Robotics Research*, Vol.13, n.3, 1994, pp.272-287.
10. Williams R.P., Tsai L.W., Azaram S., "Design of a Crank and Rocker Driven Pantograph: A Leg Mechanism for the University of Meryland's 1991 Walking Robot", *2nd National Applied Mechanisms and Robotics Conference*, Cincinnati, Vol.1, 1991, pp. 2.1-2.6.
11. FESTO, Catalogue "Programma Base", Festo Pneumatic, 1995.

Acknowledgements

A financial support of the University of Cassino through funds "Ricerche di Ateneo" is gratefully acknowledged.

ENERGETIC ANALYSIS OF THREE CLASSICAL STRUCTURES OF WALKING ROBOTS LEG

O. Bruneau and F. Ben Ouezdou
University of Paris 6, Paris, France
and
Versailles Saint-Quentin University, Vélizy, France

Key words : trajectories for impact minimization, inverse kinematics, inverse dynamics, power of actuators, energy consumption.

Abstract :

The aim of this paper is to compare the performance of several mechanical structures of legs for walking robots in transfer phase from the viewpoint of energy consumption and maximum power exerted by actuators, when the tip of leg follows a given trajectory. At first, the study deals with the construction of a cycloidal trajectories family for the leg tip depending on locomotion parameters. Secondly, three classical structures used in the legged locomotion are modelized : an anthropomorphic leg with two rotary joints, a structure with two orthogonal prismatic joints and a pantograph mechanism with two closed chains. In order to follow the trajectories described before, these three structures are controlled by the explicit inverse kinematic model. The generalized forces are then obtained by the inverse dynamic model. Finally, these three structures are compared from the viewpoint of energy consumption and maximum power exerted by the actuators.

1. Introduction

Most of the time, the construction of legged robots is made in an empiric way and the optimization of the mechanical structure is seldom taken into account. In order to avoid spending time and money on the construction of many prototypes to test their performance, a CAD tool seems to be necessary [*Ouezdou 90*]. Thanks to this kind of device, it will be possible to optimize on the one hand the mechanical structure of the legs and on the other hand the gaits which will be used by the future robot. Basic efficiency of the robot during its motion depend on the the mechanical design of the structure and specially of the propelling agents. However, the trajectory followed by the leg tip must satisfy some constraints in order to minimize the energy consumption. Three classical structures were often used to build legged robots : a leg with two rotary joints ([*Grishin 95*]), a structure with two orthogonal prismatic joints ([*Krotkov 92*]) and a pantograph mechanism with two closed chains ([*Waldron 84*], [*Lin 93*]). Thus, we discuss here the performances of these devices for kind of trajectories minimizing the impact on the ground.

2. Trajectory minimizing the impact

The impacts between the legs and the ground have a big influence on the stability and the continuity of the walking cycle and reduce the nominal robot capabilities to achieve a given task. These impacts must be avoided or at least minimized. The best way to do that is making a zero velocity of the leg tip when it touches the ground[*Lin 93*].

In animal locomotion field, a steady walking can be considered as an autonomous motion because the gait is stationary. Then, this stationary gait is changed only when the animal comes across an obstacle [*Kaneko 85*]. In order to reach a steady walking gait, it is necessary to use a cycloidal

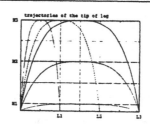

(a) Several trajectories of the tip of leg T according to the stride height H and the stride length L tested for the three mechanisms.

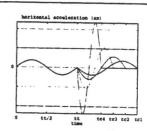

(b) Several horizontal accelerations of the tip of leg T according to the duration of the cycle, for a given trajectory.

trajectory for the leg tip. So, we prescribe the following objectives : the trajectory (with regard to the platform) must be composed by a half-ellipse or a sinusoid in the transfer phase and by a straight line in the contact phase with the ground, must be identical for each cycle n and then will satisfy repeatability boundary conditions. All the locomotion parameters (stride length **L**, stride height **H**, duration of the transfer phase **tt**, total duration of the cycle **tc**) must be changeable for this kind of trajectory. Thus, a wide range of trajectories will be usable (graphs a,b). Furthermore the trajectory must satisfy boundary conditions for positions, velocities and accelerations prescribed by the gait of the robot. In order to minimize the impact and to ensure the motion continuity during the cycles, the following conditions are required : horizontal velocity and acceleration are zero at the beginning and at the end of the transfer phase and horizontal velocity has to be maximum in the middle of the transfer phase. In the direction of the vertical axis (Oy), velocity and acceleration are zero at the beginning, in the middle and at the end of the transfer phase. The boundary conditions will be written at the following instants: beginning and end of the cycle n : $T_1 = (n-1)tc$, $T_4 = ntc$. Middle and end of the transfer phase : $T_2 = (n-1)tc + tt/2$, $T_3 = (n-1)tc + tt$. In order to satisfy all the boundary conditions and the repeatability conditions between two successive cycles, we choose composite functions for the coordinates (x,y) of the tip leg with regard to the platform, which are C^2 on the boundary points and C^∞ elsewhere :

$$\begin{cases} x(t) = \sigma_1 f_1(t) + \sigma_2 f_2(t) \\ y(t) = \sigma_3 f_3(t) + \sigma_4 f_4(t) + \sigma_5 y_0 \end{cases} \qquad \text{(eq. 1)}$$

Where $\sigma_1 = 1$ if t $\in [T_1, T_3]$ and 0 else. $\sigma_2 = 1$ if t $\in [T_3, T_4]$ and 0 else.
$\sigma_3 = 1$ if t $\in [T_1, T_2]$ and 0 else. $\sigma_4 = 1$ if t $\in [T_2, T_3]$, 0 else. $\sigma_5 = 1$ if t $\in [T_3, T_4]$, 0 else.
Functions $f_i(t)$ (i=1,..,4) must satisfy the following boundary conditions :

Cond.1 The boundary conditions for the positions are : $f_1(T_1) = f_2(T_4) = x_0$
$f_1(T_3) = f_2(T_3) = x_0 + L$ $\quad f_3(T_1) = f_4(T_3) = y_0$ $\quad f_3(T_2) = f_4(T_2) = y_0 + H$

Cond.2 The boundary conditions for the velocities and accelerations are equal to zero.
We assume that acceleration is given by : $\ddot{f_i}(t) = A_i sin(e_i t_i)$

Using Cond.1 and Cond.2, velocities and positions are also obtained :

$$\dot{f_i}(t) = -\frac{A_i}{e_i} cos(e_i t_i) + \frac{A_i}{e_i} \qquad f_i(t) = -\frac{A_i}{e_i^2} sin(e_i t_i) + \frac{A_i}{e_i} t_i + C_i \qquad \text{(eq. 2)}$$

With

f_i	e_i	A_i	t_i	C_i
f_1	$2\pi/tt$	$e_i L/tt$	$t - (n-1)tc$	x_0
f_2	$2\pi/(tc-tt)$	$-e_i L/(tc-tt)$	$t - ((n-1)tc + tt)$	$x_0 + L$
f_3	$4\pi/tt$	$2e_i H/tt$	$t - (n-1)tc$	y_0
f_4	$4\pi/tt$	$-2e_i H/tt$	$t - ((n-1)tc + tt/2)$	$y_0 + H$

Having $f_i(t)$ (i=1,..,4) (eq. 2) gives us the two coordinates of the leg tip according to the time $x(t)$ and $y(t)$ (eq. 1).

3. Simulation tool

The three structures studied have two active degrees of freedom. In order to obtain a given trajectory for the tip of leg in a cartesian frame, the time varying laws of the two active joint variables $q_i(t)$ have to be calculated. At first, we obtain the joint positions by using the inverse kinematic model : $Q(t) = F(X(t))$, where $X(t) = (x(t), y(t))^T$. Then, the joint velocities and accelerations are calculated in an explicit way by using the symbolic computation : $\dot{Q}(t) = J^{-1}\dot{X}(t)$ and $\ddot{Q}(t) = J^{-1}\ddot{X}(t) - J^{-1}\dot{J}J^{-1}\dot{X}(t)$. The generalized forces τ are obtained by the inverse dynamic model $M\ddot{Q} + S(Q, \dot{Q}) + G(Q) = \tau + D(Q)^T\Gamma$ with $D(Q)\dot{Q} = 0$. Thus, it is possible to calculate the power $P_i(t)$ exerted by each actuator i to obtain a desired trajectory for the leg tip : $P_i(t) = \tau_i(t)\dot{q}_i(t)$. The maximum power exerted during the motion (i.e. for $t \in [0, tt]$) is numerically calculated for each actuator i : $Pmax_i = max|P_i(t)|$. The energy consumption E for the whole system is calculated by numerical integration of the total power $P(t)$: $E = \int_0^{tt} \sum_i |P_i(t)|dt = \int_0^{tt} P(t)dt$.

The structures with two rotary or two prismatic joints are open kinematic chains. If the link lengths l_1 and l_2 are fixed (we choose here $l_1 = l_2$ to ensure the largest workspace), no optimization is possible from the geometrical viewpoint. However, the pantograph has two closed kinematic chains and even if the lengths of the two main links are fixed, it is then possible to optimize the mechanism by changing kinematic intrinsic parameters, such as factor k_1 (see fig. (c)). It is noteworthy that larger k_1 is, wider the workspace is [Yang 85]. The influence of k_1 on the maximum power exerted by actuators and on energy

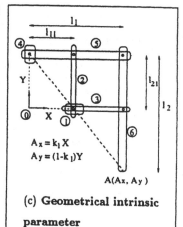

(c) **Geometrical intrinsic parameter**

$k_1 = l_1/l_{11} = l_2/l_{21}$.

consumption is then analysed for a given trajectory of the leg tip. From a kinematic viewpoint, the structure with two prismatic joints and the pantograph have a decoupling between vertical and horizontal motions, unlike the structure with two rotary joints (see fig (d)). In this paper we will show the coupling (or the decoupling) of the three structures from a dynamical point of view.

4. Results and analysis

We think that three main factors influence the energy cost : the initial configuration of the mechanism, the choice of the locomotion parameters, the structural properties of the mechanism. So, we give here some results concerning these three factors. Thanks to the simulation tool, we show that, on the one hand, we find again physical qualitative results which are intuitively foreseeable, on the other hand, we can precisely quantify the influence of the varying parameters in order to choose the best. Now we will have the following notations: **Pan i** for the pantograph structures with the amplification factor $k_1 = i$ $(i = 2, ..5)$, **RR** for the structure with two rotary joints and **PP** for the structure with two prismatic joints. For the PP and the pantograph, the horizontal actuator is called actuator 1, the other. actuator 2.

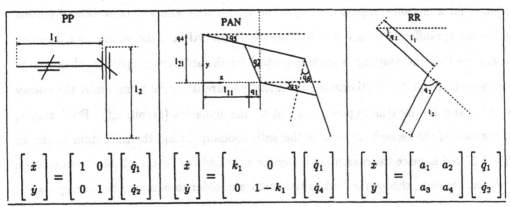

(d)Studied Mechanisms with the forward kinematic model

1) Dynamic coupling, influence of the initial configuration of the mechanism

From a dynamic point of view, it is easy to proove analytically, by using the energy consumed by the structure. that vertical and horizontal motions of the PP are not coupled. Thus, if $E_v(t)$ represents the energy consumed for a pure vertical motion and $E_h(t)$ the energy consumed for a pure horizontal motion, the energy consumed for a motion composed by the two previous motions will be always equal to the sum $E_v(t) + E_h(t)$ as shown as graph (f). Futhermore it can be observed that the initial configuration of this mechanism do not influence the power exerted by the actuators and the energy required to follow a given trajectory (graph f). This is due to the fact that the two links have only translation motions according to an absolute frame. So there is no inertia moments which influence

the dynamic behaviour of the structure. Concerning the RR, the dynamic coupling effects
are due to the kinematic coupling and to the inertia effects of the two links which must
be continuously compensated
to follow a given trajectory.
The dynamic coupling is ev-
ident for this structure (gr.
f). In the case of the pantograph mechanism, there is
kinematic decoupling between
vertical and horizontal mo-

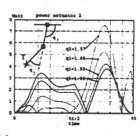

(e) Power of the second rotary
joint for the anthropomorphic leg
according to the initial angle q_1.

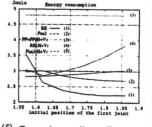

(f) Dynamic coupling. Energy
consumed for a composed motion
compared to $E_h + E_v$.

tions, however the forces exerted by the two prismatic joints are coupled. Thus, the energy
required for a motion composed by a pure vertical motion and a pure horizontal motion
will be always different from the sum of the two for the RR and the pantograph. Besides,
we can see that the coupling is more important for the anthropomorphic leg than for the
pantograph (graph f). Furthermore, the initial posture is crucial to minimize the energy
required and the maximum power exerted by the actuators (graphs e,f). For instance,
in the case of the second actuator of the anthropomorphic leg, the maximum power for
$q_1 = 1.57$ rad is twice the maximum power for $q_1 = 1.86$ rad (graph e). So the simulation
tool permits us to choose the best initial posture in order to minimize the energy cost.

2) Influence of the locomotion parameters.

a) Influence of the stride length and of the duration of the transfer phase

For the three structures, it was shown that, when the velocity of the leg tip ($v = L/tt$)
increases, the maximum power exerted by the actuators goes up in an exponential way
(graphs i,j). Furthermore, for the three structures, the energy consumption (and the
maximum power) are more influenced by the factor tt than the factors L and "initial
configuration of the mechanism" (graphs f,g,h). It can be observed that for each duration
of the transfer phase, the energy consumed by the RR is smaller than the energy consumed
by the PP and the pantograph (graph h).

b) Influence of the couple (L,tt) with a constant average velocity.

For a given leg tip velocity, we can observe that there is an optimum couple (L,tt) which minimizes the energy cost for the three structures (graphs k,l). Besides, it was shown from an energetic view-point that bigger the stride length L and the duration of the transfer phase tt are, the more performing pantograph mechanism is, compared to the RR and PP structures. However, it is obvious that for quick small strides, the pantograph is very unefficient (graph m). Finally we conclude that from an energetic viewpoint, it is more advantageous for the three structures to make long slow strides than small quick strides as shown as graphs (k), (l), (m).

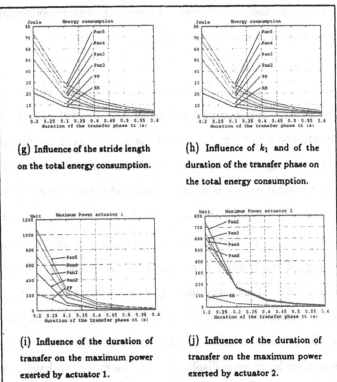

(g) Influence of the stride length on the total energy consumption.

(h) Influence of k_1 and of the duration of the transfer phase on the total energy consumption.

(i) Influence of the duration of transfer on the maximum power exerted by actuator 1.

(j) Influence of the duration of transfer on the maximum power exerted by actuator 2.

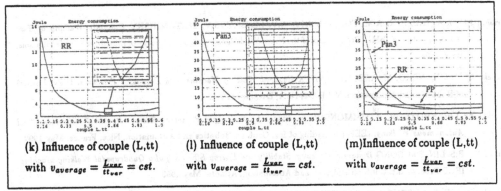

(k) Influence of couple (L,tt) with $v_{average} = \frac{L_{var}}{tt_{var}} = cst.$

(l) Influence of couple (L,tt) with $v_{average} = \frac{L_{var}}{tt_{var}} = cst.$

(m) Influence of couple (L,tt) with $v_{average} = \frac{L_{var}}{tt_{var}} = cst.$

3) Influence of the mechanism structural properties

We illustrate here the influence of this factor by studying the effects of the geometrical intrinsic parameter k_1 of the pantograph on its dynamical performances. In order to maximize the workspace, the factor k_1 has to be increased. In this case, it is shown that the energy consumption and the power exerted by the actuator 1 go up. It can be noticed that the power exerted by the actuator 2 decreases. Furthermore, when the velocity of the leg tip diminishes, the difference between the energy consumption (and maximum power) for varying k_1 is reduced (graphs h,i,j).

5. Conclusion

We have built an analysis tool for walking robots legs which is based on : the determination of trajectories minimizing the impacts, the use of the explicit inverse kinematic model to animate the structure, the use of the inverse dynamic model to obtain the generalized forces to compute, on one hand the maximum power exerted by actuators and on other hand the total energy consumed. Using this tool, we have analysed comparatively three classical structures of robot leg from the viewpoint of energy consumption and maximum power exerted by the actuators according to : the initial configuration of mechanism, the duration of the transfer phase tt, the stride length L, the couple (tt,L). We have shown numerically that an optimum couple (tt,L) exists for each structure, which minimizes the energy consumption. The dynamic decoupling of the structure with two prismatic joints and the dynamic coupling of the structure with two rotary joints and the pantograph mechanism have been displayed. Furthermore, we have observed the influence of the intrinsic geometry of the pantograph on its dynamic performances.

References

[Grishin 95] A.A. GRISHIN, A.M. FORMAL'SKY, A.V. LENSKY, S.V. ZHITOMIRSKY *Dynamic Walking of Two Biped Vehicles*, IFToMM Proceedings of the Ninth World Congress on the Theory of Machines and Mechanisms, Italy, 1995.

[Kaneko 85] M. KANEKO, M. ABE, K. TANIE *A Hexapod Walking Machine with Decoupled Freedoms*, IEEE Journal of Robotics and Automation, Vol RA-1, No 4, December 1985.

[Krotkov 92] E. KROTKOV, R. SIMMONS *Performance of a Six-Legged Planetary Rover : Power, Positionning, and Autonomous Walking*, IEEE Proceeding of Int. Conf. on Robotics and Automation, Nice, France, May 1992.

[Lin 93] B.S. LIN, S.M. SONG *Dynamic Modeling, Stability and Energy Efficiency of a Quadrupedal Walking Machine*, IEEE Proc. of Int. Conf. on Robotics and Automation, Atlanta, May 1993.

[Ouezdou 90] F.-B. OUEZDOU, V. PASQUI, P. BIDAUD, J.-C. GUINOT *Kinematic and Dynamic Analysis of Legged Robots*, RoManSy Proceedings of the Eight CISM-IFToMM Symposium on Theory and Practice of Robots and Manipulators 1990.

[Yang 85] D.C.H. YANG, Y.Y. LIN *Pantograph Mechanism as a non-traditional Manipulator Structure*, Mech.Mach.Theory, Vol.20, No 2, pp. 115-122, 1985.

[Waldron 84] K.J. WALDRON, V.J. VOHNOUT, A. PERRY, R.B. McGHEE *Configuration Design of the Adaptive Suspension Vehicle*, Int. J. of Rob. Res., Vol. 3, No 2, Summer 1984.

DYNAMIC CONTROL OF WHEELED MOBILE ROBOT USING SLIDING MODE

P. Ruaux, G. Bourdon and S. Delaplace

Pierre et M. Curie University, CNRS, Paris, France

and

Versailles Saint-Quentin University, Vélizy, France

Abstract : *In this paper, we present a new robot ROMAIN. We describe the used mechanical, hardware and software architecture. This open structure is realized with a serial high speed bus : CAN. We develop the contraint and dynamic equations and we explain the state system. To control this robot, we use a variable structure control to track desired trajectories. However, this control method is too constraining to DC motor. We adjust this controller to implement it in the view to delete the chaterring. Different simulations are performed to show the efficiency of this controller.*

1 Introduction

After somes kinematic studies and experimentations on our old robot ROMEO (classical control, neural network control, fuzzy control, multi-agent control), we focuse the dynamic effects on this type of structure. We have built a new robot ROMAIN (figure 1) (french acronymous of instrumented autonomous mobile robot). In a first part, we present the hardware, software and mechanical structure.

To control this robot, we implement the variable structure control method. This control allows to converge to the desired orders to processing switching laws imposing the change between two states when the trajectory meet the selected surface in the state space [6]. We show different simulations which allow to set futur experimentations.

2 Robot description

2.1 Structure characteristics

To make this new robot, we choose the same configuration that the last robot of our laboratory ROMEO [1]. The locomotion is realized with four wheels; two rear driving wheels and two front free wheels. The robot turns easily with a lot of dexterity and

Figure 1: ROMAIN Mobile Robot

stability nevertheless in a constraint environment. Each driving wheel is made up a DC motor (500 W), a reducer (1/16) and an optic incremental encoder (2000 pts). The available power is 60 V with automobile batteries. Two DC-DC converters pilot the motors by a control voltage with a current mode.

2.2   Hardware architecture

To realize the hardware part, we opt for a multi layer system. This conception allows to decentralize different tasks and relieve the principal system. The computation time is improved with this parallel structure. Furthermore it's easy to add or delete layer without remake all hardware architecture.

The high level is supported by an industrial 486-DX2-66 PC. It's dedicated to tracking trajectory algorithm. It generates orders to the low layer. At present, only two layers exist. The second level (low level) manages different small tasks. Many cards compose this layer, each of them controls a physical system as motors, encoders, ultrasonic sensors, communication protocol, accelerometers, etc... These cards are made around microcontroller (80C31 or 80C592). They have RAM, ROM, interface communication. Annex cards are realized to motor axis control, to manage ultrasonic sensors, etc... The sampling time to control motors is 2 ms with these cards. The middle layer is in realization progress. It will be composed of 16 bit microcontroller dedicated to many computations.

It will support different algorithms as odometer for instance. At present, this layer is enclosed in the high level.

It's necessary to communicate between this layers. We choose to take the local network : CAN (Control Area Network). This serial bus was developed during the last decade by BOSH and INTEL corporation to manage the ABS system car. This high speed bus (1Mbits/s), has a protocol allowing to add or delete nodes without modify the hardware, a non destructive collision detection by bitwise arbitration, a predictable maximum latency time message. Each level has a communication interface.the final structure is represented in the figure 2 (a). All cards have been developed in our laboratory.

2.3   Software architecture

An OS9000 operating system is installed on the board industrial PC. This real time system of Microwave corporation is the Intel target of OS9. Tasks have been developed allowing to manage several actions as :

- CAN reading task   - CAN writing task
- keyboard capture task   - putting screen task
- initialisation task   - end task
- odometer task   - control task

The task schedulling permits to send the orders at the low level every 10 ms.

3   Dynamics of wheeled mobile robot

To determinate the dynamic model of our robot ROMAIN, we choose the Lagrangian method to easely take into account the nonholonomic constraints.

3.1   Constraint equation and inertia Matrix

We only consider the dynamic of the driving wheels (free wheels are passives).

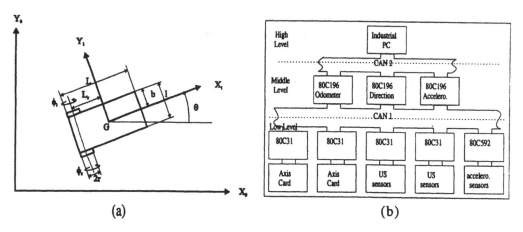

(a)   (b)

Figure 2: Hardware structure (a) and Coordinate of mobile robot (b)

Description of parameters figure 2(b):

(x_0,y_0): the world coordinate system

(x_1,y_1): the coordinate system fixed to the robot

G: the center of mass with its coordinates (x_G,y_G)

L: length of the robot

l: width of the robot

b: the distance between either driving wheels and the axis of symmetry

L_r: the distance between the center of mass and the axle of the rear driving wheels

L_{rr}: length of the plateform rear part

L_f: length of the plateform front part

ϕ_l: the angular position of the left rear driving wheel

ϕ_r: the angular position of the right rear driving wheel

θ: the angle between the two coordinate systems

r: the radius of each driving wheel

c: defined by $\frac{r}{2b}$

M_f: front part mass of the plateform

M_r: rear part mass of the plateform

M_t: total mass of the plateform

M_w: wheel driving mass

We consider the following generalized coordinates vector q :

$$q = (x_G, y_G, \phi_l, \phi_r)^t \tag{1}$$

The contact between the wheels and the ground satisfies the condition of rolling without slipping. We obtain this holonomic equation :

$$\dot{\theta} = \frac{r}{2b}(\dot{\phi}_r - \dot{\phi}_l) \tag{2}$$

And two nonholonomic equations who defined the D matrix :

$$D(q) = \begin{bmatrix} -sin(\theta) & cos(\theta) & \frac{rL_r}{2b} & \frac{-rL_r}{2b} \\ cos(\theta) & sin(\theta) & \frac{-r}{2} & \frac{-r}{2} \end{bmatrix}$$

(3)

To compute the inertia matrix, we neglect the inertia expression due to the orientation. We concider the plateform as two parallelepipeds with different lengths and mass. The front part corresponds to the batteries and the rear part to electronic cards and PC. We explain the inertia matrix on the mass center :

$$I_p = \begin{bmatrix} (l^2 + H^2)\frac{Mt}{12} & 0 & 0 \\ 0 & \frac{Mt}{12}H^2 + \frac{M_fL_f + M_rL_{rr}}{12} + \frac{L}{4M_t}(M_f^3 + M_r^3) & 0 \\ 0 & 0 & \frac{Mt}{12}l^2 + \frac{M_fL_f + M_rL_{rr}}{12} + \frac{L}{4M_t}(M_f^3 + M_r^3) \end{bmatrix}$$

(4)

We must take into account the turn part of this robot which correspond to rotor motors, reducers and wheels. Let N the reducer ratio equal to 16 and J_{rr} the inertia of reducer-motor set. Wheels are considered as cylenders.
We obtain :

$$I_t = \frac{m_w r^2}{2} + N^2 J_{cc}$$

(5)

3.2 Dynamic equations

The total kinetic energy of this robot is :

$$K = \frac{1}{2}m_t(\dot{x_G}^2 + \dot{y_G}^2) + I_{p\,3}\dot{\theta}^2 + \frac{1}{2}I_t(\dot{\phi_l}^2 + \dot{\phi_r}^2)$$

(6)

We explain the motion equation on the form :

$$A(q)\ddot{q} + B(q, \dot{q}) + C(q) = \tau - D^t\lambda$$

(7)

$$A(q) = \begin{bmatrix} M_t & 0 & 0 & 0 \\ 0 & M_t & 0 & 0 \\ 0 & 0 & I_{p\,3}c^2 + I_t & -I_{p\,3}c^2 \\ 0 & 0 & -I_{p\,3}c^2 & I_{p\,3}c^2 + I_t \end{bmatrix} \quad : \text{Inertial matrix}$$

(8)

$B(q, \dot{q}) = 0$: Coriolis matrix, τ : Torque matrix, λ : Lagrange multipliers matrix
We choose the state vector as :

$$\dot{x} = \begin{bmatrix} q \\ \nu \end{bmatrix} \quad \text{with } \nu = \begin{bmatrix} \dot{\phi_l} \\ \dot{\phi_r} \end{bmatrix}$$

(9)

We can explain the state equation [3] as :

$$\dot{x} = f(x) + gu \quad f(x) = \begin{bmatrix} -\frac{rL_r}{2b}sin\theta(\dot{\phi}_r - \dot{\phi}_l) + \frac{r}{2}cos\theta(\dot{\phi}_l + \dot{\phi}_l) \\ \frac{rL_r}{2b}cos\theta(\dot{\phi}_r - \dot{\phi}_l) + \frac{r}{2}sin\theta(\dot{\phi}_l + \dot{\phi}_l) \\ \dot{\phi}_l \\ \dot{\phi}_r \\ 0 \\ 0 \end{bmatrix} \quad g = \begin{bmatrix} 0 & 0 \\ 0 & 0 \\ 0 & 0 \\ 0 & 0 \\ 1 & 0 \\ 0 & 1 \end{bmatrix} \quad (10)$$

4   Sliding Mode Control

The ROMAIN mobile robot is a nonholonomic system. We can't control all of position and orientation components. We choose to control the following position in the world frame :

$$Z = \begin{bmatrix} x_G + L_a cos\theta \\ y_G + L_a sin\theta \end{bmatrix} \tag{11}$$

L_a is the distance between the mass center and the control point on the x_1 axis. To control this robot, we take the sliding mode method. The variable structure control allows to take into account uncertain parameters [4] [5]. It corresponds to inaccuracies on the terms actually in the model. This typical structure of a robust controller is composed of a nominal part, similar to a feedback linearizing and of additional terms aimed to deal with model uncertainty [5]. We have a nonlinear system described by the (10) equation. We write our system in the following canonical form [2] :

$$\dot{x} = f(x) + gu \quad and \quad z(x) = c(x) \tag{12}$$

The state vector x is partitioned into x_1 and x_2 where dim-x_2 = dim u = m and dim-x_1 = n-m. The reduced form of the system model takes the form

$$\begin{aligned} \dot{x}_1 &= f_1(x) \\ \dot{x}_2 &= f_1(x) + g^*u \quad where \quad f_1(x) = 0 \end{aligned} \tag{13}$$

We determine two switching surfaces as :

$$\begin{cases} s_x = \dot{\tilde{z}}_1 + \lambda_1\tilde{z}_1 \\ s_y = \dot{\tilde{z}}_2 + \lambda_2\tilde{z}_2 \end{cases} \quad where \quad \begin{aligned} \tilde{z}_i &= z_i - z_d \\ \dot{\tilde{z}}_i &= \dot{z}_i - \dot{z}_d \end{aligned} \quad i = 1, 2 \tag{14}$$

We use the reaching law approach [2] to specify the dynamic of the switching function. Let the dynamic of the switching function be specifed by the differential equation

$$\dot{s} = -Qsgn(s) - Kf(s) \tag{15}$$

Using the reaching law (15) in the system (13) yields the VSC signal

$$u(x) = -\left(\frac{\partial s}{\partial x_2}g^*\right)^{-1}\left(Qsgn(s) + Kf(s) + \frac{\partial s}{\partial x_1}f_1(x)\right) \tag{16}$$

This relation exists only if $L_a \neq -L_r$ (Inversion condition of the $\frac{\partial s}{\partial x}g^*$ matrix)

So, with this law, the control can be switched from one value to another at will, infinitely fast. In practical systems, it is impossible to achieve the high switching control that is necessary to most VSC design (presence of finite time delay for control computation, limitations of physical actuators,...). To eliminate the chattering, we replace the sign function by a saturation function. The behaviour controller is linear when $| s | < L$.

5 Simulations

To illustrate this control law, we determine two trajectories. In fact, the ROMAIN robot (RR) tracks a virtual robot (VR). On the figure 3 (a) the initial VR position is $(x_d, y_d) = (5,0)$ and its orientation is $\theta_d = 45°$. Its velocity increases during 3 seconds, remains constant during 9 seconds and decreases during 3 seconds. The RR position is $(0,0)$ and its orientation is 0. This robot catches up VR with real characteristic respect (maximum torque value : 100 Nm). The maximum velocity is 2.5 m/s. We have simulated with different λ values. If values are too great, the system can't correctely react, the system is too much solicited and quits the sliding surfaces. On the contrary, if values is too small, the system has a great response time and it converges slowly. In the other simulation (figure 3 (b)), we find the same characteristics. It exists a tracking error arround one millimeter. At the end of the straight trajectory, the VR stops, and the tracking error decreases to evolve nil. With the substitution of sign function by saturation function, the chattering is erased and the experimentation is possible (torques are compatible with motors).

6 Conclusion

In this paper, we have presented a new fast robot. It's based on an open structure to have a great flexibility. We determine dynamic equation and we establish a sliding mode

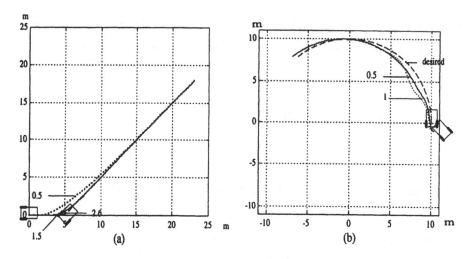

Figure 3: Straight trajectory (a) curve trajectory (b)

control to track a trajectory. In simulations, we emphased good results. We have shown the importance of λ values which permit to give dynamics of our system keeping physical values. At present, experimentations are in progress.

References

[1] S. DELAPLACE. *Navigation a cout minimal d'un robot mobile dans un environement partiellement connu.* These de doctorat de l'Universite Pierre et Marie Curie, 1991.

[2] J. Y. HUNG, W. GAO, and J. C. HUNG. Variable structure control: A survey. *IEEE Transactions on Industrial Electronics*, 40(1):2–22, february 1993.

[3] P. RUAUX, G. BOURDON, S. DELAPLACE, N. PONS, and J. RABIT. A rapid mobile robot synthesis. *IEEE International Conference on Systems, Man and Cybernetics*, 5:4073–4078, 1995.

[4] H.-S. SHIM, J.-H. KIM, and K. KOH. Variable structure control of nonholonomic wheeled mobile robot. *IEEE International Conference on Robotics and Automation*, pages 1694–1699, 1995.

[5] J. SLOTINE and W. LI. *Applied Nonlinear Control*. Englewood Cliffs, NJ:Prentice Hall.

[6] V. I. UTKIN. Variable structure systems with sliding modes. *IEEE Transactions on Automatic Control*, AC-22(2):212–231, April 1977.

SIMULATION STUDY OF A GROUP-BEHAVIOR
OF MICRO-ROBOTS ON THE MODEL OF INSECT'S BEHAVIOR

K. Matsushima, Y. Hayashibara and K. Nakamoto
Toin University, Yokohama, Japan

Abstract

Recently, the researches and developments on micro-robots are being carried out. Under such circumstances, how to control such micro-robots will become a main subject hereafter. We will not be able to expect, in general, to make the micro-robot have such higher sensing ability and control function as robots we have now. It is important to study the behavior as a group of micro-robots with only simple function under no hierarchical organization. We assume a micro-robot that only has a simple and vague visual sensor and an ability which is distinguish the same race and others. This paper proposes a method for which such micro-robots, as a group, carry out a task. We suppose the task such as a group of the micro-robots search objects in a working space and carry them to a home position. Each micro-robot works under 5 behavior rules to carry out the task. The computer simulation results show that although the micro-robot has one having only simple functions, a group of such micro-robots is able to exhibit the meaningful behavior.

1. Introduction

In a group of some kinds of insect, for example ants or bees, each individual has an assigned role respectively and they, as a group, carry out a task. They have not a hierarchical organization where an individual makes the necessary decisions and the others follow them. Recently, the researches and the developments on micro-robots are being carried out with the progress of micro-mechanism and micro-machining technology. We will not be able to expect, however, to make the micro-robot have such higher sensing ability and control functions as the robots which we have now. Therefore, it is important to study the behavior as a group of micro-robots with only simple functions like the insect's behavior as mentions above.

J.L.Deneubourg and others studied on the group behavior of ant-like robots which sort and collect foods[1]. H.Miura and others studied on the algorithm in order to make the group of micro-robots form an order. Each robot behaves as a nonlinear dynamical system in a potential field.

This paper proposes a method for which micro-robots, as a group, carry out a task. We suppose the task such as micro-robots search objects in a working space and carry them to a home position. We assume two kinds of micro-robots, one of them is named the worker which carries out a task, another one is named the guide which moves only on a fixed path in a working space from the home position.

The basic functions and the possible actions of micro-robots are described in Chapter 2.

In Chapter 3, the sets of rules under which the micro-robots behave are described.

The results of the computer simulation are shown in Chapter 4.

The computer simulation was carried out using the above rules with 30 worker micro-robots , 4 guide micro-robots and 4 objects.

The simulation result shows that although the micro-robot is one having only simple functions, a group of the micro-robots is able to exhibit the meaningful behavior.

2. Functions of Each Micro-Robot

There are two kinds of micro-robots in this study. One of them is named "worker micro-robot" and another one is named "guide micro-robot". The worker is the micro-robot which actually carries out some tasks in a work space. The guide is the micro-robot which has a role to guide the worker to a home position.

2.1 Basic Functions of Micro-Robot

We assume that the functions of the micro-robot in this study are as simple as possible.

The worker has only the following basic functions.

(1) It has only a vague visual sensor with narrow viewfield as shown in Fig.1.

(2) It can recognize the difference between workers, guides, objects to handle and its home position which they come into its viewfield.

(3) The movement of a worker is basically random walk that it's step length is constant,but it's moving direction is selected randomly within the visual angle of the visual sensor as shown in Fig.2. The worker has not any memories about it's position in the work space. The guide can only move on a path between the home position and the point fixed on the border of the working space.

2.2 Basic Actions of Worker Micro-Robot

The basic actions of a worker are possible,under the above basic functions, as follows.
(1) To go away from the home position into the work space using the moving function.
(2) When it catches another worker in its viewfield using the visual sensor,to avoid it.
(3) When it catches an object in its viewfield, to cling to it using the recognition function.
(4) After it clung to the object to carry, when it catches a guide, then it follows the guide using the moving function.

3. Behavior Rules of a Group of Micro-Robots

A set of behavior rules for which the micro-robots carry out a task as a group is constructed using the above basic functions and actions.

In this study, we suppose the task that micro-robots search objects in a working space and carry them to a home position.

A set of behavior rules that is constructed to carry out the above task is shown below.

3-1 Keeping away from Another Worker

When a worker catches,during movement, another worker or the border of the working space or the home position as obstacles in its viewfield, the worker takes an action to avoid such obstacles following to obstacle avoidance rules below.(see Fig.3)

Obstacle avoidance rules:

Rule 1: If an obstacle comes into the range "near" of the viewfield, then change the moving direction by an angle which is randomly selected between 90[deg] and 135[deg].

Rule 2: If an obstacle comes into the range "middle" of the viewfield, then change the moving direction by an angle which is randomly selected between 45[deg] and 90[deg].

Rule 3: If an obstacle comes into the range "far" of the viewfield, then change the moving direction by an angle which is randomly selected between 0[deg] and 45[deg].

3-2 Clinging to an Object

A worker catches an object in its viewfield, then the worker takes an action to cling to the object following to clinging rules as shown in Fig.4.

<u>Clinging rule</u>

Rule 1: If a number of workers already clung to the object is less than the number which is necessary to move the object, then cling to the object, otherwise execute obstacle avoidance rules.

3-4 Carrying back an Object to the Home Position

The workers clung to the object become one body and it moves randomly as a worker to follow the obstacle avoidance rules and the clinging rule, but it recognizes a guide and the home position as an object. Such an object with workers freely moves in the working space until it catches a guide or the home position.

<u>Carrying rules</u>

This set of rules is basically the same as the obstacle avoidance and clinging rule,but another rule is added for putting the object on the home position.

Additional rule: If the workers carrying the object caught a new object in the viewfield and it is the home position, then leave the object on the home position.

4. Simulation

The computer simulation was carried out under the following condition.

(1) Dimension of the working space : 570 * 330 pixels

(2) Size of the worker's viewfield : $L= 45$ pixels, $\theta_w =30°$

(3) Size of the guide's viewfield   : $L= 20$ pixels, $\theta_g = 20°$

(4) Number of workers = 30, Number of Objects = 4 and Number of guides = 4

(5) Necessary number of workers to carry an object = 3

Fig.5 shows an example of a contour that one worker followed from the home position to an object put in the working space.

Fig.6 shows the following situation .

Fig.6 (a) : the workers and the guides go away from the home position one and after another.(50 steps after)

Fig.6 (b) : two workers clung to an object.(200 steps after)

Fig.6 (c) : they with the object continue random walk until they catch a guide or the home position.(350 steps after)

Fig.6 (e) : they catch a guide in their viewfield and follow the guide toward the home

position.(700 steps after)

5. Conclusion

We examined the group behavior of micro-robots by simulation. The functions of the micro-robot were a vague visual sensor, an ability to distinguish the same race and others and a moving function. The following task that " search objects in the working space and carry them to the home position" was executed by a group of 30 workers assisted by 4 guides.

The simulation showed the meaningful result. But we do not consider about the efficiency or the optimality of the process executing the task in this study. We think that micro-robot will be not able to have so higher functions that the efficiency or the optimality of its behavior can be consider.

Acknowledgments

We would like to thank the undergraduate students Mr. N.Ogino and Mr. S.Nishino for their help on this study.

Reference

[1] J.L.Deneubourg, et.al. : The Dynamics of Collective Sorting Robot-like Ants and Ant-like Robots; Proc. of the 1st International Conference on Simulation of Adaptive Behavior, 356/363, MIT Press. 1991

[2] H.Miura : Insect-model based Robot and Group Intelligence (in Japanese); Keisoku to Seigyo, (Journal of SICE), Vol.31,No.11,1180/1184, 1992

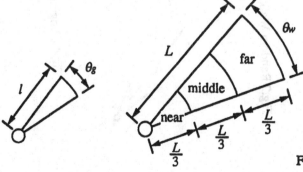

Fig.1 Viewfield of Micro-Robot

(a) Guide's Viewfield   (a) Worker's Viewfield

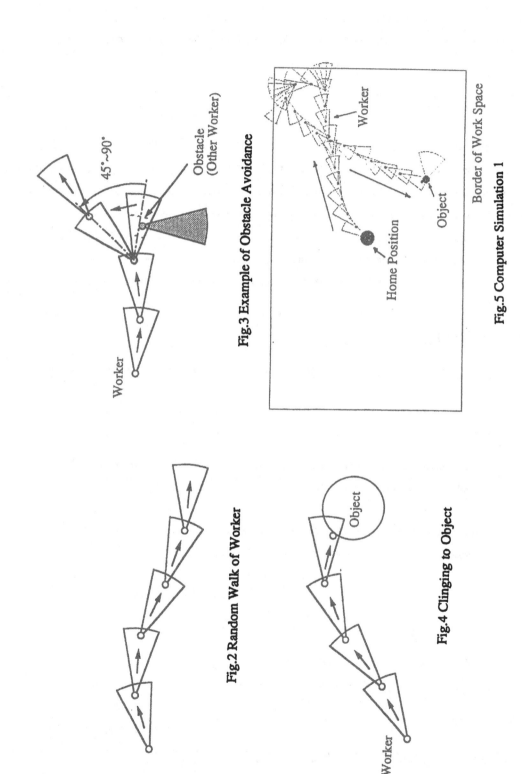

Fig.2 Random Walk of Worker

Fig.3 Example of Obstacle Avoidance

Fig.4 Clinging to Object

Fig.5 Computer Simulation 1

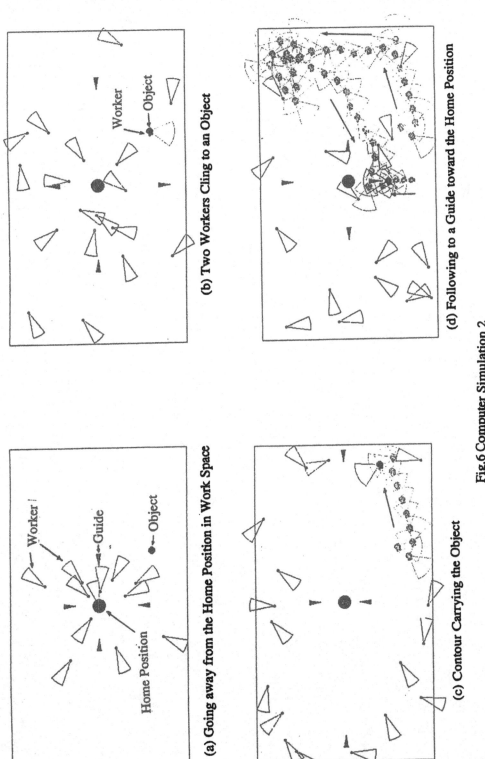

(a) Going away from the Home Position in Work Space

(b) Two Workers Cling to an Object

(c) Contour Carrying the Object

(d) Following to a Guide toward the Home Position

Fig.6 Computer Simulation 2

Chapter VI
Walking Machines
and Mobile Robots III

ACTIVE COORDINATION OF ROBOTIC TERRAIN-ADAPTIVE WHEELED VEHICLES FOR POWER MINIMIZATION

S.C. Venkataraman
The Ohio State University, Columbus, OH, USA

S.V. Sreenivasan
University of Texas at Austin, Austin, TX, USA

K.J. Waldron
The Ohio State University, Columbus, OH, USA

ABSTRACT

An active coordination scheme is developed for articulated wheeled vehicles equipped with redundant actuation, for the minimization of power consumed during locomotion on soft terrain. A detailed soil-mechanics model is used to incorporate the effects of sinkage, slip and soil resistance at the wheel-terrain contact. The coordination scheme determines the optimal actuator torques to perform vehicle guidance and is consistent with the kinematic and dynamic constraints imposed on the vehicle system due to the non-holonomic nature of the wheel-terrain contact. The key feature in the coordination scheme is the optimal allocation of the wheel plane contact forces in response to the differences in the soil condition at the different wheel contacts.

INTRODUCTION

Several wheeled vehicles that have combinations of active and passive configuration control have been studied: CARD and SAM Rovers [1]; FMC Rover [2]; Rocker Bogie [3, 4]; Wheeled Actively Articulated Vehicle, WAAV [5, 6]. Actively articulated wheeled vehicles with independently actuated wheels and fully active three degree of freedom articulations (like the WAAV), posses capabilities to control the distribution of wheel contact forces and to accommodate for terrain obstacles by varying their geometric configurations. In this respect they are most closely related to the fully terrain adaptive legged vehicles such as the Adaptive Suspension Vehicle [7] and the Ambler [8].

Active coordination schemes for legged vehicles have been extensively studied. The criteria for optimization that have been used have been maximization of static stability [9, 10], minimization of power consumption [11, 12], minimization of leg-slippage [13, 14] and elimination of interaction forces [15]. Actively coordinated wheeled vehicles have not been so intensively studied. The non-holonomic nature of the wheel contact makes steering and coordination much more difficult than for a legged system. The coordination issues of active wheeled systems have been addressed using a three-module version of the articulated chain [16, 17]. The issues of motion planning, force planning and dynamic modeling of actively articulated wheeled systems have been addressed [18]. A preliminary coordination scheme (which used the pseudo-inverse solution for contact force allocation) has been developed to perform simulation studies of such vehicles [18]. In this work, we first incorporate a detailed wheel-terrain contact model from soil-mechanics theory that simulates the off-road locomotion characteristics: wheel slip, sinkage and soil resistance. Next, the coordination scheme is extended to include an optimal allocation of the contact forces so as to minimize power consumption during locomotion.

GEOMETRY OF THE ARTICULATED WHEELED SYSTEM

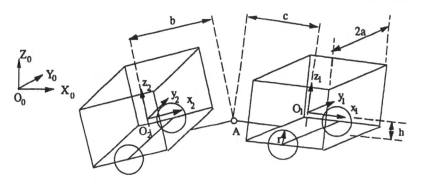

Figure 1 Two Module Articulated Wheeled System Geometry

We consider a two-module version of the articulated chain as shown in Figure 1. The frame '0' is a world fixed coordinate system. Coordinate frames '1' and '2' are introduced at the centers of mass of the modules. The coordinates of module '1' in frame '0' are given by the position vector of the module center of mass ($O_0^1 = \begin{bmatrix} 0_{1x} & 0_{1y} & 0_{1z} \end{bmatrix}^T$) and the Euler angles α, β, and γ. The coordinates of module '2' in frame '1' are given by the Euler angles ψ, θ and ϕ, which represent the joint variables of the articulation A. Let the angles θ_{Li} and θ_{Ri}, (i= 1, 2) denote the angular position of the left and right wheels of the modules.

KINEMATIC ANALYSIS OF THE ARTICULATED WHEELED SYSTEM

It can be shown that if we impose the constraints of no lateral motion at the module centers of mass and pure rolling at the contact points, the number of system degrees of freedom with respect to velocity and acceleration analysis is three. That is, the motion of the system can be determined from three velocity and three acceleration components of the master module [17, 18]. The three commanded velocity components of the main module (called commanded free rates) are chosen to be the forward velocity (v_{1xd}), the pitch rate (ω_{1yd}) and the yaw rate (ω_{1zd}).

The velocity analysis yields the commanded velocity components of the vehicle, namely: the master module roll rate (ω_{1zd}), the master module vertical velocity (v_{1zd}), the articulation angular rates ($\dot{\psi}_d$, $\dot{\theta}_d$, $\dot{\phi}_d$) and the wheel angular rates ($\dot{\theta}_{Lid}$, $\dot{\theta}_{Rid}$). It is assumed that the front module carries an inertial sensing package that provides direct measurement of the free rates. These sensed values can be compared with the commanded values to generate the three commanded free accelerations of the front module. An acceleration analysis can be performed along similar lines to the velocity analysis to obtain the commanded accelerations of the vehicle.

DYNAMICS OF THE ARTICULATED WHEELED SYSTEM

The inverse and forward dynamics analysis of articulated wheeled vehicles has been performed in reference [18] using the tools for spatial rigid-body dynamics developed in reference [19]. The inverse dynamics equations which form the basis for the contact force allocation scheme, will now be reviewed. Inverse dynamics of the wheeled system under consideration involves the calculation of the joint torques required to produce the given set of joint velocities/ accelerations and the given set of desired contact forces after compensating for soil resistance. The equations are derived in the frame of the base member (frame '1'). The recursive Newton-Euler method for performing inverse dynamics involves three steps. Step 1 involves the computation of the 6×1 spatial velocity vector ($\hat{v}$) and acceleration vector ($\hat{a}$) of all the rigid bodies from the commanded velocity/ acceleration components of the vehicle obtained from the kinematic analysis. Step 2 involves the calculation of the commanded inertia force ($\hat{f}^*$) acting on each member from its spatial velocity and acceleration. Step 3 involves the computation of the joint torques via the computation of the joint forces. Let f_{Lid} and f_{Rid} be the desired left and the right wheel contact force vectors of the ith module in frame '1'. Let r_{Lie} and r_{Rie} represent the estimates of the soil resistance vectors at the left and the right wheel contacts of the ith module in frame '1' (the soil resistance can be estimated from the tangential contact force and the wheel torque values from the previous time-step). Let c_{Li} and c_{Ri} be the position vectors of the left and the right wheel contact points of the ith module in frame '1'. The spatial vectors corresponding to the contact forces and resistances in frame '1' can be obtained as follows:

$$\hat{f}_{Lid} = [f_{Lid}^T \ (c_{Li} \times f_{Lid})^T]^T, \quad \hat{f}_{Rid} = [f_{Rid}^T \ (c_{Ri} \times f_{Rid})^T]^T$$

$$\hat{r}_{Lie} = [r_{Lie}^T \ (c_{Li} \times r_{Lie})^T]^T, \quad \hat{r}_{Rie} = [r_{Rie}^T \ (c_{Ri} \times r_{Rie})^T]^T \tag{1}$$

$$\hat{f}_{Lit} = \hat{f}_{Lid} + \hat{r}_{Lie}, \ \hat{f}_{Rit} = \hat{f}_{Rid} + \hat{r}_{Rie}, \ \text{where } i = 1, 2$$

The joint forces ($\hat{f}_{L1}$, $\hat{f}_{R1}$, $\hat{f}_{L2}$, $\hat{f}_{R2}$, $\hat{f}_{art}$, $\hat{f}_1$) are given by:

$$\hat{f}_{Li} = \hat{f}_{Li}^* - \hat{f}_{Lit}, \ \hat{f}_{Ri} = \hat{f}_{Ri}^* - \hat{f}_{Rit}, \ \text{where } i = 1, 2$$

$$\hat{f}_{art} = \hat{f}_{L2} + \hat{f}_{R2} + \hat{f}_2^*, \ \hat{f}_1 = -(\hat{r}_{L1e} + \hat{r}_{R1e} + \hat{r}_{L2e} + \hat{r}_{R2e}) = \hat{f}_{art} + \hat{f}_{L1} + \hat{f}_{R1} + \hat{f}_1^* \tag{2}$$

The joint torques are given by:

$$T_{Li} = \hat{s}_{Li}^S \hat{f}_{Li}, \ T_{Ri} = \hat{s}_{Ri}^S \hat{f}_{Ri}, \ \text{where } i = 1, 2; \ [T_r \ T_y \ T_p] = (\hat{S}_{art}^S \hat{F}_{art})^{-1} \hat{S}_{art}^S \hat{f}_{art} \tag{3}$$

T_{Li} and T_{Ri} ($i = 1, 2$) are the wheel torques. T_r, T_y and T_p are the roll, pitch and yaw torques at the articulation respectively. $\hat{s}_{Li}$ and $\hat{s}_{Ri}$ ($i = 1, 2$) are 6×1 spatial vectors that represents the screw axes of the wheel joints. $\hat{S}_{art}$ is a 6×3 spatial screw matrix that represents the screw axes of the three degree of freedom articulation joint. $\hat{F}_{art}$ is a 6×3 spatial matrix that maps the scalar articulation torques into $\hat{f}_{art}$.

CHARACTERIZATION OF REDUNDANCY
IN CONTACT FORCE ALLOCATION

It is desirable that actively coordinated vehicles posses redundant actuation, i.e. they have a larger number of actuators than the number of output degrees of freedom. This redundancy results in an underconstrained set of contact force allocation equations. The excess degrees of freedom in the force allocation problem can then be used to optimize vehicle performance. In order to address such an optimization problem it is necessary to characterize the redundancy in the contact force allocation problem. While legged systems posses complete controllability of the contact forces, wheeled systems offer only partial controllability over the contact forces. This can be explained as follows: The interaction force between the two wheel contact points (equal and opposite contact forces parallel to the axle) is not controllable for wheeled vehicles. This is due to the fact that since the axle length is fixed, the distance between the two wheel contact points cannot be actively varied. Thus, even though the lateral force components can be influenced by the articulation actuators, they cannot be arbitrarily allocated. Therefore, the redundancy in force distribution exists only in the allocation of the normal and tangential contact force components that lie in the plane of the wheel.

The force allocation problem for the actively articulated wheeled vehicle can be studied as follows: Equation 2 represents the spatial force balance for each rigid body in the system. Summation of these equations cancels all the internal joint forces and results in a relationship between the desired contact forces and the commanded inertial forces of the members:

$$\left(\hat{f}_{L1d} + \hat{f}_{R1d} + \hat{f}_{L2d} + \hat{f}_{R2d}\right) = \hat{f}_{L1}^* + \hat{f}_{R1}^* + \hat{f}_{L2}^* + \hat{f}_{R2}^* + \hat{f}_1^* + \hat{f}_2^* \tag{4}$$

which can be expanded to give:

$$[G]f_d = q_d, \quad \text{where } q_d = \hat{f}_{L1}^* + \hat{f}_{R1}^* + \hat{f}_{L2}^* + \hat{f}_{R2}^* + \hat{f}_1^* + \hat{f}_2^* \tag{5}$$

$$[G] = \begin{bmatrix} I_3 & I_3 & I_3 & I_3 \\ c_{L1} \times & c_{R1} \times & c_{L2} \times & c_{R2} \times \end{bmatrix}, \quad f_d = \begin{bmatrix} f_{L1d}^T & f_{R1d}^T & f_{L2d}^T & f_{R2d}^T \end{bmatrix}^T$$

$$c_{Li}(x_{Li}, y_{Li}, z_{Li}) \times = \begin{bmatrix} 0 & -z_{Li} & y_{Li} \\ z_{Li} & 0 & -x_{Li} \\ -y_{Li} & x_{Li} & 0 \end{bmatrix}, c_{Ri}(x_{Ri}, y_{Ri}, z_{Ri}) \times = \begin{bmatrix} 0 & -z_{Ri} & y_{Ri} \\ z_{Ri} & 0 & -x_{Ri} \\ -y_{Ri} & x_{Ri} & 0 \end{bmatrix}, \text{ where } i = 1, 2$$

For a given vector q_d, the force allocation problem involves solving Equation 5 for the contact force vector f_d. The redundancy in force allocation has been addressed in reference [17] by defining two force fields; the equilibrating force field and the interaction force field. For a given vector q_d, the equilibrating force vector f_d^{eq} is unique and is the same as the pseudo-inverse, minimum norm solution to Equation 5, i.e.

$$f_d^{eq} = [G]^+ q_d, \text{ where } [G]^+ = [G]^T ([G] [G]^T)^{-1} \tag{6}$$

The pseudo-inverse solution tends to minimize the ratios of the tangential to the normal contact force at the wheels, thereby reducing the wheel slips as well. Hence it is a good sub-optimal solution to the power minimization problem [17]. f_d^{eq} has no interaction force components [17]. Therefore, the desired lateral force at each wheel contact is obtained by taking the component of the pseudo-inverse force at that contact point along the wheel axle. Once the lateral forces have been decided, the force allocation problem in the wheel frame contact forces (f_{wd}) is given by

$[G_w] f_{wd} = q_{wd}$, where

$$G_w = \begin{bmatrix} 1 & 0 & 1 & 0 & 1 & 0 & 1 & 0 \\ 0 & 1 & 0 & 1 & 0 & 1 & 0 & 1 \\ 0 & y_{L1} & 0 & y_{R1} & 0 & y_{L2} & 0 & y_{R2} \\ z_{L1} & -x_{L1} & z_{R1} & -x_{R1} & z_{L2} & -x_{L2} & z_{R2} & -x_{R2} \\ -y_{L1} & 0 & -y_{R1} & 0 & -y_{L2} & 0 & -y_{R2} & 0 \end{bmatrix} \tag{7}$$

The redundancy in the allocation of the contact forces in the wheel plane can be characterized as follows:

$$f_{wd} = f_{wd}^{eq} + \sum_{i=1}^{3} \alpha_i n_{wi} \tag{8}$$

where n_{wi}, $i = 1, 2$ and 3 are the null space vectors of G_w. The weights α_i can be chosen so as to optimize an objective function in the contact forces.

WHEEL-TERRAIN CONTACT MODEL

Given the vehicle configuration and the local terrain characterization, the wheel-terrain contact model generates the location and magnitude of the contact forces. The terrain is assumed to be relatively soft compared to the wheels. The vertical force N acting on the wheel and the soil resistance R_s due to sinkage can be computed as follows [20]:

$$N = \frac{(3-n)}{3} (k_c + bk_\varphi) \sqrt{D} z_s^{(2n+1)/2}$$

$$R_s = \frac{3N}{(n+1)(3-n)} \left(\frac{z_s}{D}\right)^{1/2} \tag{9}$$

where n is the soil index, k_c and k_φ are the cohesive and frictional moduli of deformation, b is the wheel width, D is the wheel diameter and z_s is the sinkage due to the normal load N. Since the sinkage of a driven wheel is caused not only by the wheel load, but also by the wheel slip, there is an additional motion resistance due to wheel slip. The entire motion resistance is given as follows [21]:

$$R_{sj} = \frac{3N}{(n+1)(3-n)}\left(\frac{z_s+z_j}{D}\right)^{1/2} \tag{10}$$

where z_j is the sinkage due to tangential slip s.

The tractive force H developed at the wheel/ground interface is given as follows [20]:

$$H = H_{max}\left(1+\frac{K}{s\ell}\left(e^{-s\ell/K}-1\right)\right), \quad H_{max} = cA + N\tan\varphi \tag{11}$$

where c is the soil cohesion, A is the ground contact area, φ is the soil friction angle, ℓ is the length of the contact area and K is the soil deformation modulus. The net tangential force T that is available for locomotion is the difference between the tractive force and the soil resistance, i.e. $T = H - R_{sj}$.

The lateral force L generated at the wheel/ground interface is given by the following empirical relationship based on fitting curves to experimental data [22]:

$$L = L_{max}\left(1-e^{-B\alpha}\right), \quad L_{max} = cA + N\tan\varphi \tag{12}$$

where α is the wheel slip angle (the angle between the velocity at the wheel center and the wheel heading) and B is a constant. Equations 11 and 12 model the tractive and lateral forces independently of each other. However when both forces are present the following modification is made [22]. It is assumed that the tractive force takes precedence over the lateral force in the total force available and hence is given by Equation 11. The lateral force equation is modified as follows:

$$L = LFC * L_{max}, \quad \text{where}$$

$$LFC = LFC_{max}\left(1-e^{-B\alpha}\right), \quad \left(LFC_{max}\right)^2 + \left(COT\right)^2 = 1, \quad COT = \frac{H}{H_{max}} \tag{13}$$

CONTACT FORCE ALLOCATION FOR POWER MINIMIZATION

The objective function which is the total power consumed at the wheels and the articulation is given by

$$P = T_{L1}\dot{\theta}_{L1} + T_{R1}\dot{\theta}_{R1} + T_{L2}\dot{\theta}_{L2} + T_{R2}\dot{\theta}_{R2} + T_r\left(\dot{\psi}-s\theta\dot{\phi}\right) + T_y\dot{\theta} + T_p\left(\dot{\phi}-s\theta\dot{\psi}\right) \tag{14}$$

Equation 3 gives the wheel and the articulation torques in terms of the contact forces and the soil resistances. Our aim is to express the objective function P in terms of the contact forces alone and hence we need to relate the soil resistance to the contact forces. The soil resistance due to wheel sinkage and slip (given by Equation 10) can be approximated as a low-order polynomial function in the ratio of the tangential to the normal contact force:

$$\frac{R_{sj}}{N} = a_0 + a_1\left(\frac{T}{N}\right) + a_2\left(\frac{T}{N}\right)^2 + a_3\left(\frac{T}{N}\right)^3 \tag{15}$$

The coefficients $\left(a_0, a_1, a_2, a_3\right)$ are themselves low-order polynomial functions of the normal load N. It is assumed that such relationships are available from field tests. Such tests have been performed on the lunar rover wheels and the ELMS loop-wheel [23]. Equation 15 can now be used to express P in terms of the contact forces and hence the null space weights α_i.

Upon optimization, the optimal weights obtained give the optimal contact forces f_{wd}^{opt} for use in the inverse dynamics computation. The optimization is carried out at each time-step with the optimal weights from the previous time step as the initial guess. This results in quick convergence of the optimal solution ensuring that the coordination scheme can be implemented in real time.

COORDINATION SCHEME

Figure 2 shows the block diagram of the coordination scheme. The commanded velocity and acceleration values give the commanded inertia forces that need to be equilibrated by the contact forces (Equation 5). The redundancy in this force allocation problem is exploited to find the optimal contact forces with respect to power minimization. The optimal contact forces and the soil resistance estimates are then used along with the commanded motion of the vehicle in an inverse dynamics computation to generate the optimal actuator torques. The sensor requirements in the coordination scheme are as follows. As discussed earlier, a local inertia guidance package on the front module is necessary. The articulation joint variables, which are required to refer local vectors in module '2' to the module '1' reference frame, are assumed to be sensed. The contact force in the plane of the wheel are assumed to be sensed. The contact forces are used in the estimation of the soil resistance and the location of the contact point on the wheel.

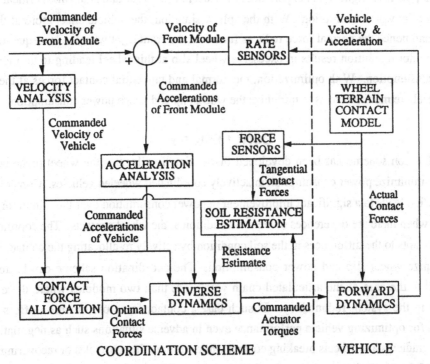

Figure 2 Block Diagram of Coordination Scheme

SIMULATION RESULTS

A simulation study was performed with the two module vehicle making a turn maneuver at a forward speed of 0.1 m/s on an even terrain that is inclined at an angle of 15^0 to the horizontal plane. The vehicle's initial heading is inclined at 45^0 to the x axis of fixed frame (X_0). The vehicle is commanded to turn up-slope by an angle of 30^0. Two cases were considered. In the first case soil strength factor k_{φ} values at all four wheel contacts are equal (this case is denoted as 'equal k' in the plots). In the second case k_{φ} is decreased by a factor of 3 at the left wheels and is increased by a factor of 1.67 at the right wheels (this case is denoted as 'split k' in the plots). The performances with the pseudo-inverse contact forces (without optimization) and with the optimal contact forces in both cases are compared. Figure 3 shows some of the simulation results. A favorable off-shoot due to optimization is the reduction of the lateral slip of the vehicle. As a result the vehicle is able to make a tighter turn up-slope, as seen from the plot of the path of the center of mass of module 1. With optimization, the overall power consumed at the wheel actuators is reduced throughout the maneuver. The maximum reduction in the instantaneous power consumed is about 7% in case 1 and about 18% in case 2 (the improvement is expected to be much more significant if compared with the performance of passive vehicles). The power consumed at the articulation was found to be negligible (within 5 Watts) in comparison with that at the wheel actuators and is not shown here. For this maneuver, the front right wheel experiences the least normal force and is the most critical wheel in terms of power consumption. With the split-k situation, the vehicle sinks more at the left wheels and hence the normal force reduces further at the front right wheel. Consequently, the minimum norm solution results in excessive wheel slip at this wheel leading to an increased power consumption. With optimization, the normal and tangential contact forces at the front right wheel are modified so as to minimize the wheel slip and hence power consumption.

CONCLUSIONS

A coordination scheme has been developed that optimally allocates the wheel plane contact forces to minimize power consumption in actively coordinated wheeled vehicles. Contact force optimization leads to a significant improvement in power consumption over the minimum norm solution when there are differences in the soil condition at the wheel contacts. The coordination scheme adapts to the differences in the soil condition by actively re-allocating the contact forces to minimize wheel slip and power consumption. The coordination scheme can be readily extended to the case of the articulated chain with more than two modules where there is an increase in the system redundancy. In such cases, contact force optimization offers great potential for optimizing vehicle performance even in adverse situations such as negotiation of extreme gradients and wheels breaking contact while surmounting obstacles or recovering from tip-over. It is also desirable that the simulator be extended to handle uneven terrain.

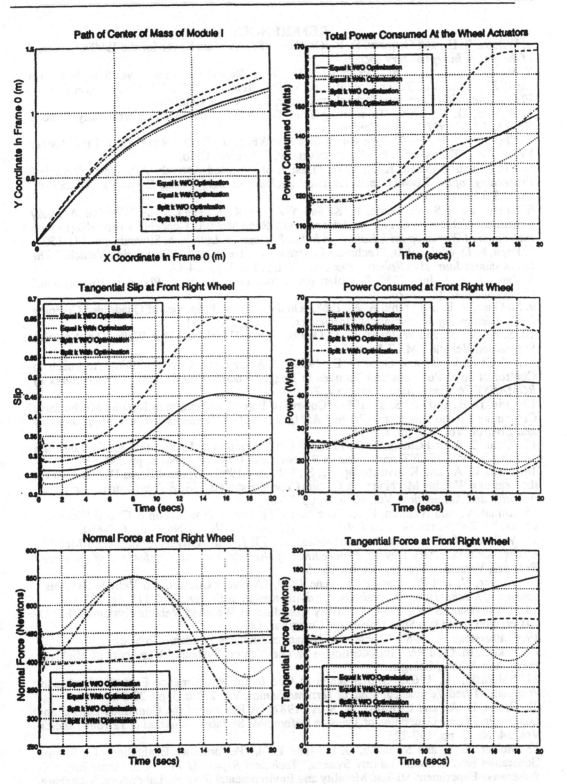

Figure 3 Simulation Results

REFERENCES

1. Wilcox, B. H., and Gennery, D. B., October 1987, "A Mars Rover for the 1990's," *Journal of the British Interplanetary Society.*
2. McTamany, L. S., *et al.*, December 1989, "Final Report for Mars Rover Sample Return (MRSR) - Studies of Rover Mobility and Surface Rendezvous," *FMC Corporation Report*, FMC Corporation Corporate Technology Center, Santa Clara, California.
3. Bickler, D. B., 1990, "Computing True Traction Forces on an Eccentrically Loaded Vehicle," *SAE Off-Highway Conference.*
4. Chottiner, J. E., 1992, "Simulation of a Six-Wheeled Martian Rover Called the Rocker Bogie", M.S. Thesis, The Ohio State University, Columbus, Ohio.
5. Yu, J., and Waldron, K. J., 1991, "Design of Wheeled Actively Articulated Vehicle," *Proceedings of the Second National Applied Mechanics and Robotics Conference*, Cincinnati, Ohio, 3-6 November.
6. Sreenivasan, S. V., Dutta, P. K. and Waldron K. J., 1994, "The Wheeled Actively Articulated Vehicle (WAAV): An Advanced Off-Road Mobility Concept," *Proceedings of 4th International Workshop on Advances in Robot Kinematics*, Ljubljana, Slovenia, pp. 141-150.
7. Pugh, R. D. *et al.*, 1990, "Technical Description of the Adaptive Suspension Vehicle," *The International Journal of Robotics Research*, Vol. 9, No. 2, pp. 24-42.
8. Bares, J. *et al.*, June 1989, "Ambler: An Autonomous Rover for Planetary Exploration," *IEEE Computer*, pp. 18-26.
9. McGhee, R. B. and Frank, A. A., 1968, "On the Stability Properties of Quadruped Creeping Gaits, " *Journal of Mathematical Biosciences*, Vol. 3, pp. 331-351.
10. Song, S. M., and Waldron, K. J., October 1989, "Machines That Walk: The Adaptive Suspension Vehicle," M.I.T Press.
11. McGhee, R. B., and Orin, D. E., 1976, "A Mathematical Programming Approach to Control of Joint Positions and Torques in Legged Locomotion Systems," *Proceedings of ROMANSY-76 Symposium*, " Warsaw, Poland.
12. Orin, D. E., and Oh, S. Y., 1981, "Control of Force Distribution in Robotic Mechanisms Containing Closed Kinematic Chains," *ASME Journal of Dynamics Systems, Measurements and Control*, Vol. 102, pp. 134-141.
13. Klein, C. A. and Chung, T. S., 1987, "Force Interaction and Allocation for the Legs of a Walking Vehicle, *IEEE Transactions on Robotics and Automation*, No. 6, pp. 546-555.
14. Klein, C. A., and Kittivatcharapong, S., February 1990, "Optimal Force Distribution for the Legs of a Walking Machine with Friction Cone Constraints," *IEEE Transactions, Journal on Robotics and Automation*, Vol. 6, No. 1, pp. 73-85.
15. Kumar, V., and Waldron, K. J., December 1988, "Force Distribution in Closed Kinematic Chains," *IEEE Transactions on Robotics and Automation*, Vol. 4, No. 6, pp. 657-664.
16. Waldron, K. J., Kumar, V., and Burkat, A., 1987, "An Actively Coordinated Mobility System for a Planetary Rover," *Proceedings of the International Conference of Advanced Robotics*, Versailles, France, October 13-15.
17. Kumar, V. and Waldron, K. J., June 1989, "Actively Coordinated Vehicle Systems", *Journal of Mechanisms, Transmissions and Automation in Design.*, Vol. 111, pp. 223-231.
18. Sreenivasan, S. V., 1994, "Actively Coordinated Wheeled Vehicle Systems", Ph. D. Dissertation, The Ohio State University, Columbus, Ohio.
19. Featherstone, R., 1987, "Robot Arm Dynamics," Kluwer Academic Publishers.
20. Bekker, M. G., 1969, "Introduction to Terrain-Vehicle Systems," University of Michigan Press, Ann Arbor, Michigan.
21. Soltynski, A., 1965, "Slip Sinkage as One of the Performance Factors of a Model Pneumatic-Tyred Vehicle", *Journal of Terramechanics*, Vol. 2, No. 3, pp. 29-54.
22. Crolla, D. A., and El-Razaz, A. S. A., 1987, "A Review of the Combined Lateral and Longitudinal Force Generation of Tyres on Deformable Surfaces", *Journal of Terramechanics*, Vol. 24, No. 3, pp. 199-225.
23. Melzer, K. J., and Swanson, G. D., June 1974, "Performance Evaluation of a Second Generation Elastic Loop Mobility System," *Technical Report M-74-7*, U.S. Army Engineer Waterways Experiment Station, Mobility and Environmental Systems Laboratory, Vicksburg, Mississippi.

PULSE WIDTH MODULATED CONTROL FOR THE DYNAMIC TRACKING OF WHEELED MOBILE ROBOTS

X. Feng and S.A. Velinsky
University of California-Davis, Davis, CA, USA

Abstract

This paper considers the control of differentially driven wheeled mobile robots through the use of a dynamic model that includes all the significant uncertainties such as external forces and tire slip in both the lateral and longitudinal directions. The influences of these uncertainties on the robot dynamics are analyzed and the matching condition for Pulse Width Modulated (PWM) control is studied. It is proven that position tracking control using the PWM method is invariant to both the external forces and tire slip. Combined position and orientation tracking control is also invariant if the desired orientation is tangent to the desired path or if the derivative of lateral slip is negligible. A multivariable PWM tracking control algorithm is developed that guarantees robustness to the influences of the uncertainties. Through experiments with a high load wheeled mobile robot developed for automating highway maintenance and construction tasks, the developed control is shown to be highly accurate, globally stable and robust to disturbances.

1 Introduction

The tracking control of differential drive mobile robots has been the subject of numerous studies, almost all of which develop algorithms based on kinematic models. However, a kinematic model based tracking control algorithm is only valid for low speed mobile robots with low payloads. Recently, a dynamic model, deduced on the assumption that the tire-ground contact is perfect and always produces sufficient traction forces, was used in control algorithm development [1]. This model is clearly not valid when slippage exists, and algorithms based on it cannot overcome all of the shortcomings of those based upon purely kinematic models.

Ideally, a complete dynamic model should serve as the basis for the control algorithm, but such a model would be of high order and computational complexity [2]. The necessity to operate in real time limits the complexity of a dynamic model used for control purposes, and additionally, some dynamic uncertainties, such as external loads and tire slippage, are difficult to accurately model. Accordingly, robustness and calculation simplicity are extremely important properties for a dynamic model based control algorithm.

The Variable Structure Control (VSC) approach has the feature of sliding mode invariance to both system perturbations and external disturbances [3]. While VSC has been studied extensively, in general, its design and implementation is not as direct nor as simple as universal Proportional-Integral-Derivative (PID) control or Pulse-Width Modulated (PWM) control for most dynamic applications.

The PWM control method is widely used in dynamic control systems, from power converters to consumer electronics to space system attitude control. PWM control is very simple in form, but its nonlinear and discontinuous nature is complex. Sira-Ramirez [4] revealed that the perturbation invariance properties of sliding regimes associated with systems regulated by the VSC is shared by the response of systems with PWM control loops. This explains the PWM approach's unique performance leading to its increased usage in robust dynamic control applications, and makes it highly suitable for the wheeled mobile robot (WMR) tracking control area.

This paper utilizes a dynamic model of a wheeled mobile robot including all uncertainties such as external forces and wheel slip. The influences of these uncertainties on the WMR dynamics are analyzed, and then, the matching condition for the dynamic system under PWM control is shown to be satisfied for the WMR tracking control problem. Finally, a multivariable PWM tracking control algorithm is developed which guarantees robustness to the influences of the uncertainties. Experiments with a high load mobile robot developed for automating highway maintenance and construction tasks show the stability, accuracy, and robustness of the PWM control approach for this application.

2 Dynamic Model of the Wheeled Mobile Robot

As noted above, a dynamic WMR model is the basis for our control approach. However, for brevity, a comprehensive derivation of the dynamic equations used herein are not included but can be found in Boyden and Velinsky [5]. The equations governing the dynamics of the differentially steered WMR shown in Fig. 1 are expressed as

$$m\dot{u} = F_{u,l} + F_{u,r} + F_{u,t} + mv\omega \tag{1}$$

$$m\dot{v} = F_{v,l} + F_{v,r} + F_{v,t} - mu\omega \tag{2}$$

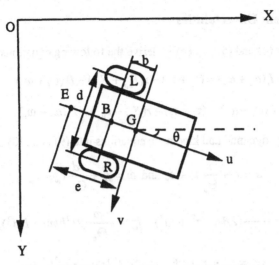

Fig. 1 WMR Structural Parameters Definition

$$I_\omega \dot\omega = \frac{d}{2}(F_{u,l} - F_{u,r}) - b(F_{v,l} + F_{v,r}) - eF_{v,t} \tag{3}$$

where u and v denote the forward and lateral WMR velocities, ω is its yaw rate, m and I_ω are the WMR mass and yaw moment of inertia, F denotes force and the subscripts u and v refer to the forward and lateral directions, and r and l refer to the right and left wheels. Forces subscripted with t refer to external forces such as those due to tooling operating on the road. Additionally, b, d, and e are geometric parameters shown in Fig. 1.

For heavily loaded WMRs, the dynamics of the wheel/tire/motors are also important and are expressed as

$$I_t\dot\omega_l = \tau_l - rF_{u,l} - B_t\omega_l \tag{4}$$

$$I_t\dot\omega_r = \tau_r - rF_{u,r} - B_t\omega_r \tag{5}$$

where I_t denotes the wheel/tire/motor inertia, r is the tire radius, ω_l and ω_r are the left and right wheel angular velocities, τ_l and τ_r are the applied motor torques, and B_t denotes the equivalent damping. With the existence of tire slip, the following kinematic equations and velocity constraints apply:

$$\dot x = u\cos\theta - (v - 2b\omega)\sin\theta \tag{6}$$

$$\dot y = u\sin\theta + (v - 2b\omega)\cos\theta \tag{7}$$

$$\dot\theta = \omega \tag{8}$$

where $v = b\omega + v^s$, $u = (u_l + u_r)/2$, $\omega = (u_l - u_r)/d$, $u_l = r\omega_l - u_l^s$, $u_r = r\omega_r - u_r^s$, and u_l^s, u_r^s, and v^s are the wheels' longitudinal and lateral slip speeds.

3 Analysis of System Uncertainties

From motor equations (4) and (5), we easily derive the following equations:

$$I_t(\dot{\omega}_l + \dot{\omega}_r) = (\tau_l + \tau_r) - r(F_{u,t} + F_{u,r}) - B_t(\omega_l + \omega_r) \tag{9}$$

$$I_t(\dot{\omega}_l - \dot{\omega}_r) = (\tau_l - \tau_r) - r(F_{u,t} - F_{u,r}) - B_t(\omega_l - \omega_r). \tag{10}$$

Substituting the system dynamic and kinematic equations into (9) and (10) gives

$$\dot{u} = f_u + \frac{r}{\Theta_u}\tau_u + \delta_u \text{ and } \dot{\omega} = f_\omega + \frac{rd}{\Theta_\omega}\tau_\omega + \delta_\omega \tag{11}$$

where

$$f_u = -\frac{1}{\Theta_u}(2B_t u - r^2 mb\omega^2), \quad f_\omega = -\frac{\omega}{\Theta_\omega}(2r^2 bmu + B_t d^2),$$

$$\Theta_u = 2I_t + r^2 m, \quad \Theta_\omega = d^2 I_t + 2r^2 I_\omega + 2r^2 b^2 m,$$

$$\delta_u = \frac{1}{\Theta_u}(r^2 F_{u,t} - B_t(u_l^s + u_r^s) - I_t(\ddot{u}_l^s + \ddot{u}_r^s) + r^2 m\omega v^s),$$

$$\delta_\omega = \frac{1}{\Theta_\omega}(2r^2(b-e)F_{v,t} - dB_t(u_l^s - u_r^s) - dI_t(\ddot{u}_l^s - \ddot{u}_r^s) - 2r^2 bm\dot{v}^s),$$

$$\tau_u = \tau_l + \tau_r, \text{ and } \tau_\omega = \tau_l - \tau_r.$$

Denoting $\delta_x = -v^s \sin\theta$, $\delta_y = v^s \cos\theta$, $f_x = u\cos\theta + b\omega\sin\theta$, and $f_y = u\sin\theta - b\omega\cos\theta$, and combining the kinematic equations with (11), the system dynamics equation can be rewritten in a compact vector form as

$$\begin{bmatrix} \dot{x} \\ \dot{y} \\ \dot{\theta} \\ \dot{u} \\ \dot{\omega} \end{bmatrix} = \begin{bmatrix} f_x \\ f_y \\ \omega \\ f_u \\ f_\omega \end{bmatrix} + \begin{bmatrix} 0 & 0 \\ 0 & 0 \\ 0 & 0 \\ \dfrac{r}{\Theta_u} & 0 \\ 0 & \dfrac{rd}{\Theta_\omega} \end{bmatrix} \begin{bmatrix} \tau_u \\ \tau_\omega \end{bmatrix} + \begin{bmatrix} \delta_x \\ \delta_y \\ 0 \\ \delta_u \\ \delta_\omega \end{bmatrix}. \tag{12}$$

Equation (12) represents the exact dynamics of the mobile robot provided that δ_x, δ_y, δ_u, and δ_ω can be exactly measured. However, these parameters represent uncertainties which consist of external disturbance inputs and unmodeled slippage dynamics. A dynamic model based control algorithm must be robust to these uncertainties in order to reach expected control performance. The sliding regimes associated with the VSC and PWM control methods are invariant to these parameters if they satisfy the so-called matching condition.

4 Matching Condition and Tracking Control

Consider the following perturbed nonlinear dynamic system controlled by the PWM method:

$$\frac{dx}{dt} = f(x) + g(x)u + \xi \tag{13}$$

where ξ can represent parametric perturbations or an external perturbation if we denote $\xi \equiv \delta f(x)$ or $h(t)$, respectively. The average PWM dynamics is said to exhibit strong invariance with respect to the perturbation signal ξ, if it is independent of the perturbation signal.

Matching Condition: The average PWM motions of the perturbed system (13) satisfy the strong invariance property with respect to ξ, if and only if the disturbance vector ξ satisfies the matching condition

$$\xi \in span\{g(x)\}. \tag{14}$$

This matching condition means that if all of the uncertainties enter the system through the input channel, then the sliding mode of the system (i.e., the average PWM motion) is invariant under the influence of the uncertainties.

Let us check the uncertainties of the WMR system in Eqn. (12). The lateral slip v^s, which appears in both δ_x and δ_y, does not enter the system through the input channel. In order to reveal the matching condition of the WMR system, we need to perform a state transformation to the system equation (12). This is easily accomplished by taking the derivatives of (6) and (7), and denoting $p = \dot{x}$ and $q = \dot{y}$, which results in the following:

$$\dot{p} = \ddot{x} = f_p + \frac{r\cos\theta}{\Theta_u}\tau_u + \frac{brd\sin\theta}{\Theta_\omega}\tau_\omega + \delta_p, \text{ and} \tag{15}$$

$$\dot{q} = \ddot{y} = f_q + \frac{r\sin\theta}{\Theta_u}\tau_u - \frac{brd\cos\theta}{\Theta_\omega}\tau_\omega + \delta_q \tag{16}$$

where
$$f_p = f_u\cos\theta + bf_\omega\sin\theta - u\omega\sin\theta + b\omega^2\cos\theta,$$

$$f_q = f_u\sin\theta - bf_\omega\cos\theta + u\omega\cos\theta + b\omega^2\sin\theta,$$

$$\delta_p = \delta_u\cos\theta + b\delta_\omega\sin\theta - v^s\omega\cos\theta - \dot{v}^s\sin\theta,$$

$$\delta_q = \delta_u\sin\theta + b\delta_\omega\cos\theta - v^s\omega\cos\theta - \dot{v}^s\cos\theta, \text{ and}$$

$$u = p\cos\theta + q\sin\theta.$$

We can then rewrite the system equation in the form of the state variables of the tracking position, (x, y), the orientation, (θ), and the derivatives of these variables, (p, q, ω), as

$$
\begin{bmatrix} \dot{x} \\ \dot{y} \\ \dot{\theta} \\ \dot{p} \\ \dot{q} \\ \dot{\omega} \end{bmatrix} = \begin{bmatrix} p \\ q \\ \omega \\ f_p \\ f_q \\ f_\omega \end{bmatrix} + \begin{bmatrix} 0 & 0 \\ 0 & 0 \\ 0 & 0 \\ \dfrac{r\cos\theta}{\Theta_u} & \dfrac{brd\sin\theta}{\Theta_\omega} \\ \dfrac{r\sin\theta}{\Theta_u} & \dfrac{brd\cos\theta}{\Theta_\omega} \\ 0 & \dfrac{rd}{\Theta_\omega} \end{bmatrix} \begin{bmatrix} \tau_u \\ \tau_\omega \end{bmatrix} + \begin{bmatrix} 0 \\ 0 \\ 0 \\ \delta_p \\ \delta q \\ \delta_\omega \end{bmatrix}. \tag{17}
$$

It is easy to verify that the following equation holds

$$
\begin{bmatrix} 0 & 0 \\ 0 & 0 \\ 0 & 0 \\ \dfrac{r\cos\theta}{\Theta_u} & \dfrac{brd\sin\theta}{\Theta_\omega} \\ \dfrac{r\sin\theta}{\Theta_u} & -\dfrac{brd\cos\theta}{\Theta_\omega} \\ 0 & \dfrac{rd}{\Theta_\omega} \end{bmatrix} \begin{bmatrix} \dfrac{\Theta_u \delta_u}{r} & -\dfrac{\Theta_u v^s \omega}{r} \\ \dfrac{\Theta_\omega \delta_\omega}{rd} & -\dfrac{\Theta_\omega \dot{v}^s}{rdb} \end{bmatrix} = \begin{bmatrix} 0 \\ 0 \\ 0 \\ \delta_p \\ \delta_q \\ \delta_\omega - \dfrac{\dot{v}_s}{b} \end{bmatrix} \tag{18}
$$

which means that as long as $\left| \dot{v}^s \right| \ll \left| b\delta_\omega \right|$ is satisfied, then the matching condition will hold and, hence, the PWM tracking control is invariant to uncertainties. However, we note that most tracking control tasks only require position tracking control and assume that the orientation lies tangent to the path. In such cases, the matching condition is always satisfied.

5 PWM Tracking Control

The principle of the PWM method for the WMR tracking control is shown in Fig. 2. The controlled WMR takes at first a pure rotation which allows it to head in the desired direction, and then it takes a pure forward motion to reach the desired position. The control inputs τ_u and τ_ω take only switching values, $\pm \tau_{u,\max}$ and $\pm \tau_{\omega,\max}$, respectively.

From Fig. 2, we have

$$
l = \sqrt{(x_d - x)^2 + (y_d - y)^2} \quad \text{and} \tag{19}
$$

$$
\alpha = a\tan 2(y_d - y, x_d - x). \tag{20}
$$

Let

$$
\alpha_d = \begin{cases} \alpha - sign(\alpha)\pi, & \text{for } abs(\theta - \alpha) > \dfrac{\pi}{2} \text{ and } abs(\theta - \theta_d) > \dfrac{\pi}{2} \\ \alpha, & \text{for all other values} \end{cases} \tag{21}
$$

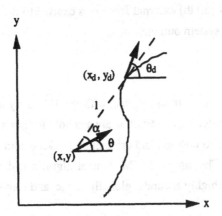

Fig. 2 Tracking Control with the PWM Method

and
$$l_d = \begin{cases} -l, & \text{for } abs(\theta - \alpha) > \dfrac{\pi}{2} \text{ and } abs(\theta - \theta_d) > \dfrac{\pi}{2} \\ l, & \text{for all other values.} \end{cases} \qquad (22)$$

Then, the time durations of the pure rotation and pure forward motion are determined as:

$$t_\omega = sat\big(abs(\alpha_d - \theta)k_\omega / T\big)T \text{ and } t_u = sat\big(k_u l / T\big)T \qquad (23)$$

where
$$sat(x) = \begin{cases} 1 & \text{for } x \geq 1 \\ x & \text{for } x < 1 \end{cases}$$

k_ω and k_u are the feedback gains, and T is the sampling period. The control action is then :

$$\tau_\omega = \begin{cases} sign(\alpha_d - \theta)\tau_{\omega,max}, & \text{for } kT < t \leq kT + t_\omega \\ 0, & \text{for all other values of } t \end{cases} \qquad (24)$$

and
$$\tau_u = \begin{cases} sign(l_d)\tau_{u,max}, & \text{for } kT < t \leq kT + t_u \\ 0, & \text{for all other values of } t. \end{cases} \qquad (25)$$

The purpose of Eqns. (21) and (22) is to ensure that rearward motion occurs, as opposed to a large turn, for the case in which the WMR is ahead of the desired position.

6 Experimental Results

Example results obtained through experiments in which the PWM tracking control was implemented on an actual high load WMR are shown in Fig. 3. The WMR design parameters are $m = 90.7$ kg [200 lb], $d = 63.5$ cm [25 in], $r = 26.0$ cm [10.25 in] and the PWM control parameters are $T = 20$ ms, $k_\omega = 30$ ms/rad, $k_u = .79$ ms/cm [2 ms/in], $\tau_\omega = \tau_u = 226$ N-cm [20 lb-in]. Both the position and orientation follow the given two sinusoidal curves very accurately and the global stability of the PWM control is also shown clearly. During the experiment, a nearly 222 N [50 lb] external force was exerted to the WMR randomly with no discernible deviation in the system output.

7 Conclusions

The perturbation invariance feature associated with the PWM dynamic tracking control of WMRs is examined. The matching condition is always hold for position tracking control. It is additionally hold for combined position and orientation tracking control if the derivative of the lateral slip is negligible. The simple PWM control algorithm developed in this paper is experimentally shown to be highly accurate, globally stable, and robust to disturbances.

References

[1] Sarkar, N., Yun, X. and Kumar, V., "Control of Mechanical Systems With Rolling Constraints: Application to Dynamic Control of Mobile Robots," *The International Journal of Robotics Research*, 13(1), 1994, pp. 55-69.

[2] Hamdy, A. and Badreddin, E., "Dynamic Modeling of a Wheeled Mobile Robot for Identification and Control," Robotics and Flexible Manufacturing Systems, Elsevier Science Publishers B. V., (North-Holland), 1992, pp. 119-129.

[3] Utkin, V.I., Sliding Modes and Their Application in Variable Structure Systems, Mir Publishers, Moscow, 1981.

[4] Sira-Ramirez, H., "Invariance Conditions in Non-linear PWM Control Systems," *International Journal of Systems Science*, 20(9), 1989, pp. 1679-1690.

[5] Boyden, D. and Velinsky, S.A., "Dynamic Modeling of Wheeled Mobile Robots for High Load Applications," *Proc. of the IEEE International Conference on Robotics and Automation*, 1994, pp. 3071-3078.

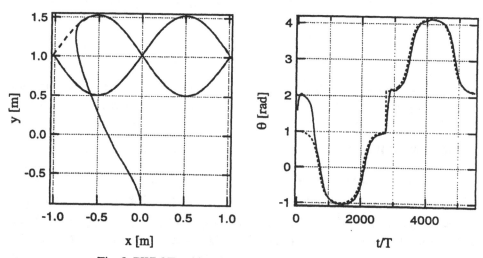

Fig. 3 PWM Tracking Control Experimental Result

MANOEUVRE ANALYSIS OF MULTILEGGED WALKING ROBOTS WITH DYNAMIC CONSTRAINTS BETWEEN LEG PROPULSORS

A.P. Bessonov, K.S. Tolmachov and N.V. Umnov
Mechanical Engineering Research Institute RAS, Moscow, Russia

The analysis of the walking robots legs on a phase of support reveals some features. All legs of a machine stand on the ground and consequently kinematicaly are connected among themselves. It results in the fact that in this case one does not manage motion of each leg independently and has to take into account all set of legs standing on the ground.

By use of propulsor of mechanism so named "insectomorphy" scheme [I], in which even for making elementary motions of a body one has to realise reasonably complex functional dependences simultaneously in all drives of all propulsors. In the more simple and economic "orthogonal" mechanisms of legs, in which the vertical motion of the basic end of the propulsor is not connected to drives, which move a machine in a horizontal direction [2, 3]. It is useful to evaluate a degree of such complication.

Transport problem of a robot or problem of manoeuvre in the majority easier, than the problem of control of positioning a manipulator. Usually at the motion of a robot as a rule there is no necessity strictly to maintain any trajectory of the motion, and it is enough to get in any point, or even any area of the space. The availability of a visual feedback with the participation of a man-operator even more reduces the requirement strictly of realisation of required motion by the robot.

Ideally, for the manoeuvre of a robot under the control of a driver in the majority of cases there would be reasonably a system of the elementary uncontrolled drives and some additional device, changing if necessary either configuration of propulsors, or characteristics of their drives, i.e. to act similarly as it takes place in a usual wheel automobile. And in the huge majority of cases the legs work independently from hands, solving their transport problem. Let us consider these questions in more detail.

We shall accept, that a robot moves in a horizontal plane, that will allow to consider its propulsors (orthogonal scheme) as planar mechanisms. Each propulsor of the robot of high mobility should have an opportunity on a phase of transfer to get in any point. Hence, by work on this phase the propulsor should have not less than two degrees of freedom. However on a phase of support such quantity of controlled drives is obviously redundant.

We shall consider the plane scheme of propulsors of an eightlegged robot, driven by "quad". In Fig. 1 a phase of the support of four propulsors, possessing 4 basic points A, B, C, D and connected with the body in four points of the suspension A'; B'; C'; D' is represented. The other 4 legs, located in given moment in phase of transfer are not shown. Two independent controlled drives of each propulsor derivate the pull force F and moment M between the body and the guide of the propulsor.

The drive can be located not only in a point of the body suspension as shown in Fig. 1, but also in the support, and then it creates a moment between the stationary support and guide. However such arrangement of a drive, equivalent from the point of view of work of a robot on a phase of support, when working on a phase of transfer creates uncertainty of positioning of a support, as far as the hinge of a point of leg suspension becomes in this case uncontrolled.

We shall consider a number of mobility of such system. Elementary count shows, that at 9 moving links and 12 pairs the system has only 3 degrees of freedom $W = 3$, and for its movement only three drives is enough. More number of drives, and in our case there are 8 of them, will require the exact co-ordination of moving force of each drive with real movement of the machine, as differently the phenomenon of dissipation, i.e. work of the drives "one against another" will take place.

Explicit decision is a procedure of switching-off part of "superfluous" drives after the ending of a transfer phase and fixing the support on the ground. Switching-off of rotary drives, creating moment M, is the most natural. Two in essence different variants are here possible: after switching-off the drive either brakes or remains an opportunity of its free rotation. In the first case, structure of the system varies (Fig. 2), as far as the hinges in the points A'; B'; C', D' disappear. In the system there will be 5 moving links and 8 pairs, that brings to $W = -1$, i.e. the system has two redundant constraints. In this case its kinematic movement without supports slipping on the ground becomes impossible. This revealed unpleasant circumstance becomes especially annoying in the light of rather logical solution of the problem of manoeuvring, repeatedly offered by various authors. The essence of it consists of that if the problem of drives in the points of suspension is guides turn by a necessary angle, dependending from radius of turn, then to ensure this turn it is possible by an additional mechanical device, functionally similar to steering mechanism of an automobile. Such decision essentially simplifies a control system of the machine, though reduces a little its opportunity at the choice of a point of leg stand. Impossibility of kinematic movement of such system, certainly, is more serious defect.

System of propulsors construction [4] is known, when some of them are operated by a steering device and as a result their position relative to the body is rigidly fixed, whereas others have an opportunity freely to turn relative to it. Such decision is shown in Fig. 3. Here front guides B

and C are, turned by steering mechanism and their position relative to the body remains stationary. Rear guides A and D always have an opportunity to rotate freely relative to the body like a "piano's" wheel and basically can in general not have any drive in a point of suspension. Incidentally, not only front or rear legs can be rotated from a steering mechanism, but also the pair of legs of one board (Fig. 4), and at shift of "quads" the position of fixed legs can simultaneously vary, i.e. "left-hand" at one quad and "right" at other. At 7 moving links and 10 pairs it gives $W = 1$, i.e. quite realisable movement. Actually in this case we have 3 redundant drives, as far as for movement it is enough only one of being present of four. In cited work [5] it was offered to connect 4 translational drives on the guides by three differential mechanisms, i.e. to impose on a system three additional constraints (Fig. 5). In such case in a system there will be only one engine, which will adjust only general speed of movement of the machine. Manoeuvring will be executed by turn of front guides with the help of steering mechanism.

The described system, despite visible simplicity, has a serious defect, in essence excluding opportunity of any its practical using, called by availability in the system of some differential mechanisms. At nonblocked differentials the loss of propulsion in one leg (for example, because of its slipping on a slippery ground) will cause immediate disappearance of the propulsion in all other legs. Vehicle completely stops.

We shall consider now another alternate approach to the control of drives at switching-off on a phase of a support all rotary drives - namely the case when the hinges of the suspension points remain free. (Fig. 6) Number of degrees of freedom in this case, as was calculated earlier, will be equally 3, that at availability of four legs with a drive in each of them results in one redundant drive.

It is interesting, that with six-legged vehicles with the similar schemes of propulsors there is no problem at all, but only in the case when the machine moves by the gaits "triples", as far as. It is easy to show, the general number of degrees of freedom with the free scheme of propulsors fastening does not depend on the number of basic propulsors and always equally $W = 3$. However small margin of static stability becoming especially small at the gait by "triples", as well as nonsymmetry of pull forces because of unequal number of basic legs on boards compel to pass to multilegged systems.

The redundancy of drives at their rigid characteristic will cause, except losses of capacity because of the work of drives "one against another" also to an opportunity of occurrence in the supports of pull forces essentially larger, than it is necessary for movement, and larger, than friction forces permit coupling supports with the ground, that will cause as a result the slipping of the support on the ground, and at impossibility of it emergencies.

It would seem for the solution of the problem it is enough to disconnect on a supporting phase one drive, having left thus an opportunity of relative movement in translational pair of a switched - off drive. But the switching-off of one drive will cause however except general decrease of propulsion also mentioned nonsymmetry of pull forces, the unpleasant consequences of which become especially noticible at rectilinear walking.

However on the scheme in Fig. 6 there is even more essential defect. The phenomenon of it was in detail considered by us in work [5] for an example of eightlegged robot with propulsors of the rhomboid-shaped scheme. In itself the scheme of the propulsor has not special significance for the problem under discussion, but its form of constraints with the body is more important. It can be either hinged, as in Fig. 6, or rigid, as in Fig. 2. We shall consider forces, affecting the body of the machine when manoeuvring and forcing it to turn. In case of rigid connection of the propulsor as in Fig. 3 or Fig. 4 - it will be reactions on the part of the ground on the guides, so and on the body, perpendicular to the guides, ensuring creation of a necessary moment. In this case on a system geometrical constraints, forcing the body to displace across longitudinal axis will be as though imposed. It is interesting, that the machine on the scheme in Fig. 3 as appear except obvious cross movement will not practically change its angular position. The system with the scheme in Fig. 4 will turn quite normally.

We shall consider movement of a robot in a mode of "circulation" - stable motion with constant circular speed V on a circle of constant radius R, and the angular speed of turn of the body is equal to V/R. In this case the vector of absolute speed of any point of the body will be stationary relative to the body, and the radius of curvature of a trajectory will coincide with an instant centre of speeds.

From the analysis of pull forces of propulsors, transmitted to the body, it is possible to conclude, that at symmetric deviation of the guides relative to the cross axis of the body sum of forces of reactions to the body will be reduced to one longitudinal force, biased relative to a centre of vehicle, whereas the projection of pull forces on across axis will completely disappear. However at movement of the vehicle in a mode of circulation the centrifugal force $F = mV^2/R$, directed along a cross axis of the body affects the body, and which will not be counterbalanced by forces of reaction from the propulsors. Hence movement of the considered system in a mode of circulation is impossible. Moreover, the numerical experiment has shown, that if a vehicle artificially to enter in a mode of circulation, i.e. to set to it speed V and angular speed, connected by necessary image through radius R, to deviate propulsors as in Fig. 6 accordingly to radius R, to apply pull forces in the propulsors and then to grant an opportunity to the machine to move, it will not be movement on the circle R, that would correspond to circulation, but will be linear movement of a centre of masses of the body and rotation of the body relative to the centre.

In the cited paper [5] it was offered "paradoxical" position of propulsors at turn (Fig. 7), at which front and rear guides of the propulsors deviate in different directions relative to longitudinal axis of the body. At such arrangement of the propulsors a constant sum longitudinal propulsion being present cross components of propulsion of each propulsor do not exclude one another, and develop, therefore there is an opportunity to compensate centrifugal force. The elementary numerical experiment has confirmed a basic opportunity of preservation in this case a mode, close to circulating.

As was already mentioned, machines with propulsors on the scheme in Fig. 6 and Fig. 7 have three degrees of freedom and, hence, can not have more than three kinematically controlled drives or drives with a rigid characteristic. However, if the characteristic of a drive is sufficiently soft, such movement is possible, as far as in this case the drive sets not movement but force, and the resulting movement is received from effect on the body of a sum of these forces and moment from them. If the forces are completely independent of movement (the ideal soft characteristic of a drive), at similarity of drives to the body 4 identical forces will be applied, the resulting movement from their action can be obviously appreciated.

It should only mean two circumstances. First, for reception of steady-state movement it is necessary to apply an external longitudinal force of resistance - analogue of a pull force on a hook of a common means of transport. Secondly, the angle of deviation of a guide does not remain constant during support phase. It is well seen from Fig. 8, in which 2 positions of a right-hand board of a machine at the moment of the beginning of the support phase (A'_1, B'_1) and at the moment of its end (A'_2, B'_2) are shown. The points of standing front (B) and rear (A) legs, are naturally thus stationary. The anle of a deviation of guides relative to the body varies depending on the extention l of legs (BB'_1) and (AA'_1). This dependence has a form:

$$\Psi_B = \pi - \arccos\frac{D}{R} - \arccos\frac{R^2 - R_B^2 + l_B^2}{\cdot 2Rl_B}$$

for front leg, and:

$$\Psi_A = \arccos\frac{D}{R} + \arccos\frac{R^2 - R_A^2 + l_A^2}{\cdot 2Rl_A}$$

for rear ones.

The angles Ψ_A and Ψ_B, as we see, are not identical, and not only as R_A and R_B different, but also as the extentions of legs vary not equally. The extention of the front leg decreases in accordance with angular motion of the body φ under the law

$$l_B^2 = R^2 + R_B^2 - 2RR_B \cos(\phi_{Bb} - \varphi)$$

Where ϕ_{Bb} corresponds to initial extention of the leg L_B at the moment of its standing on the radius R_B and is equal

$$\varphi_{Bb} = \arccos\frac{R^2 + R_B^2 - L_B^2}{\cdot 2RR_B}$$

Accordingly the extention of the rear leg increases in accordance with an increase of angular motion φ body:

$$l_A^2 = R^2 + R_A^2 - 2RR_A \cos(\phi_{Ab} - \varphi)$$

From some initial angle φ_{Ab}, corresponding to initial extention of the rear leg L_A at the moment of its standing on the radius R_A:

$$\varphi_{Ab} = \arccos\frac{R^2 + R_A^2 - L_A^2}{\cdot 2RR_A}$$

Nonsynchronism of angles change of deviation of the guides Ψ_A and Ψ_B will cause inconstancy of a sum of projections of pull forces on a cross axis of the body and, hence, imposibility of complete exeeption of influence of constant centrifugal force. Ideal circulating motion is not received.

However it is possible a little to simplify the problem, formulating it in another way: It is possible to admit, that the centre of the body for a time of support phase will deviate from the motion on the circle R of ideal circulation, having required however, that in the beginning and end of the support phase it will be strictly on it. If thus all kinematic characteristics of motion in the beginning and end of phase also coincide, there will be available a stady-state motion "on the average" on a circle of constant radius with constant average circular speed. Such mode of circulation "on the average" requires already integrated estimation of pull forces, value of their projections on the body and exposure of some average propulsor on a hook, ensuring steady-state motion.

We should take into account also one more complicating circumstance. The fact is that if, as it was mentioned before, the engines have reasonably soft characteristic $F = F_0 - kV$, a dynamic constraint [5] will be observed between drives of right and left boards. At rectilinear motion the angless of deviation of both guides are equal zero, the speeds of points of the suspension are identical, thus also identical will be and pull forces. At manoeuvring the guides of both boards will deviate by identical angle, however the speed of a point of the suspension of an external board always will be more than speed of a point of the suspension of internal one. Therefore the pull force of external legs variesin accordance with the characteristic of a drive it decreases. It will cause the change of resultant - it will also decrease, and change of a point of application of resultant and it will shift to a centre of rotation.

The problem of realisation of steady-state manoeuvre at quasi-circulation will now be formulated as follows. On known radius of turn, determined by the driver, and on known value of initial extention of the front and rear legs, determined by a vehicle design, one will find necessary angles of an initial deviation of the guides, ensuring concurrence of kinematic characteristics of motion at the beginning and at the end of a support phase. The appropriate differential equations of motion of the body have a form:

$$M\ddot{X} = \sum_{i=1}^{4} F_{Ti} \cos(\psi_i + \varphi_k) + F_{RX}$$

$$M\ddot{Y} = \sum_{i=1}^{4} F_{Ti} \sin(\psi_i + \varphi_k) + F_{RY}$$

$$J\ddot{\varphi}_k = \sum_{i=1}^{4} \left(L \cdot F_{kT_i} - H \cdot F_{kX_i} \right)$$

$$F_{Ti} = F_0 - k \cdot \overline{V}_{ki} \cdot \cos\alpha_{vi}, \qquad \overline{V}_{ki} = \overline{V}_{cm} + \overline{\omega} \times \overline{r}_3, \qquad \cos\alpha_{vi} = \frac{\overline{R}_{Li} \cdot \overline{V}_{ki}}{|R_{Li}| \cdot |V_{ki}|}$$

Complicated dependences do not permit to hope for the analytical decision of the put boundary problem. We analysed a behaviour of the system, carrying out numerical experiment on a model with reasonably real inertial and propulsion characteristics. The provisional results are indicated in Fig. 9, where dependences of initial angles of deviation of front and rear guides from the radius of turn of a vehicle, ensuring the best concurrence of final kinematic characteristics with initial ones are given. The same experiment has shown a weak critics of chosen angles to accumulation of errors of motion, that gives the good basis to hope for an opportunity of real construction of enough simple steering systems for walking robots manoeuvring, certainly under the obvious visual control of a driver.

Reference

1. Devjanin E.A. et al.: A Six-Legged Walking Robot Capable of Terrain Adaptation. Mechanism and Machine Theory, 1983,18, (4), pp.257-260.
2. Bessonov A.P., Umnov N.V.: Stabilization of Position of the Body of Walking Machines. ROMANSY-81. Proceedings. Poland, pp.356-366.
3. Hirose S., Umetani Y.: Some Consideration of Feasible Walking Mechanism as a Terrain Vehicle. ROMANSY-78. Poland.
4. Bessonov A.P., Umnov N.V. et al.: Comparative Analysis of Manoeuvring Models of Multi-Legged Mobile Taking into Account the Motion Energy., Fourth Yugoslav Soviet Symposium on Applied Robotics and Flexible Automation., Novi-Sad, Yugoslavia, 1988, pp.299-307.
5. Bessonov A.P., Umnov N.V., Tolmachov K.S.: Mechanisms with Dynamical Constraints for Walking Robots Manoeuvring., Nine World Congress on Theory Machine and Mechanisms, Milan, Italy, 1995.

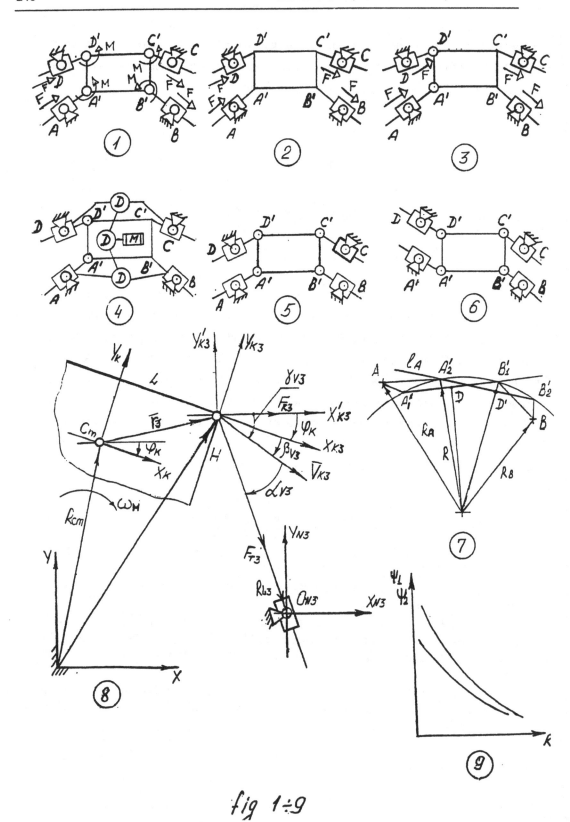

fig 1÷9

Chapter VIb
Biomechanical Aspects of Robots and Manipulators

BIOMECHANICAL ASPECT OF A MASTICATION ROBOT

H. Takanobu and A. Takanishi
Waseda University, Tokyo, Japan

Abstract —— *This paper describes a biomechanical aspect of a mastication robot, and Dental Robotics as a novel field between robotics and dental research. Since 1986 the authors have been developing the WJ (Waseda Jaws) series mastication robots on the basis of dental facts. The characteristics of mastication robots' are that these robots have similar structures with human. By using this robot we will be able to clarify and construct the quantitative engineering masticatory model for the human mastication.*

1 Introduction

The Humanoid Research Laboratory (HUREL), Advanced Research Center for Science and Engineering, Waseda University, started the "Humanoid Project" in 1992 [1]. This project has two objectives: to establish an anthropomorphic robot named humanoid, and to explore a novel field in the robotics research. The mastication robot that is described in this paper is an oral sub-project of the Humanoid Project.

The purpose of this paper is to describe the biomechanical aspect of a mastication robot and to propose a novel field of the robotics research, "Dental Robotics." Although the dental researches in the medical or dental fields have been done with methods as to measure and analyze the human masticatory jaw's motion, the quantitative and mechanical masticatory model had not been constructed. The dentists need the concrete model that is able to simulate human jaw's motion because the dentists will be able to recognize and explain the human jaw motion by using such a mechanical model. This paper describes a human jaw's me-

Fig. 2 WJ-1 (1987)

Fig. 3 WJ-2 (1988)

Fig. 1 WJ-0 (1986)

Fig. 4 WJ-3 (1992)

Table 1 Development history of mastication robots

year	robot name	DOF	contents
1986	WJ-0	1	antagonistic 2 DC motors
87	WJ-1	1	opening and unloading reflex motion
88	WJ-2	3	clenching and grinding mandibular motion
89	WJ-2R	3	measuring system using laser and gyro
90	WJ-2RII	3	adaptive motion for food characteristics
91	WJ-2RIII	3	temporomandibular joint force
92	WJ-3	3	artificial human skull model
93	WJ-3R	3	nonlinear elastic masticatory muscle
94	WJ-3RII	3	nonlinear viscoelastic masticatory muscle
95	WJ-3RIII	3	masticatory efficiency

chanical simulator based on the oral physiology and biomechanisms on the human masticatory system.

A number of studies on the masticatory system have been done mainly in the field of dental prosthesis and oral physiology. The studies in the field of engineering are classified into two kinds: those using hardware and those not using hardware. There have been only four studies using hardware. Gibbs et al. [2] developed a jaw motion replicator. Hyodo et al. [3] proposed a temporomandibular joint (TMJ) simulator using a Ni-Ti shape memory alloy. Hatcher et al. [4] developed a mechanical model to study the TMJ loading. Aoki et al. [5] proposed the concept of a self-controlled jaw-movement robot. The three works 'not using' hardware have focused on the TMJ force. Barbenel [6] analyzed the TMJ force by using linear programming. Osborn et al. [7] computed the TMJ force in the case of generating 100-1000 [N] in biting force. Throckmorton [8] examined the vector of the TMJ force by using a two dimensional and five muscle model. However, these studies only proposed a qualitative or static human masticatory model. Thus, a quantitative and dynamic masticatory model has not been proposed. The objective of this study is to construct a quantitative and dynamic engineering masticatory model by simulating the human masticatory motion with a robot similar to a human in its mechanism and control. The authors have been devel-

oping the mastication robots Waseda Jaws series
WJ-0 [9], WJ-1 [10], WJ-2 [11] and WJ-3 [12]
since 1986 as shown in Fig. 1-4 and Table 1. The
mechanical structures of these robots are similar to
those of the human masticatory system, muscle
position, sensors in muscles, and sensors under the
teeth. The mandibular movement of the mastication robots is also similar to that of the human beings such as the opening, closing, and grinding motions of the mandible (Fig. 5).

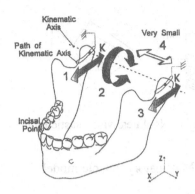

Fig. 5 Mandibular motion.

The human masticatory motion system has
similar configurations to the mechanical system driven by linear actuators such as Stewart
Platform [13]. Therefore, the authors have proposed that the mastication robot should have
a parallel mechanism driven by the masticatory muscle [14]. The nonlinearity of the muscle
parameter has been described by papers in the fields of physiology or biomechanisms.
Mackay et al. [15] measured the viscoelasticity of the human forearm. Park et al. [16], Kato
et al. [17], Kusumoto et al. [18], Shadmehr et al. [19] described that the stiffness increases
almost linearly with the contractile force. Hogan [20] pointed out the adaptive control of mechanical impedance by coactivation of antagonist muscles. On the basis of these works,
several engineering studies have focused on this characteristic and tried to simulate this nonlinearity of the muscle. Laurin-Kovits et al. [21] proposed programmable passive impedance components. Hyodo et al. [22] developed a tendon-controlled wrist mechanism using a
nonlinear spring tensioner. Goodwin et al. [23] showed the possibility of the nonlinearity
on the monkey's masticatory muscle.   This paper describes a mathematical and mechanical model based on the nonlinearity of the human masticatory muscle. The authors propose
that human control the stiffness of the masticatory muscles in chewing food, especially
crushable hard food such as rock ice. For example, humans contract the masticatory muscle
to avoid the hard contact of the teeth in the case of the real food.

This paper includes four chapters. In chapter 1, the dental research situation in recent
years and previous works on the nonlinearity of the human muscle are presented as back-
grounds of this paper. In chapter 2, feasible mathematical and mechanical models of the nonlinear viscoelastic characteristics are presented. Chapter 3 describes the mechanism and
control of the mastication robot. The concluding remarks are given in chapter 4.

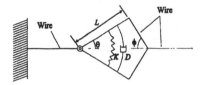

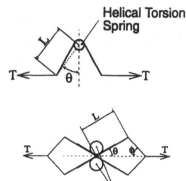

Fig. 6 Nonlinear viscoelastic model.

2 Masticatory muscle model

A mathematical model is shown in Fig. 6.

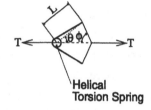

$$T = \frac{4\{K(\theta_0 - \theta) + D\dot{\theta}\}\cos\phi}{L\sin(\theta + \phi)} \quad (1)$$

$$X = L\left\{\cos\theta - \cos\theta_0 + \frac{\sin\theta_0}{\sin\phi_0}(\cos\phi - \cos\phi_0)\right\} \quad (2)$$

where T: wire tension, X: displacement of wire, K: spring coefficient of the spring, L: distance between the rotation center and the wire connecting point, $\theta\theta_0$: angle between the plate and tension direction, and its initial angle, $\phi\phi_0$: angle between the wire and tension direction, and its initial angle. As the first step toward a viscoelastic actuator, the authors constructed an elastic mathematical model. This model simulates the variable stiffness of the human muscle.

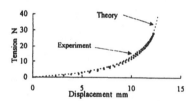

Fig. 7 Elastic model.

Some models are shown in Fig. 7. The assumption in this simulation is as follows: 1. Gravity, the mass of plates, the mass of helical torsion springs, and friction are negligible small. 2. The two plates are rigid bodies. 3. The wires never become long. As shown in this figure, the nonlinear stiffness is realized by this model.

The second step is to add the viscosity the elastic model. Simulation result is shown in Fig. 8. The tension T increases nonlinearly as the velocity and the displacement increases. The nonlinearity of this viscoelastic model is confirmed by this simulation. A mechanical model is shown in Fig. 9. This model consists of three parts, a helical torsion spring, a rotary damper, two duralumin plates. The helical torsion spring is an elastic element, the rotary damper is a viscous element of the masticatory muscle.

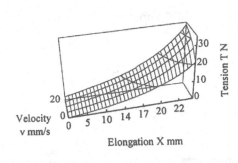

Rotary Damper

Spring

Duralumin

Fig. 8 Simulation result.   Fig. 9 Nonlinear viscoelastic mechanism.

3 Mechanism and control of mastication robot

This chapter describes the mechanism and control of the mastication robot.

The mastication robot is a mechanical simulator that is designed on the basis of the real human masticatory system. A control system for a mastication robot WJ-3RII (Waseda Jaws No. 3 Refined II) has been developed, as shown in Fig. 10, on the basis of the dental facts that is clarified in the field of the dental physiology. This robot system consists of four subsystems: a skeleton subsystem, actuation subsystem, sensory subsystem, controlling subsystem. Outlines of these four subsystems are as follows.

[Skeleton] A human skull model is adopted as a skeleton subsystem of the mastication robot. The material of this skull model is epoxy resin. A carbon rod goes through the right and the left condyles. The forward and backward motions of this carbon rod are constrained on a virtual plane inclined 40 [deg] from the horizontal plane. This inclined virtual plane is composed of two temporomandibular joints. The authors named this virtual plane as the kinematic axis plane. The temporomandibular joint for a mastication robot was designed with reference to dental facts.

[Actuators] WJ-3RII has nine artificial muscle actuators (AMA). An AMA includes a DC motor, an encoder, a tachogenerator, a wire, and a force sensor. The muscle characteristic that generates only contraction force (not pushing force) is simulated by using a wire. Two AMAs are for m. masseters (masseter muscle; exist under the cheek), two for m. temporalis (spreads both sides of the skull) anterior, two for m. temporalis posterior, two for m. pterygoideus lateralis (generate forward mandibular motion), and one for open. One end of the wire is attached to the robot mandible as a connecting point of the muscle and the other end of the wire is pulled up by the DC motor assigned to the maxilla.

[Sensors] Each AMA has an encoder, a tachogenerator, and a force sensor (a $25 \times 10 \times 1$ mm duralumin plate on which four strain gages are attached). Each of the lower and upper molar teeth has two micro pressure sensors for measuring the biting force while the robot chews a food.

[Controller] A personal computer controls the motion of the nine AMAs. In addition, this computer receives the sensory signals that are as follows: three temporomandibular joint force sensors, nine wire tension sensors, nine encoders, nine tachogenerators, four pressure sensors.

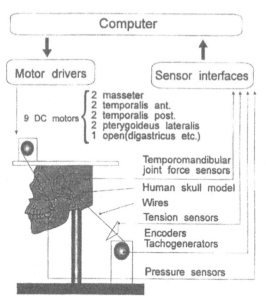

Fig. 10 Mastication robot system.

Masticatory experiments were performed. The nonlinear viscoelastic actuator that is proposed in this paper controlled the rapid closing motion of the robot jaw.

4   Conclusions

This paper is summarized by the following two points.

1) Biomechanical aspect of a mastication robot, especially, feasible mathematical viscoelastic model that is designed on the basis of nonlinearity of the human masticatory muscle is proposed and tested in computer simulations.

2) The authors made a nonlinear viscoelastic mechanism. Experimental results of the basic characteristics of this mechanism showed its nonlinearity.

The authors proposed the novel field "Dental Robotics" that is a fusion of robotics and dental research. The authors, Ohtsuki and Ohnishi *et al.* [24] of Yamanashi Medical University have started the Dental Robotics project on the mouth opening and closing machine in 1995.

Acknowledgments

   This work has been done as an oral project of the "Humanoid Project" at the Humanoid Research Laboratory, Advanced Research Center for Science and Engineering, Waseda University. The authors would like to thank Professor Kinjiro Kubota of Tokyo Medical and Dental University, Professor Shoji Kono of Niigata University, Professor Yoshio Nakamura of Tokyo Medical and Dental University. Yasushi Mori, Takeyuki Yajima, Naoki Yanagisawa made contributions to this study.

   Professor Ichiro Kato, who had been one of authors in the previous papers on the mastication robots, passed away in June 1994. The authors pray his soul may rest in peace.

Reference
[1] Kato, I., "Conception of living and life support robot," *Journal of the Robotics Society of Japan* (in Japanese), Vol. 11, No. 5, pp. 614-617, 1993.
[2] Charles H. Gibbs and Harry C. Lundeen, "Jaw movements and forces during chewing and swallowing and their clinical significance," *Advances in Occlusion*, John Wright PSG inc., pp. 2-32, 1981.
[3] Hyodo, K. et al., "Development of the joint simulators applying Ni-Ti shape memory alloy," *Seikeigeka Biomechanics* (in Japanese), Vol. 7, pp. 231-236, 1985.
[4] Hatcher, D. C., Faulkner, M. G., Hay, A., "Development of mechanical and mathematic models to study temporomandibular joint loading," *The journal of prosthetic dentistry*, Vol. 55, No. 3, pp. 377-384, 1986.
[5] Aoki, T., Hayashi, T., Nakajima, S., Kobayashi, H., Miyakawa, M., "Development of a self-controlled jaw-movement robot; trial manufacture of a jaw structure and occlusal-force sensors," *Technical report of IEICE* (in Japanese), MBE93-75, pp. 115-122, 1993.
[6] Barbenel, J. C., "The biomechanics of the temporomandibular joint: a theoretical study," *J. biomechanics*, Vol. 5, pp. 251-256, 1972.
[7] Osborn, J. W., Baragar, F. A., "Predicted and observed shapes of human mandibular condyles," *J. biomechanics*, Vol. 25, No. 9, pp. 967-974, 1992.
[8] Throckmorton, G. S., "Sensitivity of temporomandibular joint force calculations to errors in muscle force measurements," *J. biomechanics*, Vol. 22, No. 5, pp. 455-468, 1989.
[9] Kato, I., Takanishi, A., Asari, K. and Tani, T., "Development of artificial mastication system (Construction of one degree of freedom antagonistic muscle model WJ-0)," *Anatomischer anzeiger*, pp. 197-203, VEB Gustav Fischer Verlag Jena, 1988.
[10] Takanishi, A., "Development of mastication robot WJ-1," *Journal of Robotics and Mechatronics*, Vol. 1, pp. 185-191, 1989.
[11] Takanishi, A., Tanase, T., Kumei, M. and Kato, I., "Development of 3 DOF jaw robot WJ-2 as a human's mastication simulator," *Fifth International Conference on Advanced Robotics ('91 ICAR)*, pp. 277-282, 1991.

[12] Takanobu, H., Takanishi, A. and Kato, I., "Design of a mastication robot using a human skull model," *Proc. of the 1993 IEEE/RSJ International Conference on Intelligent Robots and Systems (IROS '93)*, pp. 203-208, 1993.

[13] Stewart, D., "A platform with six degrees of freedom," *The Institute of Mechanical Engineers*, Proc. 1965-66, Vol. 180, Part 1, No. 15, pp. 371-386, 1966.

[14] Takanobu, H., Kumei, M., Takanishi, A. and Kato, I., "Bio-parallel mechanism of mastication robot," *Robotics, Mechatronics and Manufacturing Systems*, pp. 75-80, Edited by Takamori, T. and Tsuchiya, K., Elsevier Science Publishers B. V., 1993

[15] Mackay, W. A., Crammond, D. J., Kwan, H. C., Murphy, J. T., "Measurement of human forearm viscoelasticity," *J. biomehanics*, Vol. 19, No. 3, pp. 231-238, 1986.

[16] Park, H. J., Kusumoto, H., Akazawa, K., "Analysis of modulation of mechanical impedance in human skeletal muscles," *Biomechanism 10* (in Japanese), pp. 45-54, 1990.

[17] Kato, A., Ito, K., "Visco-elasticity change in human muscle," *Biomechanism 11* (in Japanese), pp. 213-221, 1992.

[18] Kusumoto, H., Park, H. J., Yoshida, M., Akazawa, K., "Simultaneous modulation of force generation and mechanical property of muscle in voluntary contraction," *Biomechanism 12* (in Japanese), pp. 211-220, 1994.

[19] Shadmehr, R., Arbib, M. A., "A mathematical analysis of force-stiffness characteristics of muscles in control of a single joint system," *Biological Cybernetics*, Vol. 66, pp. 463-477, 1992.

[20] Hogan, N., "Adaptive control of mechanical impedance by coactivation of antagonist muscles," *IEEE Transaction on automation and control*, Vol. AC-29, No. 8, pp. 681-690, 1984.

[21] Laurin-Kovits, K. F., Colgate, J. E., Carnes, S. D. R., "Design of components for programmable passive impedance," *Proc. of the 1991 IEEE International Conference on Robotics and Automation*, pp. 1476-1481, 1991.

[22] Hyodo, K., Kobayashi, H., "Kinematic and control issues on a tendon-controlled wrist mechanism," *Proc. of the IMACS/SICE International Symposium on Robotics, Mechatronics and Manufacturing Systems '92 Kobe (IMACS/SICE RM2S '92 Kobe)*, pp. 67-72, 1992.

[23] Goodwin, G. M., Hoffman, D., Luschei, E. S., "The strength of the reflex response to sinusoidal strength of monkey jaw closing muscles during voluntary contraction," *J. of Physiol.*, 279, pp. 81-111, 1978.

[24] Ohtsuki, K., Ohnisi, M., Tsuzi, M., Watai, Y., Takanishi, A., Takanobu, H., "Development of a mouth opening and closing apparatus using a mastication robot -the measurement of mouth opening ability-," *J. of Japanese Society for Temporomandibular Joint*, Vol. 7, No. 3, pp. 69-78, 1995.

A STRATEGY FOR THE FINGER-ARM COORDINATED MOTION ANALYSIS OF HUMAN MOTION IN VIOLIN PLAYING

K. Shibuya and S. Sugano
Waseda University, Tokyo, Japan

Summary : This study aims to clarify the strategy of human for the finger-arm coordinated motion, and to adapt the results to the planning of anthropomorphic robots. Human motion in violin playing has been analyzed. In this paper, the results of human motion analysis and the method for planning of anthropomorphic robots based on the results is described.

1. Introduction

Human can perform many complicated tasks which need to control dexterously position, speed and force by the hands using finger-arm coordination. However, present robots which consist of redundant manipulators and fingers can not carry out such complicated tasks. One of the reasons for the fact is that strategies for the planning of the finger-arm coordinated motion to do such complicated tasks still have not been established.

To establish strategies of anthropomorphic robots for the motion, human motion can be considered as one of models for robot motion. If the strategy of human for the finger-arm coordinated motion is clarified and the results are adapted to the planning of robot manipulators, more tasks will be possible to be carried out. So, this study aims to clarify the strategy of human for finger-arm coordinated motion and to adapt the results to the planning of anthropomorphic robots.

To understand human motion, the human arm motion has been analyzed[1][2]. However, although many of the studies have dealt with arm posture, the effect of the fingers has not been dealt with . There are examples which deal finger motion of manipulators, but only tasks which need position and speed control has been proposed[3]. This study deals with the bowing in violin playing as an example of a task where force control, involving finger-arm coordinated motion is necessary.

In this paper, first, the results of human motion analysis is described. Then, a planning method for anthropomorphic robots, which is derived form the results of the motion analysis, is discussed.

2. Violin

2.1 Feature of a violin

Fig.1 shows names of each part of a violin and a bow. A violin has four strings which are called E, A, D and G string. The strings are pushed by the left fingers onto the finger-board. Both the edges of the strings are connected to pegs and a tail-piece, and supported by the bridge. A chin rest and shoulder pad is equipped on a violin which is used to support the violin.

The hair of a bow is made of tale of horses. The hair which is near the tip is called "the Point", and that which is near the frog is called "the Nut". The tension

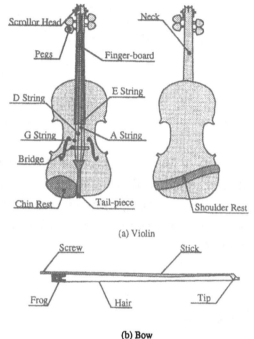

(a) Violin

(b) Bow

Fig.1 Violin and Bow

of it can be controlled by the position of frog. The position of the frog can be controlled by rolling the screw.

2.2 Bowing

The violin playing can be divided into 3 tasks. They are bowing, fingering and holding. In this paper, bowing is dealt with of the three.

Bowing is a task which the hair of a bow is rubbed with the strings. Then the sound is made. Players can control sound intensity, note value and timbre. Therefore, bowing is one of most important tasks in the violin playing.

There are two directions in bowing. They are up-bow and down-bow. Up-bow is to play the violin from the Point to the Nut, and down-bow is the reverse.

There are three important physical parameters of bowing, and they are shown below.

(1) Bow speed

(2) Bow force

(3) Sounding point

Bow speed means the relative speed of a bow to a violin. Bow force means the force to a bow from the strings. Sounding point means the position where a bow touches the strings, and it is represented by the distance from the bridge.

3. Measurement of Human Motion

3.1 Measurement System

Fig.2 shows the measurement system.

Motion of right arm, the violin and the bow was measured by 3D-Video Tracker System. They can calculate the position of the center of the makers which are equipped on subject's body from the images of some CCD cameras.

The MP and PIP joint angle of the four fingers except thumb of the right hand was measured by flexible goniometers.

Bow force was calculated from the outputs of strain gauges which were equipped to the hair of the top and bottom of the bow. All data were synchronized by the trigger, and sampled by the computers.

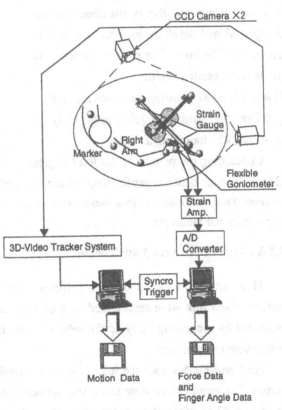

Fig.2 Measurement System

3.2 Experimental Method

Professional violinists, students of academy of music and inexperienced persons were selected as subjects. Down-bow and up-bow on "A string" which was the most basic playing in the violin playing were selected.

The subjects were instructed to play the violin with three different bow force and bow speed. The strength of bow force was instructed to the subjects by words like "strong", "middle", "weak". Bow speed was instructed to them by the time to play the violin from the Point to the Nut.

4. Experimental Results

4.1 Analysis of Arm Joint Angles

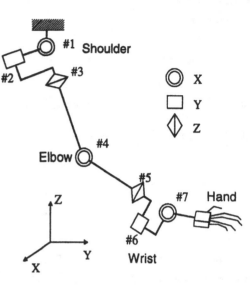

When the joint angles were calculated, the right arm were regarded as a solid link model shown in Fig.3. The standard posture of a arm is a state where the arm is stretched towards the ground, and the elbow and the palm face to the back of the body. The positive direction of joints is as described below.

(1)#1, #4 and #7 : turning the joints forward

(2)#2 and #6 : turning the joints to the body

(3)#3 and #5 : loosening a screw.

Fig.3 Link Model of Human arm

Concerning the pattern of the joint angles, clear difference between professional violinists, students and inexperienced persons can not be seen. This fact indicates that inexperienced persons can imitate the pattern of joint angles of professional violinists.

4.2 Analysis of the Arm Joint Torque

Then, arm joint torque of each subjects were calculated by using the Newton-Euler method. Each link were represented as a column, and all the parameters of the link were estimated by measuring body of the subjects. The friction between hair of the bow and A string were unaccounted.

Fig.4 and Fig.5 shows the joint torque of a professional violinist and an inexperienced person. The graphs show bow force, join torque of #1-#4, join torque of #5 and join torque of #6 and #7 in descending order. They show the results in strong, middle and weak bow force. Solid lines and broken lines in the graphs indicate the times when down-bow and up-bow begins respectively.

From Fig.4 and Fig.5, it is clear that while the peak level of professional violinist's #5 joint torque decreases according to the bow force, inexperienced person does not. Same tendency can be seen in the joint torque of #6 and #7. However, the clear tendency can not be seen in the joint torque of shoulder and elbow. This explains the reason, why the bow

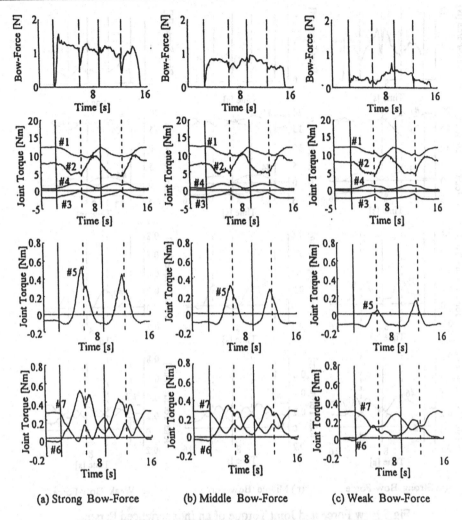

(a) Strong Bow-Force   (b) Middle Bow-Force   (c) Weak Bow-Force

Fig.4 Bow Force and Joint Torque of a Professional Violinist

force of inexperienced persons at the Point is small, while professional violinist's are not.

The above result indicate that the wrist of professional violinists control the joint torque to control bow force. The reason of this can be considered that shoulder and elbow need to use much more torque to support its self weight than it needs to control bow force. This makes it difficult to control bow force by shoulder. The wrist of professional violinists has the role of controlling the bow force by controlling its joint torque.

4.3 Analysis of the Finger Motion

Next, the finger motion of the professional violinists was analyzed. Fig.6 shows the MP and PIP joint angle of professional violinists in bowing except the thumb.

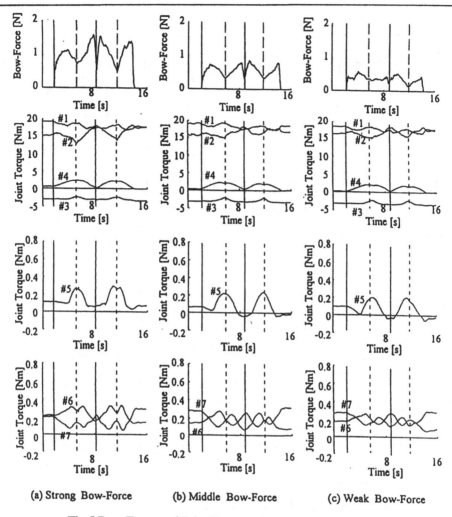

(a) Strong Bow-Force   (b) Middle Bow-Force   (c) Weak Bow-Force

Fig.5 Bow Force and Joint Torque of an Inexperienced Person

From this figure, it is clear that the MP joint angle of forefinger changes with opposite phase to PIP joint angle. However, the MP joint angle of the other 3 fingers changes with same phase to PIP joint angle.

The motion of fingers can be explained with the above results as following. As shown in Fig.7, the tip of the forefinger move straight, but that of the other 3 fingers roundly. From the above results, it is considered that forefinger control transverse direction of the bow by holding the stick, and the other 3 fingers control horizontal direction of the bow by picking up and pulling down the frog as shown in Fig.8.

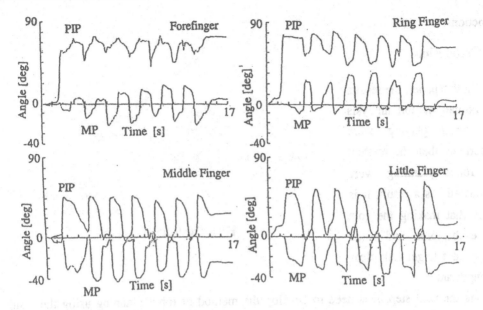

Fig.6 MP and PIP Joint Angle of a Professional Violinist

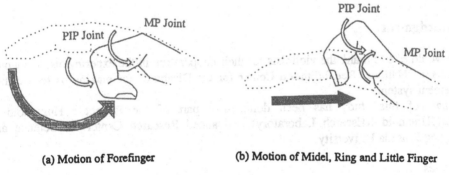

(a) Motion of Forefinger   (b) Motion of Midel, Ring and Little Finger

Fig.7 Motion of Fingers

5. Adapting Human Motion To Robot Planning

In this chapter, the method for planning of robot will be considered.

The above results of human motion analysis indicate that each joint and each finger has its clear own role. For example, the forefinger has different role form the other 3 fingers, and wrist has a role of controlling bow force in violin bowing. Human can give appropriate role to their joints and fingers by learning for a long time.

In robot planning, making the role of each joint and each finger to do a particular task is important. Because anthropomorphic robot is redundant, it needs evaluation functions to decide robot motion. It will be useful to consider each of the role to make evaluation functions.

6. Conclusion

In this paper, the results of human motion analysis in violin playing were described, then the method of robot planning were discussed. As a result, it is clear that making the role of each joint and each finger in a target task clear is important.

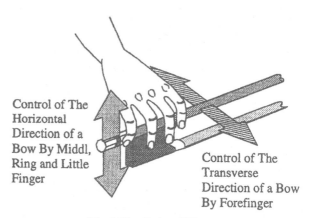

Control of The Horizontal Direction of a Bow By Middl, Ring and Little Finger

Control of The Transverse Direction of a Bow By Forefinger

Fig.8 The Role of Fingers

As the next step, it is need to develop the method of robot planning using the results much more. Then, we can plan a robot concretely. Finally, we will verify the method by a real robot.

Acknowledgment

We would like to thank the violinists for their cooperation in the experiments. We would like to thank "National Rehabilitation Center for the Disabled" for making us to use their measurement system.

A part of this study has been done as a part of the Project : Humanoid at HUREL(HUmanoid REsearch Laboratory), Advanced Research Center for Science and Engineering Waseda University.

References

[1] T.FLASH, N.HOGAN : "The Coordination of Arm Movements : Anexperimentally Confirmed Mathmatical Model", J.of Neuroscience, Vol.5, No.7, pp.1688-1703, (1985)
[2] F.A.MUSSA-IVALDI,N.HOGAN and E.BIZZI: "Neural, Mechanical, and Geometric Factors Subserving Arm Posture in Humans", J.Neuroscience, Vol.5, No.10, pp.2732-2743, (1985)
[3] S.Sugano, I.Kato: "WABOT-2:AUTONOMOUS ROBOT WITH DEXTEROUS FINGER-ARM - FINGER-ARM COORDINATION CONTROL IN KEYBOARD PERFORMANCE -", Proc. of the IEEE International Conference on Robotics and Automation, pp.90-97, (1987)

Chapter VII
Control on Motion I

HIGH PERFORMANCE CONTROL OF MANIPULATORS USING BASE FORCE/TORQUE SENSING

G. Morel

Electricité de France, Chatou, France

S. Dubowsky

MIT, Cambridge, MA, USA

Abstract

Joint friction is a major problem in accurately controlling manipulators performing small precise tasks. Previously developed solutions to this problem require the use of either complex and unreliable models of the joint friction or expensive and delicate internal joint torque sensors. Here a simple, cost-effective method for compensating the effect of joint friction is proposed. It uses a six-axis force/torque sensor mounted externally at the manipulator's base. From the base wrench measurements, the joint torques are estimated and fed back through a torque controller, that virtually eliminates friction and gravity effects. It is shown that with this high-quality torque control a simple PD position controller can provide very high precision motion control. The precision is substantially greater than for conventional methods and approaches the resolution of the Puma's encoders.

1. Introduction

Robotic manipulators are often required to perform very small, very accurate, and very slow motions, such as for micro-assembly. Precision is difficult to achieve due to the effects of nonlinear joint friction. To solve this problem three approaches been developed: model based compensation; torque pulse generation; and torque feedback control. In the first, an accurate model is needed to estimate of the friction torque [1,2,3]. The method has important limitations. It is very difficult in practice to obtain accurate and robust models that can faithfully account for many nonlinear phenomena such as Coulomb friction, dependency on joint position, influence of changes in load and temperature and nonbackdriveability.

In the torque pulse method, pulses to compensate for friction are generated using either an explicit model [1] or simple rules of qualitative reasoning [4]. While this is a practical approach the method is limited to cases where the trajectory to reach the final position is not important, since only finite displacements are controlled.

In torque feedback control the torque applied to a manipulator's joint is sensed and fed back in a joint torque loop. This method has produced good and robust experimental results; reducing the effective friction torque by up to 97% without the need for any friction model [5,6]. Unfortunately, most commercially manipulators are not equipped with joint torque sensors. Installing them in an existing manipulator would be very difficult [7]. Also, such sensors introduce number of practical problems. For example, they can add structural flexibility, that can decrease the overall performances of the manipulator, they are expensive, and they can reduce system reliability.

Here, a new approach is presented to deal with joint friction during fine motions, that overcomes the difficulties of the three discussed above. The method uses a six-axis force/torque sensor mounted between the manipulator and its base to estimate joint torques. The estimation process uses Newton-Euler equations of successive bodies. The estimated torques are used in joint torque loops. An outer position control loop provides the desired torques from measured position errors, see Figure 1.

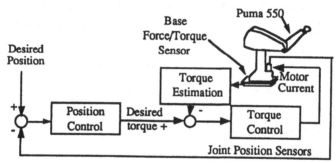

Figure 1. High precision control structure using a base force/torque sensor

2. Joint torque measurement using base sensing

Consider a manipulator mounted on a base force/torque sensor. The wrench W_b exerted by the manipulator on its supporting sensor is:

$$W_b = W_g + W_d \tag{1}$$

W_g is the wrench due to the gravity and W_d is the dynamic wrench due to the motions of the manipulator. Note, the base sensor measures wrenches only due to forces and torques effectively to the manipulator's links. Transmission and joint friction do not appear in the measured base wrench.

The first step in the process is to estimate the dynamic component, W_d, by compensating for the gravity component W_g as follows [8]:

$$W_d = W_b - W_g = W_b - \begin{pmatrix} F_g = \sum_{i=1}^{n} m_i g \\ M_g^{O_s} = \sum_{i=1}^{n} O_s G_i \times m_i g \end{pmatrix}$$

(2)

Where F_g and $M_g^{O_s}$ are the gravity force and moment at the center of the sensor, O_s, respectively, m_i and G_i are the mass and the center of mass of link i, respectively. The gravity compensated wrench (W_d) is then propagated through the successive bodies of the manipulator to yield an estimate of the dynamic joint torques, as follows:

$$\begin{cases} W_{0 \to 1} = -W_d \\ W_{1 \to 2} = W_{0 \to 1} - W_{dyn_1} \\ \vdots \\ \vdots \\ W_{i \to i+1} = W_{i-1 \to i} - W_{dyn_i} \end{cases}$$

(3)

$W_{i \to i+1}$ is the wrench exerted by the link i on the link i+1 and W_{dyn_i} is the dynamic wrench for the link i. W_{dyn_i} can be expressed at any point A in terms of the acceleration $\dot{V}_{G_i}$ of G_i, the angular acceleration, $\dot{\omega}_i$, and the angular velocity, ω_i:

$$W_{dyn_i} = \begin{pmatrix} F_{dyn_i} = m_i \dot{V}_{G_i} \\ M_{dyn_i}^A = I_i \dot{\omega}_i + \omega_i \times I_i \omega_i + G_i A \times m_i \dot{V}_{G_i} \end{pmatrix}$$

(4)

I_i is the inertia tensor of link i at G_i. Summing the equations (3) yields:

$$W_{i \to i+1} = -W_d - \sum_{j=1}^{i} W_{dyn_j}$$

(5)

Given this wrench, the torque in joint i+1 is obtained by projecting the moment vector, $W_{i \to i+1}$ acting on the ith+1 link along the z_i axis of the ith link:

$$\tau_{i+1} = -z_i^t \left[M_d^{O_i} + \sum_{j=1}^{i} \left(I_j \dot{\omega}_j + \omega_j \times I_j \omega_j + O_i G_j \times m_j \dot{V}_{G_j} \right) \right]$$

(6)

2.   Design of a torque controller for a Puma 550 using base sensing.

Experimental studies with the Puma show that its joints exhibit very large Coulomb friction [9]. In very fine motion applications this friction will severely degrade systems performance. Here a torque controller using base force/torque sensing is used to compensate for this static friction.

The torque estimation requires measurement of joint positions, velocities and accelerations. For the Puma the positions are measured with incremental encoders. This encoder data is differentiated and filtering, at a sampling rate of 2500Hz, yields the velocities and accelerations. The required mass parameters of the links are estimated from the base force/torque sensor static data [10]. The method for joint torque estimation applied to the Puma 550 is not computationally intensive. A single 68020 VME board, supporting VxWorks, can estimate the joint torques, including gravity compensation, at sampling rate of 300Hz.

The torque control law discussed here is a form of high DC gain controller implemented as an integral controller [11], but with feedforward compensation, or:

$$V_{command} = \frac{1}{K_{act}}\left[\tau_{des} + K_{int}\int_0^t (\tau_{des} - \tau_{est})\right] \qquad (7)$$

τ_{des} and τ_{est} are the desired and the actual (i.e. base-sensed) torques, respectively and K_{act} is a constant based on the amplifier, motor and transmission parameters. The gain K_{int} was set to 75% of the value that caused experimental structural oscillations.

Figure 2 demonstrates the effectiveness of base sensed torque control for the first joint of the Puma 550. Here, the desired torque is a triangular function with a maximum value of 3 Nm, while the dry friction is more than 5 Nm. Without torque feedback, the actual torque applied to the link would simply be zero, as the friction would be larger than the motor torque. However, with torque feedback, the experimental results show that the actual torque remains very closed to its desired value (Figure 2). The torque controlled motor must produce nearly 8 Nm to obtain the next 3 Nm required by the command. Results obtained in the study show that when the sign of the velocity changes (such as at approximately 3.2s, 5.35s and 6.5s in Figure 2) large torque disturbances occur. Even at these times the torque error peak remains small (±1 Nm, i.e., 20% only of the Coulomb friction) and is quickly brought under control [9].

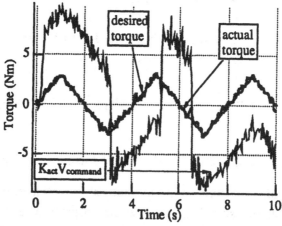

Figure 2: Joint 1 torque control experimental results

3.   Position control with base sensed torque feedback control

Experiments showed that with the based sensed joint torque feedback method the manipulator could be made to appear virtually frictionless. With such quality performance, it is easy to obtain precise position control using a simple PD loop enclosing the torque controller. Consider the simple controller (see Figure 3):

$$V_{command} = \frac{1}{K_{act}} \left[\tau_{des} + K_{int} \int_0^t (\tau_{des} - \tau_{est}) \right] \tag{8}$$

with

$$\tau_{des} = K_p(q_d - q) + K_d(\dot{q}_d - \dot{q}) \tag{9}$$

Where K_p and K_d are the proportional and derivative diagonal gain matrices, respectively.

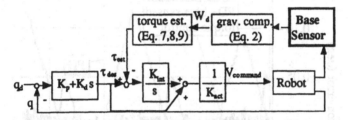

Figure 3. The precise position control scheme with base sensed torque feedback

The robustness and effectiveness of the method at the joint level is show in Figure 4. Here joint 1 is commanded to follow a very slowly a triangular wave. The magnitude of the desired motion is +/-0.1 degrees, with a period of 10 seconds. This corresponds to a desired velocity of 7 encoder counts per second.

In Figure 4 shows the performance of the Puma with base sensed torque control compared to conventional PD and PID controllers. For all three controllers, the gains have been chosen to provide a bandwidth of 5 Hz and a damping ratio of 0.5. The integral gain in the PID control has been selected to be quite high, equal to 80% of the smallest value exhibiting instability. In Figure 4 the superior performance of the base sensed torque control is clear. Conventional PD control leads to almost no motion, due to dry friction. The PID controller performs much better, and provides a zero steady state positioning error. However, when the sign of the velocity changes, the position integral compensator requires a long time (2.5 s) to compensate for the friction disturbance, resulting in lack of positioning precision. The base sensed torque feedback control method compensates rapidly for the Coulomb friction at velocity sign changes (~50 ms) and the position error remains close to zero during the task.

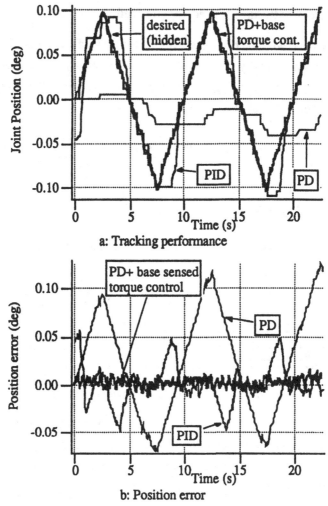

a: Tracking performance

b: Position error

Figure 4. Precise position control results for Joint 1.

Table 1 summarizes the performances of the three controllers. The base sensed torque control results show error on the order of magnitude the resolution of the encoder are reached. An encoder count corresponds to a 0.0058 degree angle, and thus the Root Mean Square error (0.0042 deg) is less than one encoder count throughout the entire task.

Table I. Summary of position control performances.

Controller	Max. Error (deg)	RMS error (deg)	Integral Sq. error (deg^{2}s)
PD	0.12	0.0590	7.7 10^{-2}
PID	0.056	0.0200	9.1 10^{-3}
PD with base sensed torque control	0.012	0.0042	4.0 10^{-4}

Figure 5 shows the ability of the manipulator with based sensed torque control to precisely tract a delicate Cartesian space trajectory. The desired end-effector trajectory is a circle with a 350μm radius. The maximum magnitude of the joint motions is 0.1 degrees. For these very fine motions the algorithm can be substantially simplified without loss of performance. First, W_g is assumed to be constant. It is simply set equal to the initial static wrench measured with the base sensor. Second, all the dynamic terms are neglected.

In the results show in Figure 5 the end-effector position is measured by a fixed 2D photodetector sensor and a laser mounted on the end-effector. The precision of the system is clearly excellent: the maximum absolute position error is less than 30μm, in spite of large frictional disturbances that occur each cycle of the motion when the sign of the velocity for each of the joint changes.

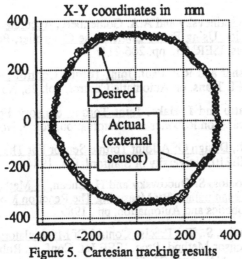

Figure 5. Cartesian tracking results

4. Summary and conclusions

A new and practical method to compensate for joint friction in fine motion control of manipulators has been proposed. It does not require either the difficult modeling of friction or the use of expensive and delicate internal joint torque sensors. It uses a 6 axis force/torque sensor mounted at the base of the manipulator. The sensor is external to the robot and hence can be easily mounted under existing manipulators.

The experimental results show dramatically fine performance. At the end-effector, during very slow displacements, the position error remains smaller than 30μm. This method is now being investigated for delicate tasks that require both fine motions and very small interaction forces.

5. Acknowledgments

Authors want to thank Electricité de France, Direction des Etudes et Recherches, Ensembles de Production - SDM - P29, Chatou, France for G. Morels financial support. We also wish to acknowledge the use of experimental equipment provided by NASA in this research.

6. References

1. B Armstrong, <u>Control Of Machines With Friction</u>, Kluwer Academic Publishers, Boston, USA, 1991

2. M.R Popovic, K.B. Shimoga and A.A. Goldenberg, <u>Model Based Compensation Of Friction In Direct Drive Robotic Arms</u>, J. of Studies in Informatics and Control, Vol. 3, No 1., pp. 75-88, March 1994.

3. C. Canudas de Wit, <u>Adaptive Control Of Partially Known Systems</u>, Elsevier, Boston, USA, 1988.

4. M.R. Popovic, D.M. Gorinevsky and A.A. Goldenberg, <u>Accurate Positioning Of Devices With Nonlinear Friction Using Fuzzy Logic Pulse Controller</u>, Proc. Int. Symposium of Experimental Robotics, ISER'95, pp. 206-211.

5. J.Y.S. Luh, W.B. Fisher and R.P. Paul, <u>Joint Torque Control By Direct Feedback For Industrial Robots</u>, IEEE Trans. on Automatic Control, vol. 28, No 1, Feb. 1983.

6. L.E. Pfeffer, O. Khatib and J. Hake, <u>Joint Torque Sensory Feedback Of A PUMA Manipulator</u>, IEEE Trans. on Robotics and Automation, vol. 5, No 4, pp. 418-425, 1989

7. Hake J.C. and Farah J., <u>Design Of A Joint Torque Sensor For The Unimation PUMA 500 Arm</u>, Final Report, ME210, University of Stanford, CA, 1984

8. H. West, E. Papadopoulos, S. Dubowsky and H. Chean, <u>A Method For Estimating The Mass Properties Of A Manipulator By Measuring The Reaction Moment At Its Base</u>, Proc. IEEE Int. Conf. on Robotics and Automation, pp , 1989

9. Morel , G. and Dubowsky, S., <u>The Precise Control Of Manipulators With Joint Friction: A Base Force/Torque Sensor Method</u>, Proc. IEEE Int. Conf. on Robotics and Automation, pp , 1989, Minneapolis, MN, April 24-27, 1996.

10. T. Corrigan and S. Dubowsky, <u>Emulating Micro-Gravity In Laboratory Studies Of Space Robots</u>, Proc. ASME Mechanisms Conf., pp. , 1994

11. R. Volpe and P. Khosla, <u>An Analysis Of Manipulator Force Control Strategies Applied To An Experimentally Derived Model</u>, Proc. IEEE/ RSJ Int. Conf. on Intelligent Robots and Systems, pp. 1989-1997, 1992

CONTROL OF A REDUNDANT SCARA ROBOT
IN THE PRESENCE OF OBSTACLES

W. Risse

Gerhard-Mercator University, Duisburg, Germany

M. Hiller
IMECH GmbH, Moers, Germany

Abstract

A redundant SCARA robot for operation in restricted work spaces is presented that performs collision-free motion while simultaneously following a specified end-effector trajectory. The control system combines resolved motion rate control with secondary goals, in particular on-line collision avoidance. The implemented control algorithms are based on artificial potential fields induced by obstacles within a model of the work space. Results of experimental studies illustrate the manoeuvrability of the robot within an environment restricted by obstacles.

1 Introduction

Robot manipulation in work spaces containing obstacles is usually limited by the manoeuvrability of the robot arm, especially for those tasks where a prescribed end-effector trajectory has to be tracked. One approach to overcome such restrictions is to improve the robots dexterity by supplying additional axes. These *redundant robots* have the ability to

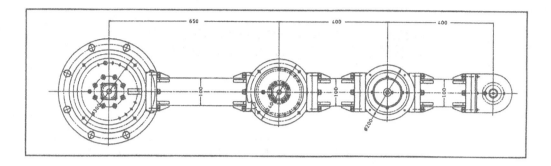

Figure 1: Redundant SCARA robot REDROB (top view)

satisfy secondary goals in addition to the specified end-effector motion, e.g. to avoid physical joint limits, singularities or collisions with obstacles in their vicinity. In the following
a planar redundant robot together with its control structure and algorithms is described
with emphasis on its behaviour in the presence of obstacles.

The manipulator under investigation is the redundant SCARA robot REDROB that was
developed and built at the *Institut für Mechatronik*. The mechanism has four revolute
joints in parallel which give the planar system one redundant degree of freedom (DOF). It
was mainly designed as an experimental setup for preliminary investigations of redundancy
control and sensor concepts to be applied to larger redundant manipulators operating in
construction or mining [1]. The modular design of REDROB allows the assembly of
different mechanisms with varying length and dimension. In the configuration discussed
(Fig. 1) the manipulator has a maximum reach of 1.54 m and can carry a payload of up
to 6 kg.

To detect obstacles located in the robot's vicinity, a configurable ultrasonic sensor system has been developed that consists of eight transducers on both sides of the three main
links [2]. Since these devices have not yet been available, a model of the planar work
space has been used to investigate on-line control strategies for operation in the presence
of obstacles. In this context obstacles can be described as lines, rectangles, circles or
combinations of these primitives. In addition the end-effector is equipped with two laser
displacement sensors allowing the tracking of unknown surfaces with respect to distance
and orientation.

2 Control Structure

The robot is controlled by means of a graphical user interface (GUI) running under X-Windows system on a UNIX workstation. It provides an interactive path editor, a teach point management system and several physical input devices for manual operation. Communication between the host and the real time controller which is implemented on a Motorola 68040 CPU is implemented via an Ethernet connection. The hierarchical control structure was developed on a workstation within an *Integrated Development Environment* and could be easily transferred to the controller by simply cross compiling the unchanged source code for the real-time target [3]. The control program consists of several processes running at different rates and priorities. Fig. 2 shows the modules and the events activating them with priority and rate increasing from top to bottom.

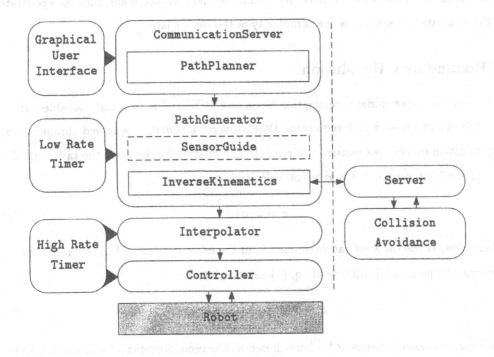

Figure 2: Extended REDROB Control Structure

The upper module is represented by a CommunicationServer managing the data exchange with the GUI to process host requests. Motion commands are handled by the PathPlanner that computes an analytical description of the corresponding trajectory, which is then evaluated by the PathGenerator at a certain rate in either joint or Cartesian coordinates. In the latter case two internal functions are called: the optional SensorGuide performing the adaption of the prescribed end-effector path depending on external sensor signals, and the InverseKinematics which is usually the most time-consuming part. For this reason the PathGenerator runs at a lower rate than the Controller, which requires the introduction of an additional Interpolator. In the current implementation the Controller and the Interpolator are called every 3 ms while the PathGenerator is evaluated every four control cycles. This ratio can be increased in case further tasks are introduced.

When operating in restricted work spaces an additional CollisionAvoidance process is activated that determines an appropriate joint motion to prevent the robot from colliding with obstacles. This process is decoupled from the main structure and runs on a separate CPU at a rate depending on the complexity of the work space.

3 Redundancy Resolution

The planar robot under investigation has $n = 4$ DOF and $m = 3$ task variables which is equivalent to $n - m = 1$ redundant DOF. Inverse kinematics is solved through local optimization by *resolved motion rate control*. The relationship between the task variables x and the joint variables θ is given on velocity level by

$$\dot{x} = J(\theta)\, \dot{\theta} \,, \tag{1}$$

where $J(\theta)$ is the $(m \times n)$ Jacobian matrix, in the following denoted as J. According to Liégeois [4] the general solution of eq. (1) can be given as

$$\dot{\theta} = J^* \dot{x} + (I - J^* J)\, z \,, \tag{2}$$

with the generalized inverse J^* – here chosen as the pseudoinverse $J^\dagger$ – of the Jacobian, the $(n \times n)$ identity matrix I and an arbitrary vector z. The resultant joint velocities comprise a *particular* solution performing the specified end-effector motion and a *homogeneous* solution representing the redundancy of the system. The latter can be utilized to match secondary goals, like the consideration of mechanical joint limits [4, 5] or collision avoidance [6], by mapping a appropriate vector z into the *null space* of the Jacobian.

To guarantee safe operation in the presence of obstacles, the manipulator arm must be protected against collisions. If the end-effector trajectory can not be completely determined in advance – because of manual operation, sensor based motion or a possibly dynamical environment – collision avoidance must be performed on-line. A suitable collision avoidance algorithm for redundancy resolution on velocity level was introduced by Maciejewski and Klein [6]. The approach is similar to the *task priority based redundancy control* [7] and uses a repelling velocity imposed on the point nearest to an obstacle as the secondary goal. To prevent the manipulator from oscillating between opposite obstacles, Maciejewski and Klein proposed to blend the respective homogeneous solutions. However, if the number of obstacles simultaneously influencing the manipulator increases, e.g. when operating in cluttered work spaces (Fig. 3), the evaluation of the homogeneous solutions may lead to high numerical costs. An efficient approach on acceleration level was proposed by Khatib [8] introducing the well known *artificial potential fields* inducing repulsive forces that lead to *artificial joint forces*. In case of redundancy these can be projected into the null space to preserve the prescribed end-effector motion.

The algorithm implemented within the REDROB controller is essentially based on Khatib's theory. However, to increase efficiency for real-time applications, inverse kinematics is resolved on velocity level and the artificial joint forces are interpreted as the vector z to be projected into the null space by eq. (2). To simplify matters only a limited number of representative points on the manipulator are considered – preferably those where the ultrasonic ranging sensors will be mounted. For each of these n_p points p_i the distance vector $d_{i,j}$ towards the obstacle O_j is determined, and a repulsive force $f_{i,j}$ is applied according to

$$f_{i,j} = \begin{cases} \left(\dfrac{\eta}{\|d_{i,j}\| + d_{ta}} - \dfrac{\eta}{d_{soi} + d_{ta}} \right) d_{i,j} & \text{for } d_{ta} < \|d_{i,j}\| \le d_{soi} , \\ 0 & \text{for } \|d_{i,j}\| > d_{soi} , \end{cases} \tag{3}$$

with the *task abort distance* d_{ta}, the *sphere of influence distance* d_{soi} and the negative scaling factor η. If the distance reaches d_{ta}, which is chosen according to the dynamic properties of the robot and the lower detection boundary of the sensors, the motion will be aborted instead of applying infinite joint forces. The total force imposed on each point is accumulated over all n_O obstacles according to

$$f_i = \sum_{j=1}^{n_O} f_{i,j} \; . \tag{4}$$

The corresponding artificial joint forces can be computed via the transpose of the Jacobian

$$\tau_i = J_i^T W f_i \; , \tag{5}$$

with a weighting matrix W taking into account the mass properties of the arm. The overall vector of the joint forces is obtained as

$$z = \sum_{i=1}^{n_p} \tau_i \tag{6}$$

and can be enlarged by further terms, e.g. considering mechanical joint limits. However, one has to keep in mind that in certain situations only one redundant DOF might already be insufficient to satisfy collision avoidance alone.

4 Experimental Results

The collision avoidance algorithm described above is utilized for operation within cluttered work spaces modelled as combinations of the above mentioned obstacle primitives (Fig. 3). The experimental setup was inspired by a welding application within double-walled ship hulls where the boundaries of the work space simultaneously represent both obstacles considered within collision avoidance, and the contour to be tracked by the end-effector. Since the work space boundaries usually do not exactly match the model – because of tolerances and bevelled edges – the end-effector path is not retrieved from the model but generated by means of sensor signals. However, these deviations are much smaller than the task abortion distance d_{ta} and thus can be neglected for collision avoidance, which allows the usage of a more efficient work space model.

Fig. 3a) shows the reconfiguration of the robot while moving on straight lines into the experimental work space. In this situation a second obstacle on the right hand side is limiting the maneouvrability and makes the robot completely fold up. In Fig. 3b) the end-effector travels along the work space boundaries in clockwise direction while collision avoidance successfully prevents the links from approaching the contour. For reasons of clarity the intermediate configurations have been reduced to lines in both figures.

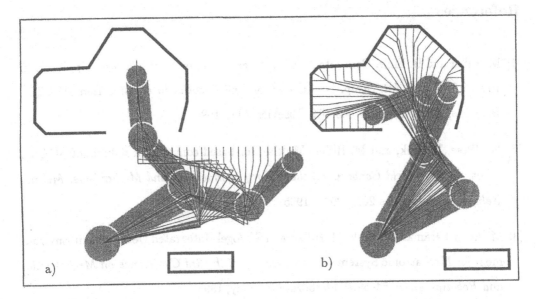

Figure 3: Collision free motion into (a) and within (b) restricted workspace

5 Conclusions

The efficient implementation of an on-line collision avoidance algorithm for a redundant SCARA robot was presented that enables safe operation in work spaces restricted by obstacles while performing a specified end-effector motion. Resolved motion rate control is applied to inverse kinematics utilizing the homogeneous solution for collision avoidance by projecting artificial joint forces into the null space of the Jacobian.

Future investigations will have to improve or replace the work space model through the application of ranging sensors. In this context the reliable determination of obstacle locations, in particular for cluttered work spaces, is the most challenging task.

Acknowledgement

This project is financed by the Ministry of Research of the Federal State North Rhine–Westphalia.

References

[1] M. Hiller and B. Fink. Concepts for a sensor-guided redundant heavy manipulator. In *Proc. of the 11th Int. Symp. on Automation and Robotics in Construction (ISARC), Brighton, England*, pages 563–570. Elsevier, May 1994.

[2] W. Risse, B. Fink, and M. Hiller. Multisensor-based control of a redundant SCARA robot. In *9th World Congress on the Theory of Machines and Mechanisms, Milano, Italy*, volume 3, pages 2037–2041, 1995.

[3] M. Anantharaman, B. Fink, M. Hiller, and S. Vogel. Integrated Development Environment for Mechatronic Systems. In *Proceedings of the 3rd Conference on Mechatronics and Robotics*, pages 54–69, Paderborn, Germany, 1995.

[4] A. Liégeois. Automatic supervisory control of the configuration and behavior of multi-body mechanisms. *IEEE Transactions on Systems, Man, and Cybernetics*, SMC-7(12):868–871, 1977.

[5] C. A. Klein and C.-H. Huang. Review of pseudoinverse control for use with kinematically redundant manipulators. *IEEE Transactions on Systems, Man, and Cybernetics*, SMC-13(3):245–250, 1983.

[6] A.A. Maciejewski and C.A. Klein. Obstacle avoidance for kinematically redundant manipulators in dynamically varying environments. *The International Journal of Robotics Research*, 4(33):109–117, 1985.

[7] Y. Nakamura, H. Hanafusa, and T. Yoshikawa. Task priority based redundancy control of robot manipulators. *The International Journal of Robotics Research*, 6(4):3–15, 1987.

[8] O. Khatib. Real-time obstacle avoidance for manipulators and mobile robots. *The International Journal of Robotics Research*, 5(1):90–98, 1986.

EXPERIMENTS OF KINEMATIC CONTROL
ON A TWO-ROBOT SYSTEM

F. Caccavale, P. Chiacchio, S. Chiaverini and B. Siciliano

Università degli Studi di Napoli Federico II, Naples, Italy

Introduction

The adoption of multiple robots for executing enhanced manipulation tasks has been receiving a great deal of attention in the robotics research community.[1-3] The key point is to achieve an effective coordination of the motion of the various robots. To this purpose, a new formulation has recently been proposed[4,5] which allows a straightforward task specification for two cooperative robot manipulators in terms of absolute and relative variables. The kinematic model of the system is conveniently expressed in terms of the direct kinematics of the two manipulators.

This paper is aimed at presenting experiments of a kinematic control scheme on a two-robot manipulator system, where inverse kinematics is solved by using a closed-loop algorithm[6] based on the differential kinematics, viz., the Jacobians, of the two manipulators. In the presence of redundant degrees of freedom, an additional constraint task can be handled according to an augmented task space technique with task priority.[7] Also, robustness of the solution in the neighborhood of kinematic singularities can be achieved by resorting to a damped least-squares inverse of a proper Jacobian.[8]

The scheme has been tested in a number of experiments on two industrial robots. The set-up available in the laboratory consists of two six-joint Comau SMART-3 S robots, one of which is mounted on a sliding track providing an extra degree of freedom. Experimental results are described to demonstrate the effectiveness of the proposed kinematic control scheme in coordinating the position and orientation of the two end effectors in several tasks with both absolute and relative motions.

Two-Robot System

Consider a system of two cooperative robot manipulators. Let p_i and R_i respectively denote the position and rotation matrix of the end-effector frame of each manipulator ($i = 1, 2$) with reference to some common base frame. According to a recently proposed task formulation,[4,5] direct kinematics of the system can be described in terms of the position

$$p_a = \frac{1}{2}(p_1 + p_2) \tag{1}$$

and rotation matrix

$$R_a = R_1 R^1_{k_{12}}(\vartheta_{12}/2) \tag{2}$$

of an absolute frame, where k^1_{12} and ϑ_{12} are respectively the unit vector and the angle that realize the rotation described by R^1_2, i.e., the orientation of frame 2 with respect to frame 1. The description is completed by specifying the relative position and orientation as

$$p_r = p_2 - p_1 \tag{3}$$

and

$$R^1_r = R^1_2. \tag{4}$$

Differential kinematics of the cooperative system can be formally obtained in terms of the stacked vectors of absolute and relative velocities $v = [\, v_a^T \quad v_r^T\,]^T$ and joint velocities $\dot{q} = [\, \dot{q}_1^T \quad \dot{q}_2^T\,]^T$ as

$$v = J\dot{q}, \tag{5}$$

where

$$J = T\bar{J} \tag{6}$$

is the overall Jacobian matrix with

$$T = \begin{bmatrix} \frac{1}{2}I & \frac{1}{2}I \\ -I & I \end{bmatrix} \tag{7}$$

and

$$\bar{J} = \begin{bmatrix} J_1 & O \\ O & J_2 \end{bmatrix} \tag{8}$$

being J_i the Jacobian of each manipulator.

Kinematic Control

A two-stage kinematic control can be adopted for the two-robot system; namely, an inverse kinematics procedure to compute the joint variables from the given cooperative task variables, which is then cascaded by two joint space controllers. This strategy has the advantage that the effort of coordination is performed at the inverse kinematics level, while conventional joint PID servos can be used to control the two robots as if they were independent.

On the basis of the above description, a closed-loop inverse kinematics algorithm can be derived which consists of computing the joint velocities as[7]

$$\dot{q} = J^\dagger(v_d + Ke) + (I - J^\dagger J)\dot{q}_0 \tag{9}$$

where v_d is the desired task velocity, K is a suitable diagonal positive matrix gain, e is an algorithmic error between the desired and actual position and orientation task variables, $J^\dagger$ denotes a pseudoinverse of J, and the operator $(I - J^\dagger J)$ projects the vector of arbitrary joint velocities $\dot{q}_0$ (aimed at utilizing possible redundant degrees of freedom of the system) into the null space of J so as not to interfere with the primary end-effector task. Notice that, in view of the transformation (7), the pseudoinverse of J can be computed as

$$J^\dagger = \bar{J}^\dagger T^{-1} \tag{10}$$

where

$$\bar{J}^\dagger = \begin{bmatrix} J_1^\dagger & O \\ O & J_2^\dagger \end{bmatrix} \tag{11}$$

leading to a considerable computational savings.

A possible choice of $\dot{q}_0$ in (9) is

$$\dot{q}_0 = k_c \left(\frac{\partial c(q)}{\partial q} \right)^T \tag{12}$$

where $c(q)$ is a constraint function of the joint variables to be locally optimized and k_c is a signed constant. It should be pointed out that kinematic redundancy of the system of two manipulators may be due either to the effective presence of additional joint variables, i.e., more than six degrees of freedom for either manipulator, or to relaxation of some task variables.

In the neighbourhood of kinematic singularities, Jacobian pseudoinversion becomes ill-conditioned and high joint velocities may occur. Numerical robustness can be gained by resorting to a damped least-squares inverse of either manipulator Jacobian close to a singularity, i.e.,

$$J_i^\# = J_i^T(J_i J_i^T + \lambda^2 I)^{-1} \tag{13}$$

where λ is a damping factor achieving a trade-off between solution accuracy (small algorithmic errors) and solution feasibility (limited joint velocities). The damping factor can be suitably tuned by defining a singular region in the neighbourhood of the singularity on the basis of the estimate of the smallest singular value of J_i

$$\lambda^2 = \begin{cases} 0 & \hat{\sigma}_m \geq \varepsilon \\ \left(1 - \left(\frac{\hat{\sigma}_m}{\varepsilon}\right)^2\right) \lambda_M^2 & \hat{\sigma}_m < \varepsilon \end{cases} \tag{14}$$

where $\hat{\sigma}_m$ is the estimate of the smallest singular value, and ε defines the size of the singular region; the value of λ_M is at user's disposal to suitably shape the solution in the neighbourhood of a singularity.[9]

Experiments

The kinematic control scheme has been tested in a number of experiments on two cooperative industrial robots. The set-up available in the laboratory consists of two six-joint Comau SMART-3 S robots, one of which is mounted on a sliding track providing an extra degree of freedom (Fig. 1). A common base frame has been located at the basis of the robot on the left with y-axis along the track towards the robot on the right and z-axis pointing upwards. Each joint is actuated by brushless motors via gear trains; shaft absolute resolvers provide motor position measurements. The kinematic structure of each robot manipulator has a non-spherical wrist whose joint axes intersect two-by-two. The native controller is the C3G 9000, a VME-based system. The robots are controlled by open versions of the controller, and bus-to-bus communication links are established with a single PC, where control algorithms can be implemented as C modules; a sampling time of 2 ms has been used.

Fig. 1. The cooperative robot manipulator set-up.

In the first case study, the cooperative system is initially placed in a configuration so that $p_a = [-0.2 \quad 1.3 \quad 0]^T$, $p_r = [0 \quad 0.4 \quad 0]^T$, $R_a = I$ and $R_r^1 = I$ with the sliding track at 0.4, where standard SI units are adopted. The given task is composed of the following phases:

⊙ an absolute displacement of 0.4 along x-axis and 0.3 along z-axis to be executed in 2.5 s;

⊙ an absolute counter-clockwise rotation of $\pi/4$ about z-axis to be executed in 1.5 s;

⊙ a relative displacement of 0.2 taking away the two end effectors to be executed in 1 s;

⊙ a relative clockwise rotation of $\pi/4$ about the normal axis of the end-effector of the robot on the left to be executed in 1 s;

⊙ return to the initial configuration while maximizing the task manipulability measure for the absolute position[10,11] to be executed in 2.5 s.

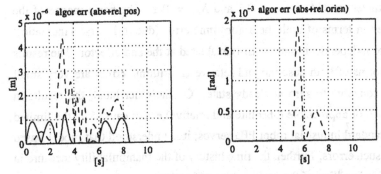

Fig. 2. Norms of absolute (solid) and relative (dashed) algorithmic errors for position and orientation in the first case study.

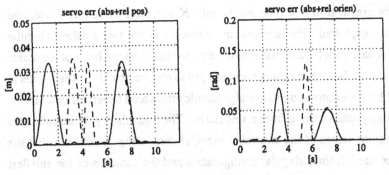

Fig. 3. Norms of absolute (solid) and relative (dashed) servo errors for position and orientation in the first case study.

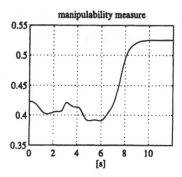

Fig. 4. Manipulability measure in the first case study.

The joint velocity solution in (9) has been utilized with the damped least-squares in (13) in lieu of the pseudoinverse. The matrix gain has been chosen as $K = 500I$ and the gain in (10) as $k_c = 10$, while the parameters in (14) are $\varepsilon = 0.1$ and $\lambda_M = 0.1$. The performance of the system has been evaluated in terms of both the algorithmic error (due to inverse kinematics) and the servo error (due to independent joint control) evaluated at the end effector. The results in Fig. 2 show that the norms of both absolute and relative algorithmic errors are very small during the transient and converge to zero at steady state. On the other hand, the results in Fig. 3 reveal the occurrence of appreciable (absolute and relative) errors caused by the limited tracking capabilities of standard industrial robot PID servos; it is understood that model-based controllers would reduce such errors. Further, the time history of the manipulability measure in Fig. 4 confirms the effective handling of the extra degree of freedom.

The second case study is aimed at studying the effectiveness of the inverse kinematics algorithm in the neighbourhood of singularities. The cooperative system is initially placed in a singular configuration and the given task consists of an absolute displacement of 0.05 along x-axis and -0.3 along y-axis to be executed in 5 s. The matrix gain K has been chosen as above, and redundancy has not been exploited. Furthermore, the estimates of the two smallest singular values of the Jacobian of each robot manipulator have been computed on the basis of a suitable algorithm[12], and the damping factor has been set accordingly as required by (14). The results in Fig. 5 demonstrate that both absolute and relative algorithmic errors tend to zero after an initial effort to track the trajectory while exiting from a singularity. The time history of the estimates of the two smallest singular values of the two robot manipulators in Fig. 6 confirms that the seven-joint manipulator starts from a singular configuration and the estimate of the smallest singular value remains anyhow within the threshold set by ε, i.e., a damped least-squares inverse is used along the whole task.

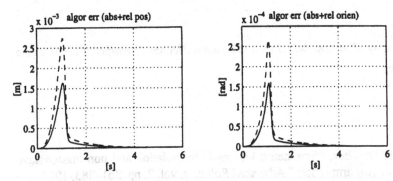

Fig. 5. Norms of absolute (solid) and relative (dashed) algorithmic errors for position and orientation in the second case study.

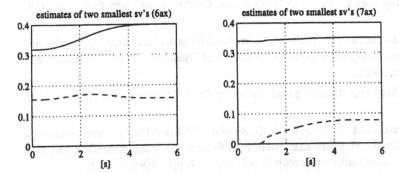

Fig. 6. Estimates of two smallest singular values of the Jacobians of the two manipulators in the second case study.

Conclusion

A kinematic control scheme for two cooperative robots has been presented in this paper. Kinematic modelling of the system has been derived on the basis of an effective task formulation which allows the absolute and relative position and orientation to be expressed in terms of the single manipulator kinematics. A closed-loop inverse kinematics algorithm with redundancy resolution and singularity robustness has been devised to generate the reference inputs to the independent joint controllers of the two robots. Experiments on the laboratory set-up have been worked out which have shown the good performance of the proposed technique for coordinated motions of the two-robot system.

Acknowledgements

This work was supported by *Ministero dell'Università e della Ricerca Scientifica e Tecnologica* under 40% and 60% funds.

References

1. M. Uchiyama and P. Dauchez, "Symmetric kinematic formulation and non-master/slave coordinated control of two-arm robots," *Advanced Robotics*, vol. 7, pp. 361–383, 1993.

2. P. Hsu, "Coordinated control of multiple robotic manipulators," *IEEE Trans. on Robotics and Automation*, vol. 9, pp. 400–410, 1993.

3. O. Khatib, "Inertial properties in robotic manipulation: An object level framework," *Int. J. of Robotics Research*, vol. 13, pp. 19–36, 1995.

4. P. Chiacchio, S. Chiaverini, and B. Siciliano, "Task-oriented kinematic control of two co-operative 6-dof manipulators," *Proc. 1993 American Control Conf.*, San Francisco, CA, pp. 336–340, 1993.

5. P. Chiacchio, S. Chiaverini, and B. Siciliano, "Direct and inverse kinematics for coordinated motion tasks of a two-manipulator systems," *ASME J. of Dynamic Systems, Measurement, and Control*, in press, 1996.

6. L. Sciavicco and B. Siciliano, *Modeling and Control of Robot Manipulators*, McGraw-Hill, New York, 1996.

7. P. Chiacchio, S. Chiaverini, L. Sciavicco, and B. Siciliano, "Closed-loop inverse kinematics schemes for constrained redundant manipulators with task space augmentation and task priority strategy," *Int. J. of Robotics Research*, vol. 10, no. 4, pp. 410–425, 1991.

8. S. Chiaverini, B. Siciliano, and O. Egeland, "Review of the damped least-squares inverse kinematics with experiments on an industrial robot manipulator," *IEEE Trans. on Control Systems Technology*, vol. 2, pp. 123–134, 1994.

9. S. Chiaverini, O. Egeland, and R.K. Kanestrøm, "Weighted damped least-squares in kine-matic control of robotic manipulators," *Advanced Robotics*, vol. 7, pp. 201–218, 1993.

10. P. Chiacchio, S. Chiaverini, L. Sciavicco, and B. Siciliano, "Global task space manipulability ellipsoids for multiple arm systems," *IEEE Trans. on Robotics and Automation*, vol. 7, pp. 678–685, 1991.

11. P. Chiacchio, S. Chiaverini, and B. Siciliano, "Kineto-static analysis of cooperative robot manipulators achieving dexterous configurations," *9th CISM-IFToMM Symp. on Theory and Practice of Robots and Manipulators*, Udine, I, Sep. 1992, in *RoManSy 9*, Lecture Notes in Control and Information Sciences 187, A. Morecki, G. Bianchi, K. Jaworek (Eds.), Springer-Verlag, Berlin, D, pp. 93–100, 1993.

12. S. Chiaverini, "Estimate of the two smallest singular values of the Jacobian matrix: Appli-cation to damped least-squares inverse kinematics," *J. of Robotic Systems*, vol. 10, pp. 991–1008, 1993.

OPTIMAL PATH PLANNING FOR MOVING OBJECTS

K. Lasinski, M. Galicki and P. Gawlowicz
Technical University of Zielona Góra, Zielona Góra, Poland

Abstract. One of the most important takes of the computer integrated manufacturing is the automatic planning of the optimal collisions-free path for many vehicles. It becomes more difficult, shou the vehicles operate in a computer work spaces including a lot of obstacles. In such cases, the methods of potential and harmonic function, mans fail to find the optimal path (an to find path, at all).In this studs a non-chemical version of Pontryagin's Maximum Principle is used to find the globally optimal vehicles paths. Unlike the penalty function method, this approach does not require an initial solution satisfy up the state equality and in equality constraints which may be trouble some to find in practice. In addition, the above approach makes it possible to handle these constraints efficiently. A computer example involving two vehicles and an obstacles being in the work space in presented.

1. Introduction

One of the most important tasks (if not the most important) of the robotic applications is the automatic planning and execution of the actions, in general by a group of intelligent machines (robots, trolleys, machine tools) to realize user specified tasks.

Problem becomes especially important in the case of repeatable tasks, which should be realized by these machines or when they have to execute operations in complicated work spaces (e.g. including many obstacles). There are everyday-life examples where these systems are inculcate e.g. one of the Siemens company has got self-propelled trolleys - robovehicles, which drive trough works and execute specified functions.

The collision-free path planning for mobile robots has been widely studied in the literature. The methods of potential and harmonic functions [1,4,5] are attractive because of real time computations. However, they suffer both from local minima (equilibrium and stagnation points for potential and harmonic functions, respectively) or lack of optimality. The first property expresses the fact that a robot may stop before its final given position is reached. A lack of optimality means that a motion from initial position to a final one may not be optimal with respect to some performance index. The repetitous processes consisting of many cycles and the complex work spaces including e.g. many obstacles cause the above methods to be inadequate. For these applications it is economically attractive to invest considerable effort in refining the gross motion characteristics of the process to improve its performance according to some criterion and to ensure the vehicles task execution. Reducing the energy required to move vehicles from their initial positions to the final ones is a good example. Here even a small energy decrease in one cycle of the process, multiplied by the number of cycles in the lifetime of the process, could imply substantial savings in manufacturing cost.

On account of the above requirements, the global path planning methods seem to be the most appropriate. The penalty functions method has been used by Gilbert et al. [9] to find minimum energetic motion for one mobile robot and by Shiller the minimum time [11], respectively. However, its application requires knowing an initial admissible vehicle paths, which satisfy collision-free conditions. This limitation may be troublesome to satysfy in practice especially when the work space includes many obstacles. Another approach, based on the discretization technique has been presented by Singh [6]. Again, the discretization problem increases considerably the dimensionality of the optimal path planning for many vehicles.

This study presents energetic-optimal path planing for many vehicles in the presence of obstacles included in the work space. It is assumed that the vehicles task (their kinematics task) is collision-free motion from the given initial positions to the given final ones. Obstacles and vehicles are modelled by means of analytic functions (used by [1]), which describe their boundaries. In addition, the vehicles controls are subject to the limitations. In that case, the Pontryagin's Maximum Principle (in its classical form) is difficult to use [3]. Instead, the negative formulation of Maximum Principle is used [8,10]. Unlike the method presented in [9], this approach does not require an initial solution (which might be difficult to find in practice). In addition, the value of performance index is usually smaller than that found by using the penalty function method. The use of Maximum Principle in this form enebles us to handle efficiently the state inequality constraints. By assuming the control inputs in a class of quasilinear functions, the above optimization problem may be reduced to linear programming problems, which standard algorithms solve efficiently. The paper is organized, as follows. Section 2 formulates the kinematic task for the vehicles as an optimal control problem. Application of the negative formulation of Maximum Principle is presented in section 3. A numerical example of optimal path planning for two vehicles moving in a work space including an obstacle is given in section 4.

2. Formulation of the vehicles kinematic task as energetic optimal control problem

Without loss of generality we consider optimal path planning for the case of two vehicles. Their simplified dynamic models [9] needed to find energetic-optimal controls (if they exist) are as follow:

$$M\ddot{q} + C\dot{q} + Fq = Hu \tag{1}$$

where: $q = (q_1 \ q_2)^T$; $q_i = (q_i^1 q_i^2 q_i^3)^T$ - vector of generalized coordinates (configuration) of the i-th vehicle (i.e. position of its masscentre and angle of rotation)

$$M = \begin{bmatrix} M_1 & O_3 \\ O_3 & M_2 \end{bmatrix}; \ C = \begin{bmatrix} C_1 & O_3 \\ O_3 & C_2 \end{bmatrix}; \ F = \begin{bmatrix} F_1 & O_3 \\ O_3 & F_2 \end{bmatrix}; \ H = \begin{bmatrix} H_1 & O_3 \\ O_3 & H_2 \end{bmatrix}; \ M_i\text{-(3×3)-dimensional}$$

inertia matrix of the i-th vehicle; C_i, H_i - (3×3) - dimensional matrixes which present dc motor drives; F_i - (3×3)-dimensional matrix of static friction; O_3-(3×3)-dimensional null matrix; $u = (u^1 \ u^2)^T$; $u^i = (u_1^i \ u_2^i \ u_3^i)^T$ - voltages in driving motors of the i-th vehicle.

While controlling the motions of the vehicles it becomes indispensable to take into consideration geometrical dimensions of these vehicles and obstacles, which exist or appear in the work space. It is assumed that vehicles geometrical models (for a fixed vector q) are described as $V^i(\mathbf{q}) = \{\mathbf{p}^i : b^i(\mathbf{p}^i, \mathbf{q}) \le 0\}$, where: $b^i : \mathbf{R}^2 \times \mathbf{R}^3 \to \mathbf{R}^1$; $\mathbf{p}^i = \left(p_1^i \ p_2^i\right)^T$ -a current point of the i-th vehicle (which depends on the vector $\mathbf{q}$): $i = 1,2$. A geometrical model of environment (obstacles) is also given by a set of analytic functions describing the equations of their boundaries which in general case may depend on time - mobile obstacles. Without loss of generality, further considerations are limited to static models of obstacles with boundary equations $b^j(\mathbf{p}) = 0$, where: $b^j : \mathbf{R}^2 \to \mathbf{R}^1$; $\mathbf{p}$ - a point lying on the boundary of the j-th obstacle model; $j = 1 : o; o$ - a number of obstacle in the work space.

It is assumed that at the initial moment, the vehicles take on the collision free configurations

$$\mathbf{q}(0) - \mathbf{q}_0 = 0. \tag{2}$$

Given vehicles final position in the work space (expressed in scalar form for convenience of further computations)

$$< \mathbf{q}(T) - \mathbf{q}_T , \mathbf{q}(T) - \mathbf{q}_T > = 0, \tag{3}$$

where: $\mathbf{q}_T$ - 6-dimensional vector representing vehicles final positions to be reached; T-a fixed final time of task execution; $<,>$ - scalar prodct of vectors.

It is natural to assume that at the initial and final moments of task execution, the velocities of vehicles are equal to 0

$$\dot{\mathbf{q}}(0) = 0 \tag{4}$$

and:

$$<\dot{\mathbf{q}}(T), \dot{\mathbf{q}}(T)> = 0, \tag{5}$$

Constant limits for controls are assumed

$$\mathbf{u}_l \le \mathbf{u}(t) \le \mathbf{u}_u \tag{6}$$

where: $\mathbf{u}_l, \mathbf{u}_u$ -lower and upper limits for control vector $\mathbf{u}(t)$, respectively.

When the motion in the work space is forced, the constraints are induced because of the fact that the vehicles should neither collide with each other nor with the obstacles. Using analytic descriptions of vehicles and obstacles, the collisions-free conditions take on the following form

$$\{ g(\mathbf{q}(t)) - \varepsilon \ge 0 \} \tag{7}$$

where: $g(\mathbf{q}) = b^j(\mathbf{p}^i)$ for each $\mathbf{p}^i \in V^i(\mathbf{q})$, for the case when the i-th vehicle should not collide with the j-th obstacle or $g(\mathbf{q}) = b^i(\mathbf{p}^k, \mathbf{q})$ for each $\mathbf{p}^k \in V^k(\mathbf{q})$, if the k-th vehicle should not collide with the i-th one; ε - a small positive number (safety smargin). The simplest way for numerical calculating the values of function $g(\mathbf{q}(t))$ is to discretize one of vehicles geometrical models. Thus, infinitely many of the above inequalities are reduced to a finite number. It is assumed, that the vehicles movement should minimise the following performance index (energy consumption)

$$I(\mathbf{u}) = \sum_{i=1}^{2} \int_{0}^{T} \langle S_i \mathbf{u}^i, \mathbf{u}^i \rangle dt \tag{8}$$

where S_i - (3×3)-dimensional positive definite, symmetric matrix.

Expressions (1) and (2)-(8) form the vehicles kinematic task as (a rather generally expressed) energetic optimal control problem. The fact that there exist collision-free inequality constraints (7) makes its solutions difficult.

The next chapter shows an approach, which makes it possible to solve the optimalization problem, using non-classical (variational) formulation of the Pontryagin's Maximum Principle.

3. Application of the negative formulation Pontryagin's Maximum Principle

To express the vehicles kinematic task in terms of the optimal control problem we use a state vector $\mathbf{x} = (\mathbf{q} \ \dot{\mathbf{q}})^T$. Hence, the equation (1) takes on the following form:

$$\dot{\mathbf{x}} = \mathbf{f}(\mathbf{x}, \mathbf{u}) \tag{9}$$

where: $\mathbf{f}(\mathbf{x}, \mathbf{u}) = \begin{bmatrix} \mathbf{O}_6 & \mathbf{I}_6 \\ -\mathbf{M}^{-1}\mathbf{F} & -\mathbf{M}^{-1}\mathbf{C} \end{bmatrix} \mathbf{x} + \begin{bmatrix} \mathbf{O}_6 \\ \mathbf{M}^{-1}\mathbf{H} \end{bmatrix} \mathbf{u}$, with initial condition given as:

$$\mathbf{x}(0) = (\mathbf{q}_0, 0)^T; \tag{10}$$

where: $\mathbf{O}_6$ - 6×6 - dimensional null matrix; $\mathbf{I}_6$ - 6×6 - dimensional identity matrix, and the performance index

$$I(\mathbf{u}) = \int_{0}^{T} \langle S\mathbf{u}, \mathbf{u} \rangle dt \tag{11}$$

where: $S = \begin{bmatrix} \mathbf{S}_1 & \mathbf{O}_6 \\ \mathbf{O}_6 & \mathbf{S}_2 \end{bmatrix}$. We notice, that the state vector $\mathbf{x}$ is functionally dependent on the control vector $\mathbf{u}$ (on account of equation (9)). Hence, the constraints (3), (5), (7) depend on the vector $\mathbf{u}$ too. Thus, they take on the following forms, which are equivalent to the prior ones.

$$\{d(\mathbf{u})\} = \mathbf{x}(T) - \mathbf{x}_T = 0 \tag{12}$$

where $\mathbf{x}_T = (\mathbf{q}_T \ 0)^T$ and :

$$\{h(\mathbf{u}) \le 0\} \tag{13}$$

where $h(\mathbf{u}) = \max_{t \in [0,T]} \{\varepsilon - g(T)\}$.

In order to use the negative form relation of the Pontryagin's Maximum Principle [7,8] it is assumed, that an admissible control $u^0=u^0(t)$ is known which satisfies only (6), but not necessarily the constraints (3), (5), (7).

In addition, the control $u^0(\ )$ does not minimise performance index (11). A trajectory $x^0(\ )$, which corresponds to this control is calculated from dynamic model (9).

Next, the increments of functionals given by the left sides of expressions (12), (13) should be determined for control $u^0(\ )$. For this purpose admissible control $u^0(\ )$ is perturbed by a small function (control variation) $\delta u = \delta u(t)$, where $\|\delta u\|_c < \lambda$; $\|\ \|_c$ - Chebyshev norm; λ a small positive number making the correctness of the presented method safe .

Hence, the value of functional $I(u)$ for control $u^0 + \delta u$ may be expressed (omitting the terms of the higher order than the first one), as follows

$$I(u^0 + \delta u) = I(u^0) + \int_0^T \langle Su, \delta u \rangle dt \qquad (14)$$

Similarly the values of functionals determined by the left sides of (12) for control $u^0 + \delta u$ are equal to

$$\{d(u^0 + \delta u) = d(u^0) + \delta d(u^0 + \delta u)\} \qquad (15)$$

where $\delta d(u^0, \delta u) = \int_0^T \langle f_u^T(t)\psi(t), \delta u \rangle dt$ - Frechet's differential of functional $d(u)$;

$$\dot\psi + f_x^T(t)\psi = 0; \quad \psi(T) = 2(x(T)\text{-}x_T);$$

$$f_u(t) = \left(\frac{\partial f}{\partial u}\right)_{(x,u)=(x^0(t),u^0(t))};$$

$$f_x(t) = \left(\frac{\partial f}{\partial x}\right)_{(x,u)=(x^0(t),u^0(t))}.$$

The increments of functionals given by the left sides of (13) require the use of Gateaux's differentials (due to non-existence of Frechet's differentials in this case). Hence, they take on the following forms

$$\{h(u^0 + \delta u) = h(u^0) + \delta h(u^0, \delta u)\} \qquad (16)$$

where: $\delta h(u^0, \delta u) = \max_{t \in S}\left\{\int_0^T \langle f_u^T(t)\psi(\tau,t), \delta u \rangle d\tau\right\}$ - Gateaux's differential of functional $h(u)$;

$S = \{t' : g(x(t')) = h(u)\};$ $\quad \dfrac{d\psi(\tau,t')}{d\tau} + f_x^T(\tau)\psi(\tau,t') = \dfrac{\partial g(x(t'))}{\partial x}\delta(\tau,t'); \quad \psi(\tau,t') = 0;$

$\delta(\ )$-Dirac's distribution.

The expressions (16) show, that Gateaux's differentials require of solving a number of Cauchy's problems depending on the profiles of functions $g(x(t))$.

In this way, the negative formulation of Pontrygin's Maximum Principle for the case considered herein, takes on the following form (by neglecting small terms $o(\delta u)$).

$$I\left(u^{o}+\delta u^{\prime}\right)=I\left(u^{o}\right)+\min_{\delta u}\left\{\int_{0}^{T}\langle Su,\delta u\rangle dt\right\}$$ (17)

with constraints :

$$d(u^{o})+\delta d(u^{o},\delta u^{\prime})=0;$$

$$\{h(u^{o})+\delta h(u^{o},\delta u^{\prime})\leq 0\};$$ (18)

$$u_{l}\leq u^{o}+\delta u^{\prime}\leq u_{u};$$

$$\left\|\delta u^{\prime}\right\|_{c}\leq\lambda$$

Due to non-optimality of control $u^{o}(\ )$, the solution of the minimisation problem (17), (18) implies the existence of small variation $\delta u^{\prime}$, such that $I(u^{o}+\delta u^{o})<I(u^{o})$. The process of minimisation is then iterated for control $u^{\prime}=u^{o}+\delta u^{\prime}$. A sequence of controls u^{k}, $k=0,1,2,...$ and their corresponding state trajectories x^{k} are obtained, as a result of successive solving the problems (17), (18). The convergence of the above method may by proved by using Arzeli's theorem [2]. To solve the optimisation problem (17), (18) numerically, the control $u(t)$ is assumed in this work to belong to a class of quasiconstant mappings. The internal $[0,T]$ is divided into $N>0$ subintervals, where in each of these subintervals the control $u(t)$ takes on a constant value. In this way, the problem of determining a sequence of controls u^{k}, is reduced to solving linear programming problems.

4. Computer example

Two vehicles shown in figure 1 are considered. There is one obstacle O_{l} in the work space.

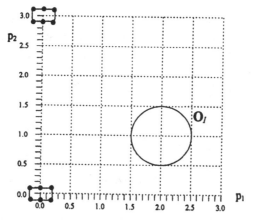

Fig 1. A scheme of vehicles and obstacles O_{l}

Data used in the numerical example are as follows:
- components of dynamic models of vehicles:

$$M = I_6, \quad C = 3I_6, \quad F = 2I_6, \quad H = I_6 \; ;$$

- geometrical models of vehicles

$$V^i(\mathbf{q}_i) = \left\{ \mathbf{p}_i : 1 - \sum_{l=1}^{2} \left(AR_i \left[\mathbf{p}_i - \binom{q_i^1}{q_i^2} \right] \right)_l^4 \geq 0 \right\}; i = 1:2 \; ;$$

$$A = \begin{bmatrix} 5 & 0 \\ 0 & 10 \end{bmatrix}; \quad R_i = \begin{bmatrix} \cos q_i^3 & \sin q_i^3 \\ -\sin q_i^3 & \cos q_i^3 \end{bmatrix};$$

notation $(\ )_l$ means the l-th component of the vector $(\ )$;
- boundary equation of obstacle O_l :

$$b^i(\mathbf{p}) = \left\langle \mathbf{p} - \binom{2}{1}, \ \mathbf{p} - \binom{2}{1} \right\rangle - 0,25 = 0 \; ;$$

- discrete models of vehicles $V^i(\mathbf{q})$, $i=1,2$, taken for the numerical simulations are shown in Figure 1 (marked dots);
- constrains (2)-(6) correspond to $\mathbf{q}_0 = (0\ 0\ 0\ 0\ 3\ 0)^T$; $T=1$; $\mathbf{q}_T = (3\ 3\ 1\ 3\ 0\ 0)^T$; $\mathbf{u}_l = (-50\ -50\ -50\ -50\ -50\ -50)^T$; $\mathbf{u}_u = (50\ 50\ 50\ 50\ 50\ 50)^T$; $N = 10$;
- initial admissible control $\mathbf{u}^0(t)$ is chosen to fail state constraints (12)-(13); $\varepsilon = 0$; was gradually decreased from 7,0 to 0,005; convergence criterion (for sequence) $I(\mathbf{u}^k)$ equals 0,05.
   Figure 2a shows the initial trajectories of the vehicles corresponding to control $\mathbf{u}^0(t)$. As it is shown from the Figure 2a, trajectory $\mathbf{x}^0(t)$ does not satisfy boundary and collision-free condition at the initial iteration of the algorithm.
   Figure 2b presents optimal vehicle trajectories obtained after 33 iterations of working the algorithm.

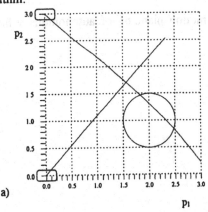

a)

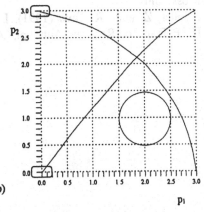

b)

Fig. 2. Initial (a) and energetic-optimal (b) vehicle trajectories

5. Conclusions

In the paper, a method of optimal path planning for multiple vehicles has been presented. It is based on the use of negative formulation of Pontryagin's Maximum Principle. An important factor that affects the velocity of determining the optimal solution is a possibility of using the parallel processors to calculate Frechet's and Gateaux's differentials. In contrast to the method of penalty function, the solution proposed herein does not require initial vehicle trajectories, which must satisfy the state inequality constraints.

6. References

[1] Khatib, O. (1986). Real - time obstacles avoidance for manipulators and mobile robots. The J. Rob. Res., vol. 5, 1, 90 - 98.

[2] Lyusternik, L. A. and V. I. Sobolev (1951). Elementy funktionalnogo analiza, Gosudarstv. Izdat., Moskva.

[3] Pontryagin, L. S. (1961). Matematiceskaya teoriya optimalnych procesov, Nauka, Moskva.

[4] Kim, J. and P. K. Khosla (1992). Real - time obstacle avoidance using harmonic potential functions, IEEE Trans. on Rob. and Aut., vol. 8, 338 - 349.

[5] Rimon, E. and D. E. Koditschek (1992). Exact robot navigation using artificial potential function, IEEE Trans. on Rob. and Aut., vol. 8, 501 - 518.

[6] Singh, S. K. and M. C. Len (1991). Manipulator motion planning in the presence of obstacles and dynamic constraints, The J. Rob. Res., vol 10, 177 - 187.

[7] Dubrovicki, A. J. and A. A. Miliutin (1965). Zadaci na ekstremum pri nalici ograniceniy, Zurnal Vycislitelnojy Mat. i Mat.-Fizyki, Tom 5, 3, 395 - 453.

[8] Fedorenko, R. R. (1978). Pribliziennye resenie zadac optimalnogo upravlenia, Nauka, Moskva.

[9] Gilbert, E. G. and D. W. Johnson (1985). Distance function and their applications to robot path planning in the presence of obstacles, IEEE J. Rob. and Aut., vol. 1, 1, 21-30.

[10] Girsanov, I. V. (1970). Lekcii po matematiceskoy teorii extremalnych zadac, Izdat. Mosk. Univ.

[11] Schiller, Z. and R. Gwo (1991). Dynamic motion planning of autonomous vehicles, IEEE Trans. on Rob. and Aut., vol, 241 - 249.

STIFFNESS CONTROL OF AN OBJECT GRASPED
BY A MULTIFINGERED HAND

K. Tanie and H. Maekawa

Mechanical Engineering Laboratory, AIST-MITI, Tsukuba, Japan

Y. Nakamura

NTT Human Interface Laboratories, Tokyo, Japan

1. Introduction

When a robotic manipulator is required to carry out tasks during which the manipulated object will receive constraint forces from the environment, adjusting the stiffness condition of the object grasped or the end effector is necessary. To achieve this function, several stiffness control laws or special accommodation devices like remote center compliance(RCC) have been developed. In robotic arms with a simple gripper, the stiffness control is done by adjusting the joint displacements according to the external forces applied to the grasped object or the end effector, and vice versa. However, when a hand with fingers which have several joints actively controlled is used for the end effector, the object stiffness control can be done by controlling the finger joints cooperatively. This paper describes this issue and propose an object stiffness control method for multifingered hands.

In order to achieve the stiffness control of an object grasped, the object has to be stably grasped by a multifingered hand. Therefore, a stiffness control law which will be implemented has to simultaneously satisfy the stable grasping condition. From this requirement, the proposed stiffness control law was implemented on the system, adding to the grasping control law previously proposed [1].

There are several research activities in this area. A typical way proposed so far by researchers for the object stiffness control is to adjust forces and torques applied to the grasped object through fingertips according to the object displacement caused by the external force and torque so that a desired object stiffness condition will be achieved[2][3][4]. In such a case, generally, each finger tip forces and torques will be explicitly controlled. In order for a multifingered hand to grasp an object stably without manipulation, at least both summations of the fingertip forces applied to the object and the moments produced by the forces have to be equal to zero, in which, of course, the fingertip forces should be generated in an appropriate way to support the gravity force applied to the object through the friction forces. If manipulating an object is required, the fingertip forces also have to be generated to produce the desired object motion through the object dynamics, keeping the above conditions.

However, to achieve the stable grasp which is robust against disturbances, controlling not only forces, but also stiffness at the fingertip, is very important. To satisfy this requirement, several multifingered hands in which the fingertip stiffness can be adjusted have been developed [5][6]. In such multifingered hands, the stiffness control or the impedance control is used for each fingertip motion control. The proposed object stiffness control was implemented on such a multifingered system. To simplify the analysis, it is assumed that each fingertip contacts a grasped object at a point with no slipping.

The following sections will describe briefly the stable grasp theory for the multifingered system with the fingertip stiffness control capability and its control system structure, and how to implement the object stiffness control law on the system. The experimental results will also be shown to confirm the effectiveness of the proposed control method.

2. Stable Grasp in a Multifingered Hand with Stiffness Adjustment Capability

When each finger will apply the force to the object through the fingertip to hold the object in the static way, the equilibrium conditions in the forces and moments applied to the object by the fingertip forces and the gravity force must be satisfied. However, satisfying only this equilibrium condition is not enough to obtain the robust stable grasp. If some external disturbance forces will be applied to the object, the equilibrium condition will not be satisfied and the grasping is not achieved. Introducing an object displacement feedback which can produce a compensatory forces, the equilibrium condition may be kept, even if under the external disturbances. However, it will be difficult to implement such a control method practically because of the difficulty of obtaining the exact model of the real system and also the inadequate accuracy of the practical force control system. In general, the object motion caused by the unstable grasp can not be exactly measured and it makes it difficult to generate the fingertip forces which exactly satisfy the equilibrium conditions. Even if the geometrical parameters could be exactly measured, the inadequate accuracy of the force control system makes it difficult to apply the fingertip forces to the object along exact directions. This also makes another reason for the difficulty of achieving the robust stable grasp in practical systems by the feedback method.

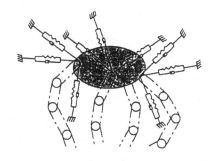

Fig. 1 Multifingered hand system with stiffness adjustable fingertip

In order to solve this problem, as shown in Fig. 1, the stiffness adjustable capability is introduced at each fingertip as well as applying the forces to the object. In such a system, the object can be considered as if it is supported by a three dimensional spring whose stiffness matrix K_{obj} is the function of the components of the stiffness matrix k_i ($i=1..n$; n is the number of finger) at each fingertip. So, if K_{obj} is always positive definite, under the application of some external disturbance forces to the object, the finger system can generate the proper restoring forces against the disturbance to keep stable grasp. K_{obj} can be written as the following form;

$$K_{obj}=K_{obj}(k^1{}_{11}.., k^2{}_{11}., ..., k^n{}_{11}.., \delta f, x_1, x_2,..) \tag{1}$$

where $k^i{}_{qr}$ is the component at q-th column and r-th raw in the i-th fingertip's stiffness matrix, δf is the force vector applied to the object which is produced by the unbalanced components in each fingertip force and is given by the summation of the fingertip force vectors, $x_1, x_2..$ are geometrical parameters of the grasping system, and n is the total number of fingers. Using equation (1), the stabile grasp condition can be written by

$$K_{obj} > 0 \tag{2}$$

The detail structure of equation (1) is given in reference[6]. To realize the stable grasp condition, a control system shown in Fig. 2 can be introduced. Suppose that a

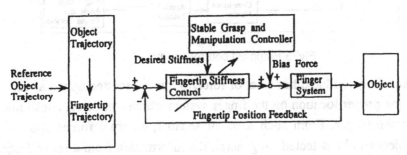

Fig. 2 Control System for the multifingered hand with stiffness adjustable fingertip

multifingered hand whose finger joints are driven by some actuators like DC servo motor. Each finger has a position controller with a fingertip position feedback loop. In such a system, the loop gain determines the fingertip stiffness. Therefore, the loop gains are adjusted to maintain the stable grasp condition (2). Simultaneously, proper bias forces which will approximately satisfy the equilibrium conditions to hold an object through the friction between the fingertip and the object are generated at the fingertip by controlling the actuators. Adjusting the reference to the position control system, also the object can be manipulated along the trajectory. The reference for the fingertip motion control can be calculated from the desired object trajectory, if the geometrical structure of the grasping system will be known. The details of this algorithm is given in reference [1][6].

3. The Proposed Object Stiffness Control Method

Though the method described in Section 2 is useful to realize the robust grasping function, it also provides another benefit for controlling non constraint motions of the finger efficiently. When the robot will move in space without any constraints, the position control must be used to control the motion. Because of this reason, the multifinger system using the explicit force control for the finger motion control will be required to switch the control mode from the position control to the force control and vise versa, according to the change of constraint situations the robot will have. Since the stiffness control described in Section 2 is based on the position control, it can be applied to both of controlling the motion in free space and constraint space. With this benefit also in mind, the proposed object stiffness control was implemented. The structure of the proposed control method is shown in Fig. 3. It will be assumed

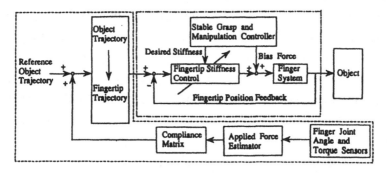

Fig. 3 Proposed control system

that the multifingered hand has a set of force/torque sensors equipped at each finger joint or some proper location on the finger system which can detect constraint forces applied to the finger. From such a set of sensors, external forces applied to the grasped object can be detected. In general, the information from a set of finger joint torque sensors include both of the grasping forces, which are corresponding to the internal forces, and the external forces. Therefore, in order to know the external forces, how to discriminate it from the internal forces has to be considered. This can be done by considering the equilibrium condition of the object applied by the internal forces and the external forces. After obtaining the external forces, it will be multiplied by a desired object compliance matrix (equal to the inverse stiffness matrix) given according to the required task. The result will be the object displacement which should be generated to realize the desired object stiffness. Adding it to the reference trajectory, finally, the object stiffness control can be achieved.

In order to formulate the reference trajectories, a grasping system shown in Fig. 4 is considered. In Fig. 4, let's set absolute coordinate system, object coordinate system and reference coordinate system which describes the desired trajectory for the object

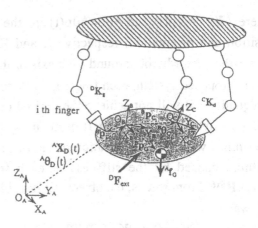

Fig. 4 Structure of the grasping
system

and call them as O_A-$X_A Y_A Z_A$, O_B-$X_B Y_B Z_B$, and O_D-$X_D Y_D Z_D$, respectively, and write $^J X$ (J=A, B or D) for a vector X defined in each coordinate system. $R(^K\theta_L) \in R^{3\times3}$ is the rotational matrix from the coordinate system L to K. Also, let's define the desired object stiffness matrix $^C K_d$ in the coordinate system O_C-$X_C Y_C Z_C$ whose location and orientation is $^B P_C \in R^3$, $^B\theta_C \in R^3$, respectively, in the object coordinate system and which is called as the stiffness coordinate system. From the equilibrium conditions of the grasped object, the external force $^D f_{ext} \in R^3$ and the external moment $^D n_{ext} \in R^3$ can be obtained by the following equations

$$^D f_{ext} = \sum_{i=0}^{n} R^{-1}(^A\theta_D)^A F_i - R^{-1}(^A\theta_D)^A f_G \tag{3}$$

$$^D n_{ext} = \sum_{i=0}^{n}\left[R^{-1}(^D\theta_B)^B P_i \times R^{-1}(^A\theta_D)^A F_i \right]$$
$$-R(^D\theta_B)^B P_G \times R^{-1}(^A\theta_D)^A f_G \tag{4}$$

where $^A F_i$, $^B P_i$, $^A f_G$, $^B P_G \in R^3$ are the force generated at the i-th fingertip ($0 < i \leq n$, n: total number of fingers), the fingertip position vector, the gravity force vectors applied to the object and the position vector for the center of gravity of the object, respectively, all of which can be assumed to be known. If a desired object stiffness matrix is defined as $^D C \in R^{6\times6}$, the required object displacement $^D\delta r = [^D\delta x^T, ^D\delta\theta^T]$ is given as follows;

$$^D\delta r = [^D\delta x^T, ^D\delta\theta^T] = {}^D C^D F_{ext} \tag{5}$$

where the transposed matrix is shown with superscript "T" and $^D F_{ext} = [^D f_{ext}, ^D n_{ext}]$. Adding the displacement shown in Equation (5) to the object reference and applying the inverse kinematic calculation, the fingertip trajectory which satisfies the desired object stiffness can be obtained as

$$^A H_{di}(t) = {}^A X_D + R(^A\theta_D)^D\delta x + R(^A\theta_D)R(^D\delta\theta)^B P_i \tag{6}$$

where $^A X_D = ^A X_D(t)$ and $^A \theta_D = ^A \theta_D(t)$ are the reference trajectories for the object position and orientation, respectively, and $R(^D \delta\theta)$ is the rotational matrix which rotates the vector by $\delta\theta$ around each axis of the reference coordinate system.

In the proposed system, each fingertip is stiffness-controlled. So, the reference given by equation (6) will not achieve the desired object stiffness $^D C$, because the motion will be affected by the fingertip stiffness. To solve this porblem, $^D C$ must be determined with the consideration of the fingertip stiffness. Let's indicate the object stiffness caused by the stiffness at each fingertip and the grasping forces as $^D K_{obj} \in R^{6 \times 6}$. From statics, the object stiffness $^D K_d$ will be given using $^D K_{obj}$ and $^D C$ as follows;

$$^D K_d = [I + ^D K_{obj} {}^D C]^{-1 D} K_{obj} \qquad (7)$$

where $I \in R^{6 \times 6}$ is the unit matrix. Therefore, $^D C$ must be determined so that $^D K_d$ satisfy the desired object stiffness matrix. The desired stiffness will be defined in the stiffness coordinate system. So, converting $^D K_d$ to $^C K_d$, the equation which shows how to set $^D C$ can be obtained as

$$^D C = \left[Q^T(^B\theta_C, {}^B P_C) {}^C K_d Q(^B\theta_C, {}^B P_C) \right]^{-1} {}^D K_{obj}^{-1} \qquad (8)$$

where

$$Q(^B\theta_C, {}^B P_C) = \begin{bmatrix} R^T(^B\theta_C) & -R^T(^B\theta_C)G(^B P_C) \\ 0 & R^T(^B\theta_C) \end{bmatrix} \qquad (9)$$

and if $^B P_C = [^B P_{Cx}, {}^B P_{Cy}, {}^B P_{Cz}]$ and $0 \in R^{3\times3}$,

$$G(^B P_C) = \begin{bmatrix} 0 & -^B P_{Cz} & ^B P_{Cy} \\ ^B P_{Cz} & 0 & -^B P_{Cx} \\ -^B P_{Cy} & ^B P_{Cx} & 0 \end{bmatrix} \qquad (10)$$

4. Experiments

To confirm the method described in the above, the experiments were carried out using a two fingered hand so far developed [7]. A finger hand used in the

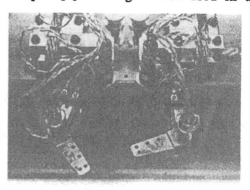

Fig. 5 Two-fingered system
using experiments

experiments is shown in Fig. 5 which moves in a horizontal plane. Each finger has three joints, each of which is driven by a geared DC motor through a tendon and pulleys. In the experiments, one joint in each finger nearest to the base was fixed and only two joints in each finger were used. At each joint, a torque sensor and an angular displacement sensor was installed. With those sensors' feedback, the fingertip stiffness control system was constructed. Also, the joint torque sensors were used to measure the applied forces to the grasped object. There were two kinds of experiment planned. The first experiment was to measure the stiffness ellipsoid of the object under various stiffness conditions. To do it, a constant force(2.94 N) was applied to a point at the center of the object along various directions and the displacements of the object were measured through the joint angular sensors. The second experiment was relating to a peg-in-hole task. A rectangular solid object with the width 25.0mm was grasped by the hand and was inserted into or drawn from a rectangular shape hole with the width 25.6mm, after a proper stiffness matrix has been set on the object. The results of the first experiments are shown in Fig. 6. In the

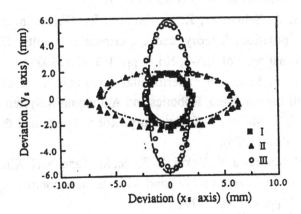

Fig. 6 Experimental results

experiments, the stiffness coordinate system was set at the center of the line which connects with the contacts between the fingertip and the object. As the object was not moved, the object coordinate system coincides with the stiffness coordinate system. For $^DK_{obj}$, diag(1500N/m, 1500N/m, 0.88Nm/rad), in which diag(a,b,c...) means the diagonal matrix with components a, b, c,..., was used, and diag(10^9N/m, 10^9N/m, 10^9Nm/rad), diag(750N/m, 10^9N/m, 10^9Nm/rad) and diag(10^9N/m, 750N/m, 10^9Nm/rad) were used in experiments (I), (II) and (III), respectively, for $^DK(=^DC^{-1})$. The dotted lines show the calculated ellipsoids. The measured ellipsoids were approximately same as the calculated ones. From those results, it was confirmed that the proposed method worked well. Also, in the second experiments, the peg-in-hole tasks with several stiffness conditions were carried out, and it was confirmed that each trial generated proper behaviors of the object motions.

5. Conclusions

The paper proposed a method of controlling the object stiffness grasped by a multifingered had. The method was implemented together with the stable grasp control law previously proposed. Through the experiments using a planer two fingered hand, the effectiveness of the method was confirmed. For future works, expansion of the method from two dimensional to three dimensional case will be necessary.

References

[1]H. Maekawa, K. Yokoi, K. Tanie, M. Kaneko, N. Kimura, N. Imamura, Position/stiffness based manipulator by three-fingered robot hand, Proc. of International Symposium on Advanced Robot Technology, Tokyo, pp.597-603(1991).

[2]Y. Nakamura, K. Nagai, T. Yoshikawa, Mechanics of coordinative manipulation by multiple robotic mechanisms, Proc. of 1987 IEEE International Conf. on Robotics and Automation, pp.991-998(1987)

[3]S. A. Schneider, R. H. Cannon, Jr., Object impedance control for cooperative manipulation: Theory and experimental results, IEEE Trans. on Robotics and Automation, Vol. RA-8, No. 3, pp. 383-394(1992).

[4]G. P. Starr, An experimental investigation of object stiffness control using a multifingered hand, Robotics and Autonomous Systems 10, pp. 33-42(1992).

[5]M. T. Mason, J. K. Salisbury, Robot Hands and the Mechanics of Manipulation, MIT Press(1985).

[6]H. Maekawa, K. Yokoi, K. Tanie, M. Kaneko, N. Kimura, N. Imamura, Development of a three-fingered robot hand with stiffness control capability, Mechatronics, Vol. 2, No. 5, pp. 483-494(1992).

[7]N. Imamura, M. Kaneko, K. Yokoi, K. Tanie, Development of a two-fingered robot hand with compliance adjustment capability, Proc. of 1990 Japan-U.S.A. Symposium on Flexible Automation, pp. 997-1004(1990).

Chapter VIII
Control of Motion II

PRACTICAL STABILISATION OF ROBOTS INTERACTING WITH DYNAMIC ENVIRONMENT BY DECENTRALISED CONTROL

D. Stokic

ATB - Institute for Applied System Technology, Bremen, Germany

M. Vukobratovic

M. Pupin Institute, Belgrade, Yugoslavia

1. Introduction

Although a high number of papers[1-8] has been devoted to the problem of the so-called constrained motion control of robots, the model uncertainties, representing the crucial problem in control of robots interacting with the dynamic environment, still have not been appropriately addressed. Especially the uncertainties in a dynamic model of the environment in different technological tasks may have high influences, due to difficulties in the identification/prediction of the environment parameters and behaviour of the environment. Therefore, it may be difficult to achieve the asymptotic (exponential) stability of the system. It is of practical interest to require more relaxed stability condition, i.e. to consider the so-called practical stability of the system. An approach to analysis of the practical stability of manipulation robots interacting with the dynamic environment has been presented in our previous papers[9-10]. In this paper a new approach is presented following the basic idea of a decomposition/aggregation method for the stability analysis of large-scale systems in which the system is decomposed into 'subsystems', but without ignoring dynamic interactions among these 'subsystems'. The objective of the paper is to establish for the first time less conservative conditions for the practical stability of the robots which are suitable for the analysis and synthesis of decentralised control laws.

2. Model of Robot Interacting with Dynamic Environment

The complete dynamic model of the robot with n degrees of freedom and the dynamic model of the environment (described by the second-order differential equations) are considered in Cartesian space. The model of dynamics of the robot can be written in the form:

$$\Lambda(p, d)\ddot{p} + \rho(p, \dot{p}, d) = J^{-T}(p, d)\,\tau + F \qquad (2.1)$$

where $p = p(q)$ is the nx1 vector of the robot Cartesian coordinates, while q is the nx1 vector of the robot internal coordinates, $\Lambda(p, d)$ is the nxn inertia matrix, $\rho(p, \dot{p}, d)$ is the nx1 non-linear vector function of Coriolis, centrifugal and gravity moments, d is the $l \times 1$ vector of parameters which belongs to the constrained set $\mathbf{D}$, $J(p, d)$ is the nxn Jacobian matrix, τ is the nx1 vector of driving torques (inputs), F is the mx1 vector of Cartesian forces, generalised interaction forces (forces and moments) acting upon the end-effector of the robot. For the sake of simplicity we shall consider the second order models of actuators, which are assumed to be included in the robot dynamic model. In this paper we shall consider the case n = m.

The model of dynamic environment can be written in the form:

$$M(p, d)\ddot{p} + L(p, \dot{p}, d) = -F \qquad (2.2)$$

where M is the nxn matrix, L is the nx1 non-linear vector function. It is assumed that all mentioned matrices and vectors are continuous functions of their arguments. The model of the robot in the state space can be defined in the following form:

$$\dot{x} = f(x, d) + B(x, d)\,\tau + G(x, d)\,F \qquad (2.3)$$

where $x = (p^T, \dot{p}^T)^T$ is 2n x 1 state vector, $f(x, d)$ is 2nx1 vector function, $B(x, d)$ is 2nxn matrix, $G(x, d)$ is 2n x m matrix.

3. Definition of Control Task

Let us assume that in m_1 directions ($m_1 < m$) desired force trajectories $F^{01}(t)$ are specified, where $F^{01}(t)$ is m_1 x1 vector, while in n_1 directions desired trajectory $x^1(t)$ is specified, where $x^1(t)$ is n_1x1 and where $n_1 + m_1 = n$. Note that the meaning is following: in some directions only forces are specified, in other directions only desired position trajectories are specified. Let us introduce the following notations: $p^0(t) = (p^{01T}(t), p^{02T}(t))^T$, where $p^{02}(t)$ is $(n - n_1) \times 1$ vector of the nominal trajectories of the Cartesian coordinates in the directions in which force trajectories are specified. Note that the trajectories $p^{02}(t)$ are not specified in advance, but have to be determined based on the environment model. Similarly the vector of desired force of the nominal force trajectories in directions in which forces are acting upon the robot, but the nominal trajectories of the Cartesian space coordinates are specified. The force trajectories $F^{02}(t)$ are not specified in advance, but have to be calculated based on the dynamic model of environment. The nominal trajectories of the forces and of the Cartesian coordinates must satisfy the model of the environment i.e.:

$$M (p^0(t)) \ddot{p}^0(t) + L(p^0(t), \dot{p}^0(t)) = - F^0(t) \tag{3.1}$$

Let us introduce the following notations:

$$M = \begin{bmatrix} M_{11} M_{12} \\ M_{21} M_{22} \end{bmatrix} \quad \text{and} \quad L = [L_1^T, L_2^T]^T \tag{3.2}$$

where dimensions of matrix M_{11} are $m_1 \times n_1$, of M_{12} are $m_1 \times (n-n_1)$, of M_{21} are $(m-m_1) \times n_1$, M_{22} are $(m - m_1) \times (n-n_1)$, while L_1 and L_2 are vectors of dimensions $m_1 \times 1$ and $(m-m_1) \times 1$ respectively. The desired trajectory in the state space is denoted by $x^0(t) = (p^{0T}(t), \dot{p}^{0T}(t))^T$.

Now, the control task is specified in the following form, as a task of practical stability of the robot around the nominal trajectory $x^0(t)$: The control of robot has to ensure that $\forall x(0) \; \varepsilon \; X^I$ and $\forall d \; \varepsilon \; D$ imply $x(t) \; \varepsilon \; X^t(t)$, $\forall t \varepsilon T$, where X^I and $X^t(t)$ are the finite regions in the state space around the prescribed nominal trajectory $x^0(t)$ and $T = (t, t \varepsilon \; (0, \; t_1))$, t_1 is the pre-defined time period. It is assumed that $x^0(0) \; \varepsilon \; X^I$, $x^0(t) \; \varepsilon \; X^t(t)$, $\forall t \; \varepsilon \; T$, $X^t(0) \supset X^I$. This formulation of the control task can be interpreted in the following way. Given the desired trajectory $x^0(t)$, an initial error around the nominal trajectory is allowed in such a way that the state $x(0)$ must belong the pre-defined region X^I around $x^0(0)$; the control has to ensure that the robot state follows desired trajectory $x^0(t)$ with an allowed error which is constrained by requirement that the state of the robot $x(t)$ must belong to the pre-defined region $X^t(t)$ around the nominal trajectory $x^0(t)$, within the pre-defined time period T. This has to be guaranteed for each allowable values of parameters d. Due to (3.1), and as shown in[6-7], a fulfilment of the specified control task also guarantees tracking of the desired force trajectories $F^0(t)$, i.e. it guarantees that $F(0)\varepsilon F^I$ and $\forall d\varepsilon D$ imply $F(t)\varepsilon F^t(t)$, $\forall t\varepsilon T$, where F^I and $F^t(t)$ are regions in the $m \times 1$ space around the nominal trajectory $F^0(t)$. Note that regions F^I and $F^t(t)$ must correspond to the regions X^I and $X^t(t)$, respectively. In order to simplify stability analysis let us consider specific forms of the finite regions X^I and $X^t(t)$: $X^I = \{x (0): \|\Delta x(0)\| < \underline{X}^I\}$, $X^t(t) = \{x (t): \|\Delta x(t)\| < \underline{X}^t exp(- \alpha t)\}$, $\forall t \; \varepsilon \; T$, $F^I = \{F(0): \|\Delta F(0)\| < \underline{F}^I\}$, $F^t(t) = \{F(t): \|\Delta F(t)\| < \underline{F}^t exp(- \beta t)\}$, $\forall t \; \varepsilon \; T$, where $\underline{X}^t > \underline{X}^I > 0$, $\underline{F}^t > \underline{F}^I$, $\alpha > 0$, $\beta > 0$. Here $\underline{X}^t$, $\underline{X}^I$, $\underline{F}^t$, $\underline{F}^I$, α, β denote

is 2n x 1 vector of the state deviation around the desired nominal trajectory $x^0(t)$, i.e. $\Delta x(t) =$ $x(t) - x^0(t) = (\Delta p^T(t), \Delta \dot{p}^T(t))^T$, while $\Delta F(t)$ is n x 1 vector of the force deviation around the desired force trajectory.

4. Control Law

Dynamic position/force control law is considered:

$$\tau = U^*(p, \dot{p}, \ddot{p}_c, F) \tag{4.1}$$

where $U^*(p, \dot{p}, \ddot{p}, F) = J^{*T} (\Lambda^* \ddot{p} + \rho^* - F)$, where J^*, Λ^*, ρ^* denote matrices and vector corresponding to J, Λ, ρ from the model (2.1) but with the assumed parameters values $d = d_0 \varepsilon D$. This means that we assume that the parameters values are not accurately known.

$$\ddot{p}_c = \begin{bmatrix} \ddot{p}^{01} + P_1(\Delta p^1, \Delta \dot{p}^1) \\ W^*(p, \dot{p}, \ddot{p}^{01} + P_1(\Delta p^1, \Delta \dot{p}^1), F^{01} + \Delta F^1) \end{bmatrix}$$

$$W^*(p, \dot{p}, \ddot{p}^{01}, F^{01}) = M_{12}^{*-1}(-F^{01} - M_{11}^* \ddot{p}^{01} - L_1^*)$$

$$\Delta F^1 = K^{1F} \int (F^1(t) - F^{01}(t)) dt, \quad P_1(\Delta p^1, \Delta \dot{p}^1) = K_1^1 \Delta p^1 + K_2^1 \Delta \dot{p}^1$$

where M_{11}^*, M_{12}^*, L_1^* denote matrices and vector corresponding to M_{11}, M_{12}, L_1 from the model (3.2) but with the assumed parameters values d_0. F^1 is m_1x1 sub-vector of force vector F for which the nominal trajectories are specified in advance, $F = (F^{1T}, F^{2T})^T$, $\Delta p^1 = p^1(t) - p^{01}(t)$ is n_1x 1 vector, K^{1F} is $m_1 \times m_1$ matrix of force feedback gains, K_1^1 and K_2^1 denote n_1x n_1 matrices of the position and velocity feedback gains, respectively. For the sake of simplicity we shall assume that the both matrices are diagonal. Note that contrary to the so-called classical hybrid control schemes[1,2], the control law (4.1) takes into account complete dynamic models of the robot and environment as well as interaction among directions in which position is controlled and directions in which force is controlled. This means that both position and force feedback loops are used in all directions. The control law may be considered in a more general form, i.e. P_1 and ΔF^1 can be defined in a more general form.

5. Stability Conditions for Proposed Control Law

The closed loop model of the robotic system (model of deviation around the desired nominal trajectory $x^0(t)$) in the state space is obtained by combining the robot model in the state space (2.3) and the corresponding control law (4.1):

$$\text{...} \tag{5.1}$$

where $\Delta f\,(\Delta x, x^{0*}, d)$ is 2nx1 vector and $\Delta G(\Delta x, d)$ is 2nxn matrix. In the previous papers[6,7] it has been shown that application of the control law (4.1) ensures desired tracking of the prescribed nominal force trajectory (i.e. desired transient behavior of the force F(t)), if an ideal model of the system used in the control law is assumed. Since the model of the system (robot + environment) in (4.1) is not ideal (due to parameters uncertainties), the desired force transient process cannot be perfectly achieved. However, it can be shown that by appropriate selection of the force feedback gains, and under assumption of the limited deviations of the assumed model parameters from the actual values, the control law (4.1) can guarantee that the force transient process satisfies the conditions of the practical stability, i.e.

$$\|\Delta F(t)\| < \underline{F}^t \exp(-\beta t) \tag{5.2}$$

Starting from the assumption that the force transient process satisfies (5.2), it has to be examined whether the control law (4.1) can ensure the practical stability of the over-all system. This means that the conditions, under which the proposed control law fulfils the specified control task, i.e. the conditions of the practical stability of the robotic system have to be derived. It can be shown[11-13] that the system is practically stable with respect to $(X^I, X^t(t), T)$ as defined above, if there exist a real valued continuously differentiable function $v(t, x)$ and a real valued function of time $\Psi\,(t)$ which is integrable over the time interval T such that

$$\dot{v}\,(t, x) \leq \Psi\,(t), \; \forall x \varepsilon \tilde{X}^t\,(t), \; \forall t\,\varepsilon\,T \tag{5.3}$$

$$\int_0^t \Psi(t')\,dt' < v_m^{\partial x^t}\,(t) - v_M^{\partial x^t}\,(0), \; \forall t\,\varepsilon\,T \tag{5.4}$$

$\partial X(t)$ denotes the boundary of the corresponding region and $\tilde{X}^t\,(t) = X^t(t) - \underline{X}^I \exp(-\alpha t)$. In (5.3) $\dot{v}$ denotes the time derivative of the function $v(t, x)$ along the solution of the closed-loop system. v_m and v_M denote minimum and maximum values of v at the corresponding boundaries of the regions, respectively. Note that the conditions (5.3) and (5.4) are sufficient but not necessary conditions for the practical stability of the system. The proof of the above stated method for the testing of practical stability is provided in previous papers[11-13].

Following the aggregation/decomposition approach to system stability analysis we may decompose the system (5.1) into two 'subsystems': the position controlled part and force controlled part:

$$\Delta \dot{x}^i = A_{ii}\,\Delta x^i + \Delta f_i(\Delta x, x^{0*}, \Delta F, F^0, d) + \Delta G_i\,(\Delta x, d)\,F, \; i = 1,2 \tag{5.5}$$

where $\Delta x^i(t) = x^i(t) - x^{0i}(t) = (\Delta p^{iT}(t), \Delta \dot p^{iT}(t))^T$ and Δf_i and ΔG_i are vectors and matrices of the appropriate dimensions, while the matrices A_{ii} are given by:

$$A_{11} = \begin{bmatrix} 0 & I_{n1} \\ \overline{Q}_{11} K_1^1 & \overline{Q}_{11} K_2^1 \end{bmatrix}, \quad A_{22} = \begin{bmatrix} 0 & I_{n2} \\ \overline{Q}_{22} L_1^P & \overline{Q}_{22} L_1^V \end{bmatrix}$$

where $\overline{Q}_{11}$ and $\overline{Q}_{22}$ are $n_1 \times n_1$ and $n_2 \times n_2$ estimates of $Q_{11} = \Lambda_1^{-1} J^{-T} J^* \Lambda_1^*$ and

$Q_{22} = \Lambda_2^{-1} J^{-T} J^* \Lambda_2^*$ (where $\Lambda_1^{-1}, \Lambda_2^{-1}, \Lambda_1^*, \Lambda_2^*$ are appropriate submatrices of the matrices Λ^{-1}

and Λ^*). L_1^P and L_1^V represent estimates of the stability factors in the dynamic model of the

environment, i.e. $M_{12}^{*-1} L_1^* - M_{12}^{0-1} L_1^0 = L_1^P \Delta p^2 + L_1^V \Delta \dot p^2 + L_1^R$.

Let us consider the practical stability of the decoupled subsystems $\Delta \dot x^i = A_{ii} \Delta x^i$. We may

assume that the regions of practical stability can be presented in the form

$X^I = X^{I(1)} \times X^{I(2)}$ and $X^t = X^{t(1)} \times X^{t(2)}$, where $X^{I(i)} = \{x^i (0): \|\Delta x^i(0)\| < \underline{X}^{I(i)}\}$, $X^{t(i)}(t) = \{x^i$

(t): $\|\Delta x^i(t)\| < \underline{X}^{t(i)} \exp(- \alpha_i t)\}$, $\forall t \; \epsilon \; T$, where $\underline{X}^{t(i)} > \underline{X}^{I(i)} > 0$, $\alpha_i > 0$. Let us select the

functions $v_i(t, x)$ (i = 1,2) in the form: $v_i(t, x) = (\Delta x^{iT} H_i \Delta x^i)^{1/2}$, i =1,2, where H_i are the

positive definite matrices of the appropriate dimensions. The derivative of the functions v_i

along the solutions of the decoupled subsystems can be written as:

$$\dot v_i(t, x) = (\text{grad} v_i)^T A_{ii} \Delta x^i \leq - \gamma_i v_i \leq - \gamma_i' \|\Delta x^i\| \tag{5.6}$$

where $\gamma_i = - \min |\lambda(A_{ii})|$, $\lambda(A_{ii})$ denotes eigen-values of the corresponding matrix, $\gamma_i' = \gamma_i \cdot$

$\lambda_m(H_i)$. The conditions (5.6) are valid under assumption that the matrices H_i are selected to

satisfy $H_i A_{ii} + A_{ii}^T H_i \leq - 2 \gamma_i H_i$. The coupling members in the subsystems (5.5) can be

estimated in the following form:

$$(\text{grad} v_i)^T \Delta f_i (\Delta x, x^{01}, \Delta F, F^{01}, d) < \sum_{j=1}^{2n_1} \xi_{ij}^1 |x_j^{01}(t)| + \sum_{j=1}^{m_1} \xi_{ij}^2 |F_j^{01}(t)| + \sum_{j=1}^{2n_1} \xi_{ij}^3 |\dot x_j^{01}(t)|$$

$$+ \sum_{j=1}^{2n} \xi_{ij}^4 |\Delta x_j| + \sum_{j=1}^{n} \xi_{ij}^5 |\Delta F_j| \tag{5.7}$$

$$(\text{grad} v_i)^T \Delta G_i (\Delta x, d) F < \sum_{j=1}^{2n} \xi_{ij}^6 |\Delta x_j| + \sum_{j=1}^{n} \xi_{ij}^7 (|F_j^0(t)| + |\Delta f_j|), \; i=1,2 \tag{5.8}$$

where ξ_{ij}^k, k = 1, 2, ...n_k (n_k denotes appropriate number corresponding to k) are the real

numbers (note that these numbers may be also negative). Inequalities (5.7), (5.8) have to be

valid for $\forall x \; \epsilon \; X^t$ (t), $\forall t \; \epsilon \; T$ and $\forall d \; \epsilon \; D$. The practical stability conditions of the over-all

system can be established by considering the derivative of the function v_i along the solutions

of the coupled subsystems (5.5) (aggregation principle). Based on (5.6)-(5.8), candidates for

$$\Psi_i(t) = -\gamma_i \, \underline{X}^{I(i)} \exp(-\alpha_i t) + \Sigma \, (\xi_{ij}{}^4 + \xi_{ij}{}^6) \, \underline{X}^t \exp(-\alpha t) + \Sigma \xi_{ij}{}^1 |x_j{}^{01}(t)| + \Sigma \, \xi_{ij}{}^3 \, |\dot{x}_j{}^{01}(t)| +$$

$$+ \Sigma \, \xi_{ij}{}^2 \, |F_j{}^{01}(t)| + \Sigma \xi_{ij}{}^5 \lfloor \underline{E}_j{}^t \exp(-\beta \, t) | dt + \Sigma \xi_{ij}{}^7 \, |F_j{}^0(t) + \underline{F}_j{}^t \exp(-\beta t)|, \quad i=1, 2 \qquad (5.9)$$

By substituting (5.9) into (5.4) we obtain the practical stability test for the robotic system interacting with the dynamic environment when the control law (4.1) is applied.

6. Conclusion and Discussion of Stability Test

The paper presents for the first time an innovative test for the practical stability of the robotic system interacting with the dynamic environment. The test enables to explicitly examine stability of the position and force controlled part of the system taking into account dynamic coupling between these parts. The test enables to examine practical stability of the different control laws, and it is specifically convenient (i.e. less conservative) to consider so-called decentralised control law in which each direction is stabilised independently of other directions. The test enables to study influences of different model uncertainties upon the system behavior as one of the most relevant aspect for practical application of robots in the considered class of robotic tasks. The obtained test may also be used to define under which conditions the stabilisation of the force (5.2) ensures a practical stabilisation of the over-all system. Specifically, if we assume that all parameters are perfectly known (i.e. that $d = d_0$) the obtained stability test represents generalisation of the sufficient conditions for the asymptotic stability given in the previous papers[6, 7] to the case of a practical stability. Contrary to the tests obtained in the previous papers [9, 10], this test explicitly examines stability of the position control subsystem (i=1) and the force controlled subsystem (i=2). The stability test of the subsystem 2 (force controlled subsystem) explicitly shows under which conditions it is possible to ensure the practical stability if the force is practically stabilised. However, the test takes into account dynamic coupling among the position and force controlled subsystems, both through the dynamics of the robot and the dynamics of the environment. The considered control law (4.1) represents a generalisation of the so-called computed torque method[14, 15] to the case when the robot is interacting with the dynamic environment. Obviously, if the parameters of the models are perfectly known, and if the environment model possesses stability properties, this control law can practically stabilise the robotic system. However, if the approximate model of the environment is used in the control loop (e.g. where dynamic coupling among different directions is ignored, $M^* = \text{diag} \, (m_i^*)$), the control law is simplified, but it has to be tested (using the established test) whether such control can ensure practical stability. If the

the control law (4.1) practically is reduced to the so-called classical hybrid control[1,2], and the established stability test can be used to test under which conditions it is possible to stabilise the robot with such hybrid control law. The test can be effectively used to synthesise the control laws which are based on decentralised approach but which take into account different dynamics aspects of the robot and of the environment.

References

[1] M. T. Mason, 'Compliance and Force Control for Computer Controlled Manipulators', IEEE Trans. on Systems, Man and Cybernetics, Vol. SMC-11, No.6. pp. 418-432, 1981.

[2] M. H. Raibert, J. J. Craig, 'Hybrid Position/Force Control of Manipulators', Trans. ASME J. of Dynamic Systems, Measurement and Control, Vol. 102, No. 3., pp. 126-133, 1981.

[3] O. Khatib, 'A Unified Approach for Motion and Force Control of Robot Manipulators: The Operational Space Formulation', IEEE Journal of Robotics and Automation, Vol. RA-3. No. 1., pp. 43-53, 19987.

[4] A. De Luca, C. Manes 'Modelling of Robots in Contact with Dynamic Environment', IEEE Trans. on Robotics and Automation, Vol. 10. No.4., pp. 542-548, August 1994.

[5] N. Hogan 'Impedance Control: An Approach to Manipulation: Part I - Theory, Part II - Implementation, Part III - Applications', Trans. ASME J. of Dynamic Systems, Measurement, and Control, Vol. 107., No.3., pp. 1-24, 1985.

[6] Vukobratovic, K.M., Ekalo, Yu., 'Unified Approach to Control Laws Synthesis for Robotic Manipulators in Contact with Dynamic Environment ', In Tutorial S5: Force and Contact Control in Robotic Systems, IEEE Int. Conf. on R&A - Atlanta, pp. 213-229, 1993.

[7] Vukobratovic, K. M., Ekalo Yu., ' New Approach to Control of Robotic Manipulators Interacting with Dynamic Environment', Robotica, Vol. 14, pp. 31-39, 1996.

[8] An, H.C., Hollerbach, M.J., 'Dynamic Stability Issues in Force Control of Manipulators', IEEE International Conference on R&A, Raleigh, NC, pp. 890-896, March 1987.

[9] Stokic, M.D., Vukobratovic, K.M., 'Contribution to Practical Stability Analysis of Robots Interacting with Dynamic Environment', Proc. of the First ECPD International Conference on Advanced Robotics, Athens, September, pp. 693-699, 1995.

[10] Stokic, M.D., Vukobratovic, K.M., 'Practical Stability of Robots Interacting with Dynamic Environment', submitted to the International Journal of Robotics Research.

[11] Michel, N. A., 'Stability, Transient Behaviour and Trajectory Bounds of Interconnected Systems', Int. Journal of Control, Vol. 11, No. 4., 1970.

[12] Stokic, M.D., Vukobratovic, K.M., 'Practical Stabilisation of Robotic Systems by Decentralised Control', Automatica, Vol.20., No.3., 1984.

[13] Vukobratovic, K.M., Stokic, M.D., Kircanski, N., Non-Adaptive and Adaptive Control of Manipulation Robots, Springer-Verlag, Berlin, 1985.

[14] Paul, R.C., Modelling, Trajectory Calculation and Servoing of a Computer Controlled Arm, A.I. Memo 177, Stanford AI Laboratory, Stanford University, Sept. 1972.

[15] Pavlov, A.V., Timofeyev, V.A., 'Calculation and Stabilisation of Programmed Motion of a Moving Robot-Manipulator', Teknicheskaya kibernetika, No. 6., pp. 91-101, 1976.

CONTROL OF REDUNTANT ROBOTS AT SINGULARITIES IN DEGENERATE DIRECTIONS

E. Malis, L. Morin and S. Boudet

Direction Etudes et Recherches d'E.D.F., Chatou, France

Abstract

An algorithm for controlling redundant robots at singularities, when the desired motion is in a degenerate direction, is proposed. This case is not supported by other solutions, using only Jacobian information, proposed in the literature. Our algorithm uses Jacobian and Hessian matrices and it allows to choose the configuration of the robot when it exits the singularity. The algorithm was tested on a 7 degree of freedom robot.

Introduction

Let q be the current joint coordinates vector, of size $[n,1]$, of a robot, and x a $[m,1]$ vector of end-effector Cartesian coordinates, then the forward kinematics is:

$$x = f(q) \tag{1}$$

The Cartesian velocity is obtained by deriving equation (1) with respect to time:

$$\dot{x} = J(q)\dot{q} \tag{2}$$

where J is the $[m,n]$ Jacobian matrix. The inverse kinematics problem is solved by finding the joint vector $\tilde{q} = q + \delta q$ so that $\delta x = f(q + \delta q) - f(q)$. The state of the art relies on the assumption that the displacements δx and δq are small enough so that the following relationship holds from equation (2):

$$\delta x = J(q)\delta q \tag{3}$$

If $\mathbf{J}$ is a m × n matrix, equation (3) generates a linear system of m equations and n unknown. If $n > m$, the robot is redundant and we have an infinity of solutions. However, if we are at a singular joint position, equation (3) can be solved only if $\mathrm{rank}([\mathbf{J} \quad \delta\mathbf{x}]) = \mathrm{rank}([\mathbf{J}])$. When this equation is not true, an approximate solution is found using the Jacobian pseudo-inverse[1] or the damped least-squares Jacobian inverse method[2,3]. At a singularity, there may be a set of directions $\delta\mathbf{x}$ that are "degenerate". A direction $\delta\mathbf{x}$ is said to be degenerate if $\mathbf{J}^T\delta\mathbf{x} = \mathbf{0}$. Suppose the robot is at a singularity and the desired motion is in a degenerate direction. Solutions proposed in the literature using only Jacobian information are not capable of handling this particular case. Nielsen[4] and Egeland[5] have shown that the Cartesian acceleration term provides the information required to move along the degenerate direction. The Cartesian acceleration is obtained by deriving equation (2) with respect to time:

$$\ddot{\mathbf{x}} = \mathbf{J}(\mathbf{q})\ddot{\mathbf{q}} + \dot{\mathbf{J}}(\mathbf{q})\dot{\mathbf{q}} \tag{4}$$

where:

$$\dot{\mathbf{J}}(\mathbf{q},\dot{\mathbf{q}}) = [\mathbf{H}_1(\mathbf{q})\dot{\mathbf{q}} \quad \mathbf{H}_2(\mathbf{q})\dot{\mathbf{q}} \quad \cdots \quad \mathbf{H}_m(\mathbf{q})\dot{\mathbf{q}}]^T \tag{5}$$

The m Hessian matrices $\mathbf{H}_k$ of size [n,n] are symmetric. We proposed in another paper[6] an algorithm to compute in real-time Hessian matrices for solving the inverse kinematics problem and so how accomplish motion in degenerate direction at singularities without trajectory generation error. This paper is organized as follows. In the first section we illustrate the algorithm for inverse kinematics at singularities in degenerate directions. In the second section we discuss the experimental results.

Inverse Kinematics

The algorithm proposed in this paper minimizes the following set of functions:

$$\min_{\delta\mathbf{q}} \quad \Phi = \|\delta\mathbf{q}\|^2 \tag{6}$$

$$\min_{\delta\mathbf{q}} \quad \Psi = \|\delta\mathbf{x} - (\mathbf{f}(\mathbf{q}+\delta\mathbf{q}) - \mathbf{f}(\mathbf{q}))\|^2 \tag{7}$$

Our idea is to take a second order Taylor expansion of the norm of the Cartesian error:

$$\hat{\Psi} = \delta x^T \delta x - 2\delta x^T J(q)\delta q + \frac{1}{2}\delta q^T \left[2J^T(q)J(q) - \sum_{i=1}^{m} \delta x_i H_i(q) \right]\delta q \tag{8}$$

We define C, **B** and **E** in the following way:

$$C = \delta x^T \delta x, \quad B = -2\delta x^T J(q), \quad E = 2J^T(q)J(q) - \sum_{i=1}^{m} \delta x_i H_i(q) \tag{9}$$

Equation (8) gives a better approximation of the Cartesian error thanks to the term using the Hessian matrices. We test the vector **B** of equation (9). If $B \neq 0$, we can use any of the existing algorithms. If $B = 0$ and $\delta x \neq 0$, δx is a degenerate vector, we cannot use those algorithms, and we have to solve the set of equations (6) and (7) written hereafter. The minimum value of Ψ is 0, thus our problem is equivalent to minimize equation (6) with the following constraint:

$$\hat{\Psi} = C + \frac{1}{2}\delta q^T E\delta q = 0 \tag{10}$$

The Lagrange operator of equations (6) and (10) is:

$$L = \delta q^T \delta q + \lambda\left(C + \frac{1}{2}\delta q^T E\delta q \right) \tag{11}$$

The necessary conditions for solving this set of equations are:

$$\frac{\partial L}{\partial \delta q} = \delta q + \lambda E\delta q = 0 \tag{12}$$

$$\frac{\partial L}{\partial \lambda} = C + \frac{1}{2}\delta q^T E\delta q = 0 \tag{13}$$

If $\lambda = 0$, equation (12) implies $\delta q = 0$ and equation (13) doesn't hold when $\delta x \neq 0$. So λ is not null and μ is defined as $\mu = -1/\lambda$. Equation (12) can be expressed as follow: $E\delta q = \mu\delta q$. Therefore δq is an eigen vector of **E** and μ is its eigen value. We choose a normalized vector v and a scalar α so that $\delta q = \alpha v$. By plugging this equation into equation (13), we get:

$$C + \frac{1}{2}\alpha^2\mu = 0 \tag{14}$$

C is positive, therefore we have to choose a negative eigen value of **E** to verify equation (14). The solution to equation (14) is:

$$\alpha = \pm\sqrt{-2C/\mu}$$
(15)

The bigger is the norm of the eigen value μ, the smaller is α. Since **v** is normalized the norm of α is the norm of $\delta\mathbf{q}$. We want the smallest vector $\delta\mathbf{q}$ to satisfy equation (6), so we have to choose the lower eigen value of **E**. The sign of α determines the configuration of the robot when it exits the singularity.

Experimental results

To simplify the explanation of the algorithm in this paper, we blocked 4 axes of the seven axis robot Mitsubishi PA-10 used in our laboratory, and obtained a three joint planar robot. In our calculations, we only consider the Cartesian position of the end-effector, therefore the robot for our example is "redundant" (see Figure 1). The links' lengths are those of the PA-10: $l_1 = 0.45$ m; $l_2 = 0.5$ m; $l_3 = 0.08$ m.

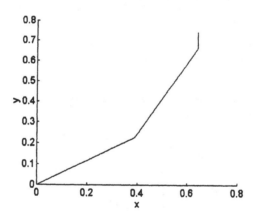

Figure 1: Three joint planar robot

If the robot is at the joint position $\mathbf{q} = \begin{bmatrix} 0 & 0 & 0 \end{bmatrix}^T$ the Jacobian is of rank 1:

$$\mathbf{J(q)} = \begin{bmatrix} 0 & 0 & 0 \\ 1.03 & 0.58 & 0.08 \end{bmatrix}$$

We want to compute the joint motion needed to move the robot along the x axis, for example $\delta x = [-0.001 \quad 0]^T$ meters. With the methods using only Jacobian information, the result is a null joint displacement, because δx is a degenerate vector ($J^T \delta x = 0$). In this case, we use the Hessian matrices. E, B and C are computed from equation (9). Then the eigen vectors and eigen values of E are found: $\mu_1 = -1.9665\,10^{-4}$; $\mu_2 = -6.7167\,10^{-5}$; $\mu_3 = 2.8060$ and $v_1 = [0.4932 -0.8509 -0.1808]^T, v_2 = [0.0310 -0.1905\,0.9812]^T, v_3 = [0.8694\,0.4895\,0.0675]^T$ As explained earlier, we choose the eigen vector whose eigen value is the smallest (it also has to be negative). In this case, we find $\mu_1 = -1.9665\,10^{-4}$.

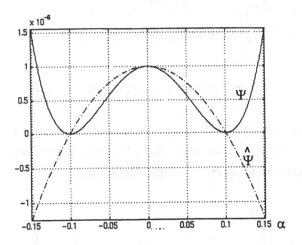

Figure 2: Cartesian error norm and its second order approximation

Figure 2 plots the functions $\Psi = \Psi(\alpha v)$ of equation (7) (continuous line) and its approximation $\hat{\Psi} = \hat{\Psi}(\alpha v)$ of equation (8) (dashed line) versus α. At $\|\delta q\| = 0$, we are at the initial position, and the first derivative is null, as expected. In this example, the Hessian in the y direction is null, therefore, $\hat{\Psi} = \hat{\Psi}(\alpha v)$ and $\Psi = \Psi(\alpha v)$ are null for the same α: $\alpha = \pm\sqrt{-2C/\mu_1} = \pm 0.1008$. The absolute value of α is the norm of δq in radians, it is equivalent to 5.77 degrees.

Generally, at a given singularity, Ψ and $\hat{\Psi}$ will not intersect at the value 0, like in Figure 2. However, it is still possible to command a motion along the vector associated with a negative eigen value of E, so that the norm $\hat{\Psi}$ decreases (i.e. Ψ decreases). If there are not strictly negative eigen values, the Taylor expansion needs to be taken to a higher order, and a new algorithm is needed. For the seven degree of freedom Mitsubishi robot arm, our results confirmed that there is at least one negative eigen value at each singularity.

Figure 3 shows the two configurations of the robot for $\delta q = \pm[2.8496 \quad -4.9169 \quad -1.0455]^{\mathrm{T}}$ degrees.

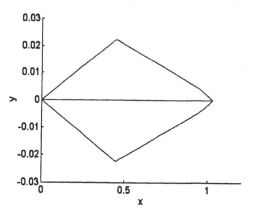

Figure 3: Movement in degenerate direction

If we choose the eigen vector corresponding to the second smallest eigen value (which is also negative, therefore it is suitable), we would find a δq with a bigger norm. We found another pair of symmetric solutions to the inverse kinematics problem.

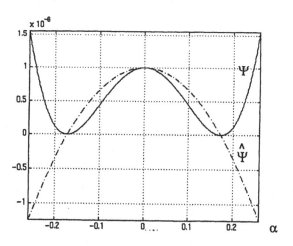

Figure 4: Cartesian error norm and its second order approximation

Figure 4 plots $\Psi = \Psi(\alpha v)$ of equation (7) (continuous line) and its second order approximation $\hat{\Psi} = \hat{\Psi}(\alpha v)$ of equation (8) (dashed line) versus α. In this case too, $\hat{\Psi} = \hat{\Psi}(\alpha v)$ and $\Psi = \Psi(\alpha v)$ are null for the same α: $\alpha = \pm\sqrt{-2C/\mu_2} = \pm0.1726$. The absolute values of α

is the norm of δq in radians, equal to 9.89 degrees. Figure 5 shows the two joints positions of the robot for $\delta q = \pm[0.3068 \quad -1.8930 \quad -9.7011]^T$ degrees.

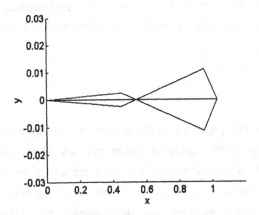

Figure 5: Movement in degenerate direction

Generally, it is better to perform the Cartesian motion with the smallest possible joint displacement. But if there is an obstacle and we cannot move the robot as in Figure 3, we can choose the second eigen value and move the robot like in Figure 5. It is therefore possible to move away from a singular configuration in a degenerate direction with little tracking error, but as the robot gets closer to the singularity, the commanded velocity should get smaller. Since $\dot{x} = J(q)\dot{q}$, the robot will always pass the singularity at a null speed in the degenerate directions.

The algorithm proposed in the paper was tested on the PA-10 from Mitsubishi, a 7 degree of freedom robot. The Singular Value Decomposition algorithm was used to invert the Jacobian. To know if the wanted motion is along a degenerate direction, we test if $\|J^T\delta x\| \le \varepsilon_1$ and $\|\delta x\| \ge \varepsilon_2$ (ε_1 and ε_2 are thresholds). There are two limitations on the norm of δq:

- a Taylor expansion is used, so it is valid only for small δqs. Since we take the expansion to the second degree, the range of δq is bigger than the other methods which take at best an incomplete Taylor expansion to the second order.

- the result of the computation is used on a real robot. The algorithm is therefore iterative in the sense that a new δq is calculated at every sampling time. During that time, the robot can only move a limited amount of degrees. This is imposed by the maximum acceleration and speed of the robot.

Because of these two constraints, the norm of δq is limited by a maximum value. This value was chosen during simulation runs, because there is not enough time to compute it in real time. In the case of the PA-10, we chose the limit on δq based upon the maximum joint speed of the robot. The algorithm taken at the second order was implemented on the 7 degree of freedom robot PA-10. The computation period is 10 milliseconds on a MVME 167 processor card.

Conclusions

The algorithm described in this paper solves the problem of moving a robot in a degenerate direction from a singularity of its workspace. Other methods which solve motion through singular points using only Jacobian information are not capable of handling this particular case. The basic idea is based on taking a Taylor expansion of the norm of the function to minimize. To illustrate the algorithm, a second order expansion was performed in this paper. There may be robots for which higher orders are needed. It is, however, unlikely that this should occur: taking a second order expansion means that there is at least one pair of joints (among all the possible pairs) that can move the robot in the degenerate direction if these two joints are actuated together. By comparison, the Jacobian only calculates the differential movement due to each joint, one at a time, and then performs a linear combination of the movement.

References

1. Whitney D.E.: "The Mathematics of Coordinated Control of Prosthetic Arms and Manipulators", *Trans. ASME J. Dynamic Systems, Measurement and Control*, 94 (Dec. 1972), 303-309.
2. Nakamura Y., Hanafusa H.: "Inverse kinematics solutions with singularity robustness for robot manipulator control", *Trans. ASME, J. Dyn. Syst., Meas. and Contr.*, 108 (Sept. 1986), 163-171.
3. Wampler C.W.: "Manipulator Inverse Kinematics Solution Based on Damped Least-squares Solutions", *IEEE Trans. Syst., Man and Cybernetics*, 16(1) (Jan. 1986), 93-101.
4. Nielsen L., Canudas de Wit C., Hagander P.: "Controllability Issues of Robots near Singular Configurations", *Proc. of the Int. Workshop on Nonlinear and Adaptive Control: Issues in Robotics*, Grenoble, France, (Nov. 1990), 307-314.
5. Egeland O., Spangelo I.: "Manipulator Control in Singular Configurations - Motion in Degenerate Directions", *Proc. of the Int. Workshop on Nonlinear and Adaptive Control: Issues in Robotics*, Grenoble, France, (Nov. 1990), 296-306.
6. Malis E., Morin L., Boudet S.: "Two new Algorithms for Forward and Inverse Kinematics under Degenerate Conditions", *W.A.C.*, Montpellier, France, (May1996).

APPLICATION OF NEURAL NETWORKS FOR CONTROL OF ROBOT MANIPULATORS-SIMULATION AND IMPLEMENTATION

T. Uhl, M. Szymkat, T. Bojko, Z. Korendo and J. Ród
St. Staszic Technical University, Cracow, Poland

1.Introduction

Several approaches to using neural networks for solving the kinematics, dynamics, motion planning and control problems in robotics applications have been proposed in past years. In the literature some applications of neural networks in kinematics are presented [3, 6, 7, 14, 17] and motion planning [7,14,17]. The use of neural networks for control of robotic manipulators motion has been proposed in the number of papers [1, 5, 7, 9, 10, 11, 12, 19, 22]. In this paper the problem of the trajectory tracking is considered. The desired trajectory for the motion of a manipulator is generated by the path planner. Having the desired trajectory with the initial and final positions of the centerpoint of the end-effector, controllers need to be constructed for the actuators that make the end-effector follow the specified trajectory as closely as possible.This is achieved by determining the torques acting on the joint shafts or inputs to the joint actuators so that the system follows the desired trajectory with minimal tracking error. The dynamic model of the general manipulator represents a multiple-input-multiple-output (MIMO) system, in which equations are nonlinear and coupled. We consider two alternatives of control of such systems: the classical control and the intelligent one. One class of intelligent control is control scheme based on neural networks.

2.Neural networks based dynamic control of robotic manipulators

The new trend in control of nonlinear systems tries to exploit neural networks as advanced nonlinear controllers. Some neural networks properties, such as their inherent computational parallelism, fuzzification of the knowledge represented by particular patterns, relative parameters insensitivity, together with the effective neural net training algorithms made this new tool attractive for the control of robotic manipulators. The most often reported control techniques based on neural networks for manipulation robots are: nonlinear control - controller parameters tuning via neural look-up table [9], adaptive control with neural parameters identification [12], inverse dynamics - neural model [11], inverse dynamics and PD controller realised as neural network [20], model predictive control scheme- neural output prediction [11], neural compensation of unknown parameters [22].

The first approach mentioned above is presented in this paper. The scheme of the control system under consideration is shown in figure 1. The design process of neural network based controller contains two main steps [1]: choice of a neural network architecture and training of the neural network, by adjustment neuron weights. The weight updating law in many cases of robot manipulators minimises the objective function E(e) [2] which is a quadratic function of the tracking error. This methodology is divided into two steps; in the first stage of training the neural network learns the output of the PD controller, in the second stage of training the PD controller and

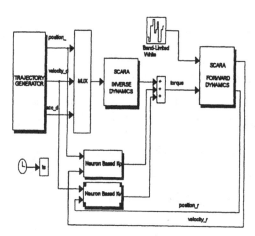

Fig. 1 Scheme of the controller with tuning parameters via neural networks look-up table.

the neural network operate in parallel. The neural network is then further trained by the output of the combined controller. Stage two is iteratively repeated until the neural network achieves desired accuracy. Neural network based controllers can be implemented in few ways [15]:

- using special chips e.g. Intel's ETANN,

- using special board with processor,which simulates the a neural networks,

- using signal processor to speed up the neural networks simulation.

The third possibility is applied in the presented investigation, because of the high computational performance of the signal processor and availability of the software and hardware integrated with input/output interfaces.

3.Control system design methodology based on mechatronic approach

Mechatronic design methodology is intensively developed and applied in automotive, robotics and aviation industry. The design process implemented here contains the following steps [8]: modeling of a system, identification of the model parameters, simulation of model behaviour, fast prototyping procedure, implementation of the controller according to industrial standards.

Fast prototyping procedure has three phases [16]: design of controller, analysis of control system properties, implementation of the prototype of controller using the DSP technology. These steps are realised employing computer assisted methods, with software used as an integration tool which transform designers ideas across all steps of the design process. The modelling phase has been realised using computer algebra system described in detail in [21]. The identification of the model parameters has been done employing model adjustment techniques formulated in [18]. This paper is focussed on the problem of controller simulation and implementation. The dSPACE products are used to realise the fast prototyping method. The products currently supplied by the dSPACE can be directly used for fast prototyping of the control system [8]. The major steps in this approach are the following:

- preparation for real-time test at the real plant with the I/O specified in the SIMULINK block diagram by drag and drop from I/O block library and textual specification of details.
- generation of C-code for the target DSP system, completed with I/O function calls as specified, embedded into timer-interrupt driven real-time software structure, compiled linked and finally downloaded to the target DSP system.
- switching the control mode, tuning the controller parameters, observing system behaviour, acquisition of process data during the closed loop system operation in order to improve control system structure.

The software for use with dSPACE board employed in this paper consists of the following programs: TRACE (MTRACE), for on-line system state visualisation, COCKPIT (MLIB) for on-line parameter tuning, RTI- SIMULINK library containing block diagrams with board drivers. This software is completely integrated with MATLAB/SIMULINK environment, so the graphical user interface of this software can be directly used.

4. Neural networks based control system design

The mechatronic design approach has been employed for the design of neural networks based controller. The idea of the controller consists in adaptive controller gains tuning using trained neural network. The parameters K_p and K_v for both joints are adaptively changed on the base of the tracking error value. The controller scheme in SIMULINK block diagram format, applied for system behaviour simulation is shown in figure 1.

This neural network in the format of SIMULINK block diagram is shown in figure 5 . The applied neural network has two layers; input layer with radial basis neurons and output layer with linear neuron structure. For the neural network training a set of different trajectories has been generated employing trajectory generator. Using these trajectories sets of optimal K_p and K_v are obtained. Values of K_p and K_v are limited due to limited current in the drives. These data was a base for supervised training of the chosen neural network. Applied neural networks approximate the mapping of the tracking errors to the values of gains in the controller. During the training process, which is based on backpropagation algorithm, the neural network weights were tuned. The system behaviour with neural network based controller has been tested using off-line simulation. The simulation results are shown in Figure 2.

a)   b)

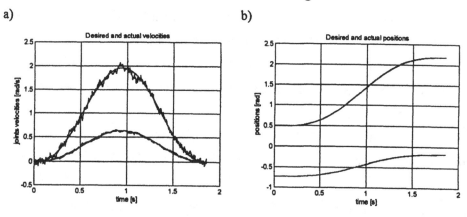

Fig.2. Results of simulation of neural network based controller performance

The controller performance has been tested with relatively big torque disturbance, the ratio between nominal torque and torque disturbances is close to one. The obtained results show the efficiency of the neural network based controller for accurate trajectory tracking.

5.Experimental results

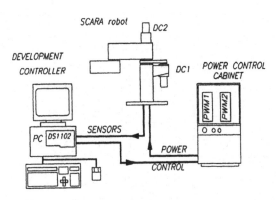

Fig. 3 Scheme of experimental setup.

During the experiment validation of the controllers has been carried out. The scheme of the experimental setup used is shown in figure 3 . Desired trajectories for each joint have been generated off-line in the 'trajectory generator' block. The actual trajectories have been measured by two encoders mounted at the axis of each joint, the actual velocities have been numerically computed and together with controller gain profile have been monitored during the experiment. The SIMULINK scheme of neural based controller with RTI blocks for direct implementation with DS1102 board is shown in figure 4. The applied neural networks architecture is shown diagramatically in figure 5. Results of the experimental investigation of neural based controller are shown in figure 6.

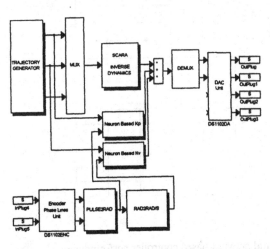

Fig. 4. SIMULINK block diagram scheme for experimental investigation of designed neural based controller on DS1102 board.

The controller gains have been changing during the neural networks based controller action (fig.7). For the controller performance comparison an integral square errors for given trajectory tracking has been computed. These factors for investigated controllers have the following values;

- for classical controller;

$$\varepsilon_{1\theta_1} = 123.56 \quad \varepsilon_{1\theta_2} = 231.12$$

$$\varepsilon_{1\dot{\theta}_1} = 200.61 \quad \varepsilon_{1\dot{\theta}_2} = 94.37$$

- for neural based controller:

$$\varepsilon_{1\theta_1} = 89.32 \quad \varepsilon_{1\theta_2} = 78.35$$

$$\varepsilon_{1\dot{\theta}_1} = 145.98 \quad \varepsilon_{1\dot{\theta}_2} = 82.34$$

From above results one can easily notice that the performance measured as the square error between desired and obtained trajectories is better for neural based controller. To improve the controller performance the more complete training set for learning of the controller should be generated.

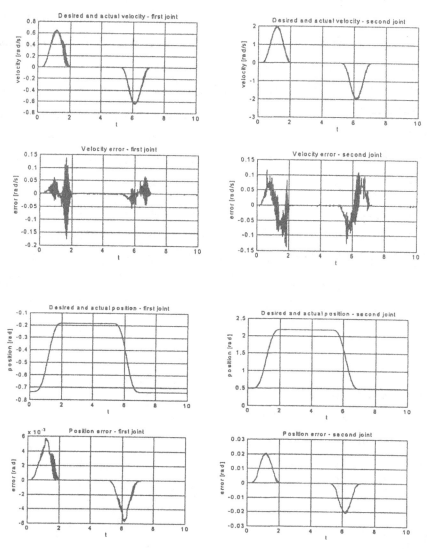

Fig.6.Results of experimental investigation of the neural network based controller performance

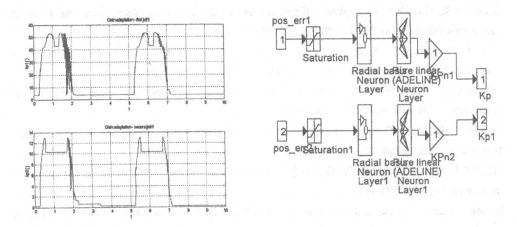

Fig.7.Controller gains changes for given trajectory   Fig. 5 Scheme of applied neural networks
tracking

The paper presents the possibilities of prototyping neural based controllers using digital signal processors board integrated with MATLAB/SIMULINK computing environment. It should be mentioned here that applied methodology is suitable for prototyping and testing of neural networks based controllers but final industrial implementation should be done with employing specialised neural network silicon chips or for less complex problems may be accomplished with DSP technology but in specialised version. The cost of the solution, applied in this work is high, but the most significant advantage of this solution is the relative ease of use freeing the designer from the burden of the manual programming of the DSP board.

7. Literature

1.Antslakis P., Special issue on neural networks in control, IEEE Control System Magazine, vol.12, no2, 1992, pp.8- 57.

2.Barto A.G., Connectionist learning for control - An overview, Neural Networks for Control, Ed. Miller W.T., Sutton R.S., Paul J., MIT Press, Cambridge, 1990, pp.5-59.

3.Cavalieri S., Martini M., Petrone F., Sinatra R., A neural network approach for position-error optimisation problem in a redundant robot, Proceedings of IX World Congress on the Theory of Machines and Mechanisms, Milan, 1995, pp.1684-1689.

4.Chen F.C., Back propagation neural networks for nonlinear self-tuning adaptive control, IEEE, Control System Magazine, pp.1448- 1453, 1989.

5.Gehlot N.S., Alsina P.J., A comparison of control strategies of robotic manipulators using neural networks, International Conference on IECON, 1922, pp. 688-693.

6.Gou J., Cherkavsky V., A solution to the inverse kinematics problem in robotics using neural network processing, IEEE Int. Conf. on NN, Washington D.C., Vol.II, pp 299-304, 1989.

7.Hamavand Z., Schwartz M., Trajectory control of robotic manipulators by using a feedback-error-learning neural network, Robotica, vol.13, 1995, pp.449-459.

8.Hanselman H., DSP in Control: The Total development Environment, Proc. of Int. Conf. on signal processing Applications & Technology, Boston, 1995. pp.107- 114.

9.Hunt K.J., Sabraro, Zbikowski R., Gawthrop, Neural networks for control system - a survey, Automatica, vol.28, no.6, 1992, pp. 1083-1112.

10.Jung S., Hsia T.C., A new neural network control technique for robot manipulators, Robotica, vol.13, 1995, pp.449-459.

11.Kawato M., Wada Y., A neural network model for arm trajectory forming using forward and inverse dynamics model, Neural networks, vol.6, no.7., 1993,pp.919- 933.

12.Khemaissia S., Morris A.S.,Neuro-adaptive control of robotic manipulators, Robotica, vol.11, no.5, pp.465-473, 1993.

13.Sciavicco L., Siciliano B., Modeling and Control of robot manipulators, The McGraw-Hill Companies,Inc., 1996.

14.Simon D., Neural network based robot trajectory generation, IEEE Transaction on Neural Networks vol.5, pp.540- 545, 1993.

15.Sontag E.D., Neural networks for control, in Essay on Control: Perspective in theory and its application , Ed. Trentelman H.L., Birkhauser, Boston, 1993.

16.Szymkat M, Ravn O., Turnau A., Kolek K., Pjeturson A., Integrated mechatronic modelling environments, Int. Conf. on Recent Advances in Mechatronics, pp. 767- 772, Istanbul, 1995.

17.Szymkat M., Uhl T., Biennier F., Neural network support for simultaneous optimisation of torque and deviation in robot path planning, Proc. of the IEEE Int. Symp. on Intelligent Control, pp. 241 - 252., Chicago IL, 1993.

18.Uhl T., Lisowski W., Bojko T., Ród J., Identification of robotic joints models using model adjustment techniques, MMAR'95, Miedzyzdroje, 1995, pp.659-664.

19.Uhl T., Szymkat M., Application of neural nets for control of robotic manipulators, Proc. of ROMANSY 10 :Theory and practice of robots and manipulators, Springer Verlag, 1994.

20.Uhl T., Szymkat M., A comparison of the classical and neural-based approach to control of manipulation robots, XVII ISMA, Leuven, 1992. pp. 255- 284.

21.Uhl T., Rod J.,Modeling and identification of robotics systems using MATLAB/SIMULINK environment, In ICRAM, Istanbul, August 1995.

22.Yabuta T., Yamada T., Possibility of Neural Networks Controller for robot manipulator, Proc. of IEEE Int. Conf. on Robotics and Automation, 1990, pp 1686-1691.

Chapter IX
Control of Motion III

ACTIVE FORCE CONTROL OF PNEUMATIC ACTUATORS

H. Busch

University of Hannover, Hannover, Germany

J. Hewit

University of Dundee, Dundee, UK

Abstract

The method of control known as Active Force Control (AFC) has been shown previously to be an effective way of using feedforward to cancel the effects of disturbances even when these disturbances are unmeasurable. By using extra instrumentation to measure the applied forcing and the resultant acceleration, the disturbances can be implicitly measured. Here AFC is applied to a pneumatic system consisting of an air cylinder with considerable Coulombic friction between the piston and the cylinder wall. It is shown that AFC can greatly reduce the damaging tendency to stick-slip motion which bedevils pneumatic controls.

KEYWORDS - Pneumatic control, disturbance rejection, robotics, mechatronics

1.   Introduction

Pneumatic actuators are used extensively in automation. They are relatively cheap to install and operate in comparison with electric servomotors and hydraulic drives and have a high power-to-weight ratio. However accurate positioning and motion control of pneumatically actuated systems can pose severe problems due to non-linear friction forces between the piston and the cylinder walls which can lead to stick-slip movements of the piston [1],[2].

Mechatronic design, which requires extra sensor-based feedback, can effect the cancellation of these Coulombic friction forces and so achieve an apparently friction-free system. Active Force Control (AFC) is used to overcome the unknown non-linear disturbances by estimation of the disturbance forces and then approximate compensation through a feed-forward controller. Then a conventional control strategy is applied to complete the position controller design.

2.   The Pneumatic Cylinder - Equations of Motion

Figure 1 shows a schematic illustration of the pneumatic cylinder.

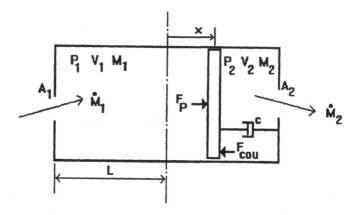

Figure 1   Model of the Pneumatic Cylinder

The orifices of the pneumatic cylinder are controlled through proportional solenoid valves with independent closed loop PI-controller circuits. The valves maintain a proportional relation between an desired cylinder pressure P_{des} which is set by an input voltage, and the output pressure P.   The valves are equipped with two equally sized outlet orifices, one opening to atmospheric pressure P_0 and the other to the pneumatic actuator. A block diagram of the solenoid valve is given in figure 2.

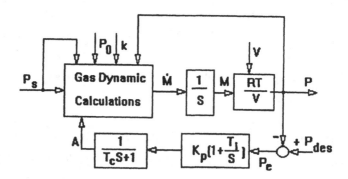

Figure 2   Proportional solenoid valve with closed loop PI-controller

The equations describing the generation of the piston force F_p are shown below [3]. It is assumed that the air flow through the orifices is an adiabatic process. The quantities involved are defined in figure 1.

$$P_1 V_1 = M_1 RT \qquad 1$$

$$P_2 V_2 = M_2 RT \qquad 2$$

$$V_1 = A_{piston} (L + x) \qquad 3.$$

$$V_2 = A_{piston} (L - x) \qquad 4$$

$$F_p = A_{piston} (P_1 - P_2) \qquad 5$$

Dependent on the desired cylinder pressures P_{des1} and P_{des2} there is a mass flow of air charging one end of the cylinder and discharging from the other cylinder chamber. The gas flows through the valve orifices are described by,

$$\dot{M}_j = - K A_j P_1 E(P_E, P_1) \qquad 6$$

$$\dot{M}_i = K A_j P_E E(P_1, P_E) \qquad 7$$

with $i, j = 1, 2$; and the constant K defined by

$$K = C \left(\frac{2\gamma}{RT(\gamma - 1)} \right)^{\frac{1}{2}} \qquad 8$$

The function $E(.,.)$ is defined for air flow into the cylinder chambers (charge) as,

$$E(P_E, P_1) = \sqrt{\left(\frac{P_1}{P_E}\right)^{\frac{2}{\gamma}} - \left(\frac{P_1}{P_E}\right)^{\frac{\gamma-1}{\gamma}}} \qquad for \qquad P_1 > .528 P_E \qquad 9.$$

i.e. subsonic flow, and as

$$E(P_E, P_1) = .259 \qquad for \qquad P_1 \leq .528 P_E \qquad 10.$$

i.e. sonic or choked airflow. Here we take $\gamma = 1.4$.
For air flow out of the cylinder chambers (discharge) $E(.,.)$ is calculated as,

$$E(P_1, P_E) = \sqrt{\left(\frac{P_E}{P_1}\right)^{\frac{2}{\gamma}} - \left(\frac{P_E}{P_1}\right)^{\frac{\gamma-1}{\gamma}}} \qquad for \qquad P_E > .528 P_1 \qquad 11.$$

and

$$E(P_1, P_E) = .259 \quad for \quad P_E \le .528\, P_1 \qquad\qquad 12.$$

Equations 1 through 12 describe the relationship between the orifice areas of the valves and the output function F_p, the force applied to the piston, which is seen to be highly non-linear. It may be written as,

$$Fp = h(A_i,\, A_j) \qquad with\ i,\, j = 1,2 \qquad\qquad 13.$$

where A_i and A_j are the valve exhaust and outlet orifice areas of both valves. However due to the use of closed loop proportional valves a significant simplification of the input output relationship is achieved. The result may be written as,

$$F_p = g(P_{des,\,i}) \quad with\ i = 1,2 \qquad\qquad 14.$$

where $P_{des,\,i}$ is the desired pressure input at the two valves. This relationship contains the non-linear gas dynamics in the form of the valve response given for a desired cylinder pressure $P_{des,}$.

Use of these equations allow the control for F_p to be generated via the pressures P_1 and P_2 according to equation 5. The movement of the piston in the cylinder is described by the equation of motion

$$M\ddot{x} = F_p - F_{cou} - c\dot{x} \qquad\qquad 15$$

3.   The Pneumatic Cylinder - Feed Forward via Active Force Control

In order to apply feed forward or Active Force Control (AFC) [3], it is necessary to find an inverse function so that given a desired force F_p we can calculate the desired input pressures $P_{des,\,i}$ for the solenoid valves [4]. Since there are two input functions $P_{des,1}$ and $P_{des,2}$ and only one output function F_p it is necessary to impose an additional constraint. Thus we need to find a suitable inverse pair of functions of the form,

$$P_{des,\,i} = u_i\,(F_p) \qquad\qquad with\ i = 1,2 \qquad\qquad 16.$$

This can be done by introduction of logic which distinguishes between possible conditions in which the piston may find itself within the cylinder.

Four cases, dependent on the position error $ex = x - x_{bar}$ and the velocity dx/dt as listed in the table below, are shown in figure 3. *(with $x_{bar} = 0$)*

	dx/dt	ex	Feed Forward ?
CASE 1	$>= 0$	> 0	yes
CASE 2	> 0	< 0	no
CASE 3	< 0	> 0	no
CASE 4	$<= 0$	< 0	yes

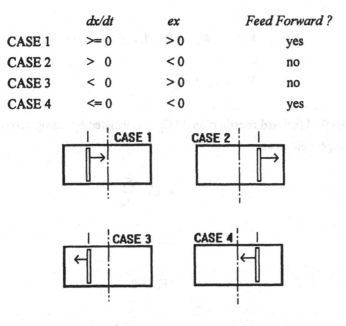

Figure 3   The four cases

The Coulomb friction force F_{cou} resists motion and its direction is therefore opposite to that of dx/dt as shown in figure 3.   We may now adopt the following strategy :-
In cases 2 and 3 the effect of F_{cou} is **beneficial**, i.e. F_{cou} is acting in the correct direction to bring the position error ex to zero; therefore AFC or feed forward **is not** applied.   In cases 1 and 4 however, F_{cou} is **detrimental**, i.e. F_{cou} is acting in the wrong direction and is tending to prevent the error ex being brought to zero. It is therefore beneficial to cancel the effect of the Coulomb friction by means of AFC and then solve the resulting control problem by a conventional controller.

To push the piston to the right requires the supply pressure P_s be applied to the left and atmosphere pressure P_0 to the right hand orifice of the cylinder.

The actual control of values of the orifice areas of the valves as functions of the position error ex is executed by the closed loop solenoid valves.   We are now in the position to obtain the inverse relationship indicated in equation 16.   We need only concern ourselves with case 1 and case 4 since in the other two cases no feed-forward is applied.

Neglecting the dynamic behaviour of the valves, it is straightforward to obtain the inverse relationship as follows. An estimate F_e^* of the disturbance force F_e is given by,

$$F_e^* = F_p - M\ddot{x} = A_{pist}(P_1 - P_2) - M\ddot{x} \qquad 17.$$

with:

$$F_e = F_{cou} + c\dot{x} \qquad 18.$$

Then the feed forward contribution AFC_i is calculated by taking account of the actual operating condition as,

Case 1

$$AFC_1 = \frac{F_e^*}{A_{pist}} \qquad 19$$

$$AFC_2 = 0 \qquad 20.$$

Case 4

$$AFC_1 = 0 \qquad 21.$$

$$AFC_2 = \frac{F_e^*}{A_{pist}} \qquad 22.$$

The feed-forward therefore cancels the estimated disturbance force F_e^* and a controller contribution to solve the remaining positioning problem can be processed by a conventional controller algorithm. The controller action depends on the position error ex. If ex is positive (cases 1 and 3), the left cylinder chamber is charged and the right one discharged. In cases 2 and 4 the controller acts the other way round. The resulting valve input pressure $P_{des,i}$ is then evaluated by adding the controller output $CONT_i$ and the feed forward output AFC_i as,

$$P_{des,i} = AFC_i + CONT_i \qquad 23$$

The complete block diagram of the system is shown in figure 4.

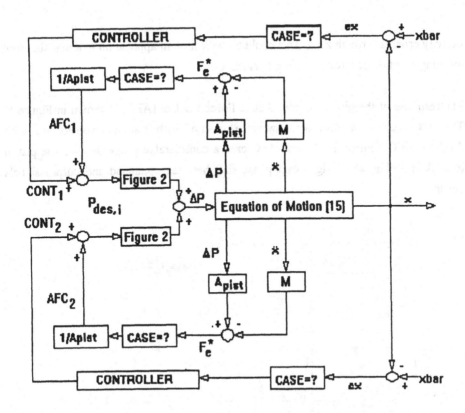

Figure 4   Block Diagram of the Pneumatic Cylinder

4.   Results

Experimental investigations are at present under way to determine the practicability of the Active Force Control in this application.   Here we present only preliminary simulation results which support the theoretical base of the proposed controller structure.

A simulation programme has been written in ACSL (Advanced Continuous Simulation Language) incorporating the mathematical model described above and an accurate model of Coulomb friction.   This has a conditional, non-analytic form and is the major source of difficulty for analytic-type control algorithms.   The Coulomb friction forces in the simulation were set to values between 50N and 75N.

To test the efficiency of the feed-forward control action, a sine wave input command for the desired piston position X_{bar} is used. The piston is commanded to move from the central position of the cylinder to one extreme, then to the other extreme and then back into mid position in a smooth sinusoidal fashion. Due to the very slow movements the piston tends to stick at the extreme positions of its trajectory where the velocity crosses through zero. This could be typical of an application in a machine tool indexing function or a robot cycling operation.

The response of the piston without Active Force Control (AFC) is shown in Figure 5. The best results in control accuracy were obtained with non-AFC control and with AFC-plus-PID control. In the non-AFC case a considerable propensity for the piston to stick is visible, and a significant phase shift between command and output signals occur.

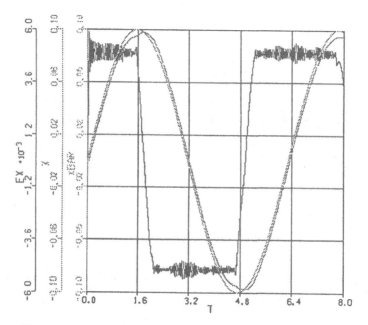

Figure 5   PID-Controller without Active Force Control.

The response with PID-controller and AFC displayed in figure 6, shows that the phase shift is almost eliminated and the sticking phenomenon is reduced to a large extent.

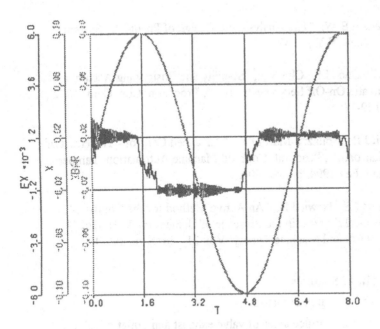

Figure 6 PID-Controller with Active Force Control (AFC).

The improvement in control using Active Force Control in conjunction with self controlled closed loop solenoid valves is obvious. Work is continuing to obtain experimental confirmation of the simulation results and to optimise the system.

5. Conclusion

It has been shown how the application of advanced control can lead to the cancellation, to a large extent, of the effects of Coulombic friction in pneumatic actuators.

First the mathematical model of the actuator and the driving valves was derived. Next the feed forward part of the controller was developed and placed in front of the highly non-linear, non-analytic plant in order to cancel the Coulombic friction effects. Finally an overall control structure was added.

Simulation results have demonstrated the ability of the controller to avoid the stick/slip phenomenon prevalent in conventionally controlled pneumatics.

6. References

1. Anderson B.W., "The Analysis and Design of Pneumatic Systems" John Wiley, N.Y., 1967.

2. Eun T., Cho H.S., Cho Y.J., "Stability and Positioning Accuracy of a Pneumatic On-Off Servomechanism.", Proc. Am. Con. Conf. 1982 , pp 1189-1194.

3. Hewit J.R., Bouazza-Marouf K., "Advanced Control - The Heart of Mechatronics", Proc. Int. Conf. on Machine Automation, Tampere, Finland, Feb.1994, pp 393-406.

4. Burdess J.S., Hewit J.R., "An Active Method for the Control of Mechanical Systems in the Presence of Unmeasurable Forcing", Mechanism and Machine Theory, Vol. 21, No. 5, pp 393-400.

Appendix - List of Symbols

A_{pist}	=	area of piston
A_i	=	orifice areas of valve exhaust and outlet of valve 1
A_j	=	orifice areas of valve exhaust and outlet of valve 2
c	=	coefficient of viscous friction.
ex	=	displacement error
F_{cou}	=	Coulomb friction force
F_p	=	force due to pressure difference
k	=	combined coefficients of valve orifice
L	=	cylinder length
M	=	mass of piston
M_1	=	mass of air in chamber 1
M_2	=	mass of air in chamber 2
P_1, P_2	=	gas pressures in cylinder
P_L, P	=	low, high pressure across orifice
P_s, P_0	=	supply, exhaust pressure
$P_{des,i}$	=	desired input pressure of valve 1 and 2
R	=	gas constant
T	=	temperature
V_1	=	volume chamber 1.
V_2	=	volume chamber 2.
x	=	piston position
x_{bar}	=	required piston position

USING BACKPROPAGATION ALGORITHM FOR NEURAL ADAPTIVE CONTROL: EXPERIMENTAL VALIDATION ON AN INDUSTRIAL MOBILE ROBOT

P. Henaff and S. Delaplace

University of Paris 6, Vélizy, France

and

Versailles Saint-Quentin University, Vélizy

Abstract : *This paper presents an original method in the use of neural networks and backpropagation algorithm to learn control of robotics systems. The originality consists to express the control objective as a criterion of which the gradient is backpropagating through the network instead of the classical quadratic error used in standard backpropagation. This technic allows on-line learning that is impossible to do with standard backpropagation. Experimental validation is realised by the position and the orientation control of a faster industrial mobile robot. Results show the feasability of the method, and particularly establish that on-line learning scheme permit to refine the weights of the network in front of the kinematics constraints of the robot.*

1   Introduction : problems of direct backpropagation

Multi-layered networks are very interesting in control of robotic systems because of their known generalization capabilities. A standard neural control scheme is build on the academic feedback control approach where the neural net replace the classical corrector.

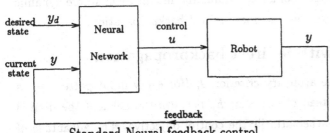

On right figure, net input is a combination between the current state of the system and a desired state. Net outputs are the control parameters (joint torques for example).

Standard Neural feedback control

Multi-layered networks are usually trained with the well known backpropagation learning algorithm we named direct backpropagation. The goal of direct backpropagation is to minimize the quadratic error E (addition of quadratic errors E_p for every input pattern p) between the net and a desired output (see Fig. 1). Consequently, it requires an a priori knowledge of the desired outputs (e.g. the desired controls parameters) corresponding to the net input.

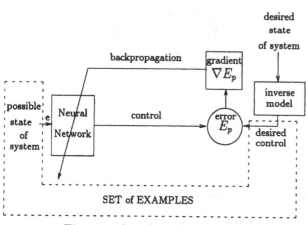

Figure 1: Standard Backpropagation

Determination of desired outputs is the main problem of using standard backpropagation in learning control of robotic applications, because either a reference model is use (PID, fuzzy controller [5], human teacher [2],...) and it does not exploit the whole learning capabilities of neural nets (except for the human teacher); or a inverse model of the robot is use, and the inversion can not, in general, easily be solved on-line.

In the case of non holonomic mobile robots (independent driving wheeled robots) the inverse model can not easily be obtained because of the non-integrability of the equations. Then it is impossible to learn on-line with Direct Backpropagation.

To make allowance for these remarks and since backpropagation is an optimization algorithm, we propose a learning method that doesn't need desired outputs, but only a criterion specifying the robotics task.

The first idea of this approach has been introduced before in [6] for a simulated arm cartesian control problem. Our method has been presented in [8] for a dynamic reflex control problem in which the network learns to control the dynamic equilibrium of a simulated planar biped. In [10], we have presented experimentations on a slow mobile robot that show the feasibility of our learning approach.

In this paper, we presents experimental results on this approach (we named it indirect backpropagation) applied to the control of a faster industrial mobile robot where dynamic effects are very important. Next section describes indirect backpropagation.

2   Solution : learning with indirect backpropagation

The basic idea is to minimize the arbitrary criterion J_p (for every input pattern p) as a function of the net output, instead of the error E_p between the net and the desired output. This criterion explain mathematically the control objective. The advantage of learning with criterion is that we need not any desired output, but only a cost function which specify the control objective and its constraints. So, the neural network determines itself the way to achieve this objective.

In order to minimize this criterion, we use also a gradient descent method :

$$\frac{\partial J_p}{\partial w_{ji}} = \frac{\partial J_p}{\partial net_{pj}} \frac{\partial net_{pj}}{\partial w_{ji}} = \frac{\partial J_p}{\partial net_{pj}} o_{pi}$$

where w_{ji} is the weight from neuron i to neuron j, o_p is the net output vector, $net_{pj} = \sum_i w_{ji} o_{pi}$ denotes the activation of neuron j and f_j its function (sigmoïde).
We set $\delta_{pj} = -\frac{\partial J_p}{\partial net_{pj}}$ and obtain:

- for a hidden neuron : $\delta_{pj}^{hid} = f_j'(net_{pj}) \sum_k \delta_{pk} w_{kj}$
- for an output one : $\delta_{pj}^{out} = -\frac{\partial J_p}{\partial o_{pj}} \frac{\partial o_{pj}}{\partial net_{pj}} = -\frac{\partial J_p}{\partial o_{pj}} f_j'(net_{pj}) = \nabla J f_j'(net_{pj})$

Instead of the error vector E_p, we backpropagate the gradient vector of the criterion $\nabla J_p = -\frac{\partial J_p}{\partial o_{pj}}$ related to the corresponding output. Therefore, the partial derivates of the criterion must be analytically calculable with respect to the control parameters. The computation of δ_{pj}^{hid} remains unchanged.

2.1 Learning Off-line and on-line with indirect backpropagation

Figures 2 and 3 illustrate the new learning algorithm (compares with figure 1). It can be use to train the net off-line with a model of the robot and on-line with the real robot.

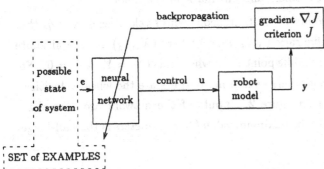

Figure 2: Learning off-line with criterion

Our learning structure is very simple. The on-line structure allows to refine the neural controller in front of real behavior of the robot. In comparison, D. H. Nguyen and B. Widrow proposed in [3] to use a network to emulate the system. Then, they can backpropagate a criterion through the emulator net and trained the controller without a gradient. Nevertheless, their method allows not to control the robot during the learning because the criterion cannot be propagated through it. Consequently, they cannot refine their neural controller to the real behavior of the robot.

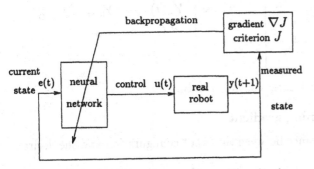

Figure 3: Learning on-line with criterion

We presents in next section experimental results of on-line learning because they are much more pertinent than results of off-line learning. Results establish that on-line learning scheme adapt the weights of the neural controller in order to solve the control objective.

3 Experimental validation of indirect backpropagation

3.1 Robot position control problem

The problem is to control the cartesian position and the orientation of a two independent
driving wheeled industrial mobile robot. Kinematic equations are done by equation (1)
on figure 1. The main characteristic of this robot is its non-holonomy (see [7] and [4]).

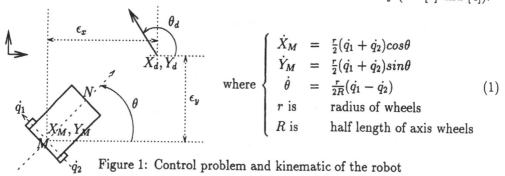

$$\text{where} \begin{cases} \dot{X}_M &= \frac{r}{2}(\dot{q}_1 + \dot{q}_2)\cos\theta \\ \dot{Y}_M &= \frac{r}{2}(\dot{q}_1 + \dot{q}_2)\sin\theta \\ \dot{\theta} &= \frac{r}{2R}(\dot{q}_1 - \dot{q}_2) \\ r \text{ is} & \text{radius of wheels} \\ R \text{ is} & \text{half length of axis wheels} \end{cases} \quad (1)$$

Figure 1: Control problem and kinematic of the robot

The neural controller has to determine the instantaneous wheel velocities $\dot{q}_1$ and $\dot{q}_2$ that
drive the robot at the desired configuration (X_d, Y_d, θ_d) where (X_d, Y_d) are the absolute
desired cartesian coordinates of the middle point M of wheel axis $(X_d = Y_d = \theta_d = 0)$. So,
net inputs are cartesian and orientation errors and net outputs are the wheel velocities.
The control structure is described on figure 2. Inputs of the network are normalized
between -1 et $+1$ with respect to their maximum value ($D = 4$ meters is the half-longer
of experimentation room).

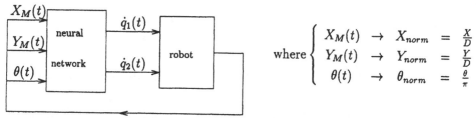

$$\text{where} \begin{cases} X_M(t) &\rightarrow X_{norm} &= \frac{X}{D} \\ Y_M(t) &\rightarrow Y_{norm} &= \frac{Y}{D} \\ \theta(t) &\rightarrow \theta_{norm} &= \frac{\theta}{\pi} \end{cases}$$

Figure 2: Neural control structure

3.2 Control criterion and learning gradient

The objective is to minimize the distance between the robot configuration and the desired
one. Then the criterion is :

$$J = \alpha_1 X_M^2(t+1) + \alpha_2 Y_M^2(t+1) + \alpha_3(\theta(t+1) - \theta_s)^2 \quad (2)$$

where α_1, α_2, α_3 are normalization coefficients and $\theta_s = arctg(2\frac{Y_M(t)}{D})$ is a strategy
constraint that helps the network on the singular configuration $X_M = 0$ and $Y_M \neq 0$.
By considering the time step Δt and the kinematic equations (1), the criterion becomes:

$$J = \alpha_1(X(t) + \Delta X(t+1))^2 + \alpha_2(Y(t) + \Delta Y(t+1))^2 + \alpha_3(\theta(t) + \Delta\theta(t+1) - \theta_s)^2 \quad (3)$$

$$\text{where}\begin{cases} \Delta X(t+1) &= \Delta U(t+1)cos(\theta(t)+\frac{\Delta\theta(t+1)}{2}) \\ \Delta Y(t+1) &= \Delta U(t+1)sin(\theta(t)+\frac{\Delta\theta(t+1)}{2}) \\ \Delta\theta(t+1) &= \frac{r}{2R}(\dot{q}_1(t)-\dot{q}_2(t))\Delta t \\ \Delta U(t+1) &= \frac{r}{2}(\dot{q}_1(t)+\dot{q}_2(t))\Delta t \end{cases}.$$

The gradient that trains the network is calculated with respect to the net outputs :

$$\frac{\partial J}{\partial \dot{q}_i(t)} = 2(\alpha_1 X(t+1)\frac{\partial \Delta X(t+1)}{\partial \dot{q}_i(t)} + \alpha_2 Y(t+1)\frac{\partial \Delta Y(t+1)}{\partial \dot{q}_i(t)} + \alpha_3(\theta(t+1)-\theta_s)\frac{\partial \Delta\theta(t+1)}{\partial \dot{q}_i(t)})$$

where:

$$\begin{cases} \frac{\partial \Delta X(t+1)}{\partial \dot{q}_i(t)} &= \Delta t\frac{r}{2}cos(\theta(t)+\frac{\Delta\theta(t+1)}{2}) - \frac{1}{2}\frac{\partial\Delta\theta(t+1)}{\partial\dot{q}_i(t)}\Delta U(t+1)sin(\theta(t)+\frac{\Delta\theta(t+1)}{2}) \\ \frac{\partial \Delta Y(t+1)}{\partial \dot{q}_i(t)} &= \Delta t\frac{r}{2}sin(\theta(t)+\frac{\Delta\theta(t+1)}{2}) + \frac{1}{2}\frac{\partial\Delta\theta(t+1)}{\partial\dot{q}_i(t)}\Delta U(t+1)cos(\theta(t)+\frac{\Delta\theta(t+1)}{2}) \\ \frac{\partial \Delta\theta(t+1)}{\partial \dot{q}_i(t)} &= \frac{r}{2R}\Delta t \ (i=1) \ or \ -\frac{r}{2R}\Delta t \ (i=2) \end{cases}$$

3.3 On line indirect backpropagation

We see in section 2 that our learning method permits to train the network off-line and on-line. The off-line learning method has been tested with good results. We solely presents in this section the results of the on-line learning method (Figure 3) because they are much more interesting. We argued that on-line learning scheme made feasible adaptive control.

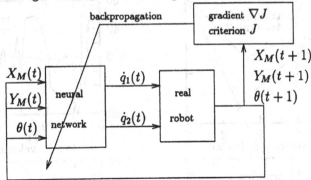

Figure 3: On-line learning scheme to solve our problem

As an illustration of that affirmation, we proposed to experiment on-line training with an **untrained network** which is updating after each control time step.

Protocol of experimentation

We have used a one hidden layer (of 9 neurons) network. The robot velocity is limited at 0.8 m/s during training. Only three training lessons have been necessary to train the network. At the beginning of each lesson, the robot is at an initial configuration far from the desired one (between 3 and 4 meters). The protocol is as follows:

- **Lesson 1:** The network is untrained (all the weights are randomly initialized). Initial robot configuration is $X_M = 3m$, $Y_M = 2m$, $\theta = 0°$.

- **Lesson 2:** Network has been trained by the first lesson. Initial robot configuration is still $X_M = 3m$, $Y_M = 2m$, $\theta = 0°$.
- **Lesson 3:** Network has been trained by the firsts lessons and Initial robot configuration is $X_M = 3m$, $Y_M = 0m$, $\theta = 180°$.

We stop each training when the robot completely stops.

First and second training

Right figure shows evolution of the training criterion. The learning duration is three times faster for the second training. Figure 4 shows cartesian trajectory, orientation of the robot, error and velocities of the wheels (i.e net outputs). For the first lesson, the robot trajectory is hazardous during ten minutes but we see clearly that the robot is attracted to the desired configuration.

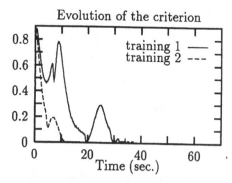

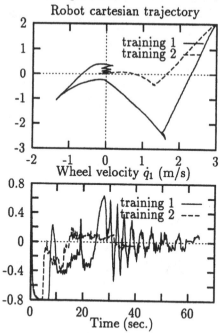

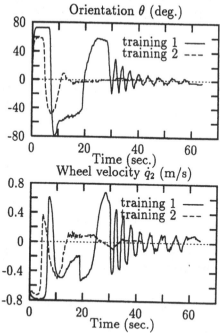

Figure 4: Trajectory (M point), orientation and control parameters of the robot

The trajectory is more direct and shorter in lesson 2 than in lesson 1. Final errors are reasonably good (about 1cm and 5 degrees) in regards with the robot size (1.025 m × 0.68 m) and are smaller after lesson 2. Wheel velocities are more smooth in lesson 2 and the average velocity greater.

The third lesson used a different initial configuration $(X_M = 3m, Y_M = 0m, \theta = 180°)$. The results are similar to the previous and the robot reaches the desired configuration in 40 seconds. minutes (results are not plotted).

These results shows that indirect backpropagation allows the neural-controller to identify the real behavior of the robot and determine a control law to drive the robot to the desired configuration and increase the quality of the control. Finally, we can argue that there is an knowledge acquisition of the robot kinematics and an adaption of the control.

Control of the real robot after on-line training

The trained network controls now the robot without learning from an initial configuration which has never been used for the learning: $X_M = 4m, Y_M = -2m, \theta = -90°$. The maximum velocity of the robot is 1.25 m/s. Right figure and figure 5 shows the net ability to control the robot with small final errors.

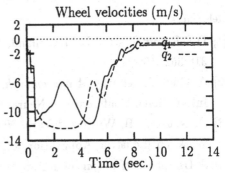

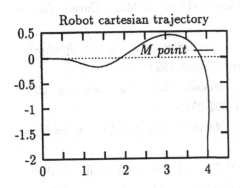

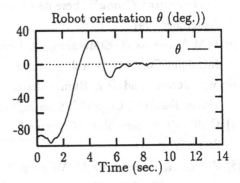

Figure 5: Control of the robot after three on-line trainings : trajectory and orientation

Others tests establish that the neural net controls the robot at any configuration in the experimentation room while $-3\ m \le X_M \le 3\ m$ and $-3\ m \le X_M \le 3\ m$.

These results show clearly that indirect backpropagation allows neural networks to learn control laws by especially specifying the control objective.

4   Conclusion

In this paper, we have presented experimentation results of an original approach to the Neural Network learning architecture for the control and the adaptive control of mobile robots. Our goal was to use multi-layer-networks and the backpropagation algorithm in order to learn control objective without desired output. The principle of our approach is to express the control objective as a criterion. The gradient of the criterion is backpropagating through the network instead of the classical quadratic error. The technic permits the neuro controller to learn off-line or on-line. Experimental validation is realized by the position and the orientation control of a fast industrial mobile robot. To show the feasibility of the method, we trained a random neural network on the real robot with the on-line learning scheme. Results show clearly the very good and very fast adaption of the neural-network in front of the kinematics constraints and the dynamics effects of the robot.

References

[1] D.E Rummelhart, J.L. McClelland, *"Parallel Distributed Processing"*, MIT Press, pp. 318-362, 1986

[2] A. Guez, J. Selinsky, *A neuromorphic controller with a human teacher*, Proc. of IEEE International Conference on Neural Networks, pp 595-602, 1988

[3] D. Nguyen, B. Widrow *"Neural Networks for Self-Learning Control Systems"*,IEEE · Work. on Industrial Applications of Neural Networks, pp 18-23,1991

[4] S. Delaplace, *"Navigation à Coût Minimal d'un Robot Mobile dans un Environnement Partiellement Connu'"* thèse de l'Université Paris 6 (Pierre et Marie Curie), France, November 1991

[5] D.M.A. Lee, W.H. ElMaraghy, *"A Neural Network Solution for Biped Gait Synthesis"*, Int. Joint Conf. on Neural Networks, Vol II, pp. 763-767,1992

[6] M.I. Jordan and D.E Rummelhart, *"Forwars Models: Supervised Learning with a Distal Teacher"*, Cognitive Science, Vol 16, pp. 307-354, 1992

[7] C. K. Ait-Abderrahim, *"Commande de Robots Mobiles"*, Thèse de l'Ecole des Mines de Paris, 1993.

[8] P. Hénaff, H. Schwenk,M. Milgram *" A Neural Network Approach of the Control of Dynamic Biped Equilibrium"*, International Symposium on Measurement and Control in Robotics,1993

[9] P. Hénaff, *"Mises en Œuvre de Commandes Neuronales par Rétropropagation Indirecte: Application à la Robotique Mobile"*, thèse de l'Université Paris 6 (Pierre et Marie Curie), France, june 1994

[10] P. Hénaff, *"Adaptive Neural Control In mobile Robotics: Experimentation for a Wheeled Cart"*, IEEE Int. Conf. on Systems, Man and Cybernetics, pp 1139-1144, Oct. 1994

HOPFIELD'S ARTIFICIAL NEURAL NETWORKS IN MULTIOBJECTIVE OPTIMIZATION PROBLEMS OF RESOURCE ALLOCATIONS CONTROL

J. Balicki and Z. Kitowski

The Naval Academy, Gdynia, Poland

Summary

Some robots cooperating each other to perform connected operations can be considered as a system. This system should be controlled to take advantages of different robots assigning to several operations. Indeed, a control of robot-operation allocations as a sequence of many static optimization task of resource allocations with different input parameters can be formulated. If some robot control problems are transformed to this resource allocation problem, then it is possible to use the proposed below methods. In this paper, analog Hopfield's artificial neural networks are used by genetic algorithms for solving NP-hard binary multiobjective optimization problems, which can be considered in modeling of resource allocations control. This problem can be solved for improving the efficiency of a few connected robots during their activities. Moreover, another neural approach for dynamic optimal control is elaborated. Finally, an example of two-layer feed-forward network in the adaptive control system of the underwater vehicle motion is submitted.

1. Introduction

The control of resource allocations as a sequence of many static optimization task with different input parameters can be discussed. In some cases, robots control problems can be transformed to this resource allocation problems, and then it is possible to use the considered method. Many multiobjective optimization problems with zero-one decisions are NP-hard [5,7,14]. Recently, interest has risen in the application of genetic algorithms to solving combinatorial optimization problems [3]. Genetic algorithms are the alternative approach compare to standard operational research methods, simulated annealing, Hopfield artificial neural networks [6,16], the Boltzman machine [8], and elastic nets [1,4]. Genetic algorithms can be combined with recurrent neural networks. This approach is very efficient because of the massively parallel processing.

From the other point of view investigations into the applications of Hopfield's artificial neural networks (HANN) for solving optimization problems proved that they can solve some optimization problems in a real time [11,13].

In this paper, genetic methods and artificial neural networks for solving binary multiobjective optimization operations allocations problem have been proposed. HANN based on nonnegative convex combination method algorithm for finding Pareto-optimal solution has been presented [3]. Designed HANN family in the genetic algorithm has been used. Results for resource allocations control are considered. Moreover, another neural approach for dynamic optimal control is elaborated. Finally, an example of two-layer feed-forward network in the adaptive control system of the underwater vehicle motion is submitted.

2. Problem formulation

In complex economics systems, technical systems and defense systems is necessary to assign operations to serve elements (processors) such, that the quality criterion is minimized. But properties of operations and serve elements can be changed, that re-optimization of operations is needed. Finally, the sequence of operation assignment optimization problems are solved (fig. 1).

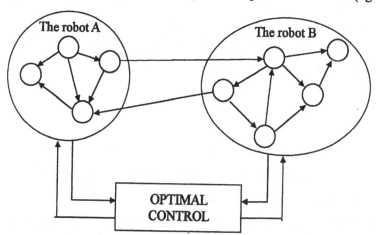

Fig. 1. Optimal control in a system of two robots and some operations.

For each phase when attributes are constant, the following multiobjective optimization model [2,9,12] for operations assignments can be considered.

1) X - a feasible solution set

$$X = \{x \in B^{2V+J} \mid x = (x_{11}, \ldots, x_{vi}, \ldots, x_{V2}, x_{11}^{\pi}, \ldots, x_{ij}^{\pi}, \ldots, x_{2J}^{\pi})^{T}; B = \{0,1\};$$

$$\sum_{i=1}^{2} x_{vi} = 1 \;, \; v = \overline{1, V}; \quad \sum_{j=1}^{J} x_{ij}^{\pi} = 1 \; i = \overline{1,2}; \}$$

2) F - a quality criterion   (1)

$$F : X \rightarrow R^{2}$$

$$F(x) = [F_{1}(x), F_{2}(x)]^{T} \times X$$

$$F_{1}(x) = \sum_{i=1}^{2} \sum_{j=1}^{J} \delta_{j} x_{ij}^{\pi}$$

$$F_{2}(x) = \sum_{j=1}^{J} \sum_{v=1}^{V} \sum_{i=1}^{2} t_{vi} x_{vi} x_{ij}^{\pi} + \sum_{v=1}^{V} \sum_{u=1}^{V} \sum_{i=1}^{2} \tau_{vu} x_{vi} (1 - x_{ui})$$

3) P - Pareto relationship [2].

In above problem the following denotations have been used:

J - a number of feasible robots types,

V - a number of operations,

δ_{j} - costs of robots type π_{j},

t_{vj} - execution costs for the operation m_{v} on robots type π_{j},

τ_{vu} - interoperations reference costs between operations m_{v} and m_{u}, which are processed on

     different robots,

Decision variables are as follows:

$$x_{ij}^{\pi} = \begin{cases} 1 \text{ if robots type } \pi_{j} \text{ is assigned to the node } w_{i} \\ 0 \text{ in the other case} \end{cases}, \; v = \overline{1, V}; \; j = \overline{1, J};$$

$$x_{vi} = \begin{cases} 1 \text{ if operation } m_{v} \text{ is assigned to the node } w_{i} \\ 0 \text{ in the other case} \end{cases}, \; v = \overline{1, V}; \; i = \overline{1,2};$$

3. Neural approach for solving optimization problems

HANN can represent one Pareto-optimal solution in an equilibrium point. In the analog model of Hopfield's ANN, the behavior of neurons is described by the differential equations [15,17]:

$$\frac{du_m}{dt} = -\frac{u_m}{\eta_m} + \sum_{i=1}^{M} w_{im} g_i(u_i) + I_m \tag{2}$$

where

M - a number of neurons,

η_m - a passive coefficient for neuron x_m,

w_{im} - a synaptic weight between neurons x_i and x_m,

I_m - an external input to the neuron x_m.

On the fig. 2 the trajectory of the activation level $u_1(t)$ for different values of η_1 passive coefficient for neuron x_1 is presented. The network called JHANN/3/5 for satisfying the constraint

$$\sum_{j=1}^{5} x_{ij}^{\pi} = 3$$ is considered. For $\eta_1<0$ network does not obtain the equilibrium point and the

continues increase of the activation level is observed. For $\eta_1<0$ network obtains the equilibrium point at value $u_1(t)=\eta_1$.

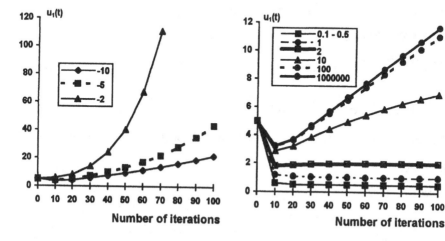

Fig. 2. Trajectories of neural activation level $u_1(t)$ in network JHANN/3/5 for different values of parameter η.

The synaptic weights are symmetric. The activation function is a sigmoid function as follows:

$$g_m(u_m) = \frac{1}{2}\left[1 + \tanh(\alpha\, u_m)\right], \qquad m = \overline{1, M} \tag{3}$$

where α the gain coefficient ($\alpha > 0$).

From the formula (3) we have $g_m(u_m) \le \varepsilon$ for $u_m \le -u_{gr}(\varepsilon)$ and $g_m(u_m) \ge 1 - \varepsilon$ for $u_m \ge u_{gr}(\varepsilon)$. If $\varepsilon \to 0+$, then we can assume that $g_m(u_m) \approx 0$ for $u_m \le -u_{gr}(\varepsilon)$ and $g_m(u_m) \approx 1$ for $u_m \ge u_{gr}(\varepsilon)$. The width of the interval $(\varepsilon, 1-\varepsilon)$ is equal to $2u_{gr}(\varepsilon)$. There is the following linear formula for finding value of the width $2u_{gr}$ and the gain coefficient α:

$$u_{gr}(\alpha, \varepsilon) = \frac{u_{gr}(1, \varepsilon)}{\alpha} \tag{4}$$

where $u_{gr}(1, \varepsilon)$ is the threshold activation value for $\alpha = 1$.

On the fig. 3 relationship between the width $2u_{gr}$ and the gain coefficient α for different values of accuracy ε is presented. If external inputs in Hopfield's ANN are constant over time, then Hopfield's ANN with feasible parameters reaches the equilibrium point. Hopfield proposed an energetic function as follows:

$$E(u) = -\frac{1}{2}\sum_{k=1}^{M}\sum_{m=1}^{M} w_{km} g_k(u_k) g_m(u_m) - \sum_{m=1}^{M} I_m g_m(u_m) + \sum_{m=1}^{M}\int_{0}^{g_m(u_m)} g_m^{-1}(\xi_m)\,d\xi_m \tag{5}$$

where g_m^{-1} - a reverse function for g_m, $u_m = g_m^{-1}(x_m)$.

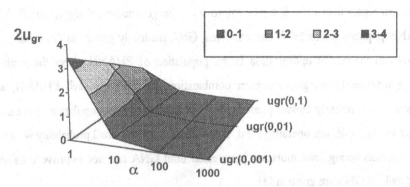

Fig. 3. Relationship between the width $2u_{gr}$ and the gain coefficient α for different values of accuracy ε.

4. Artificial neural network for optimal control of operation assignments

The energetic function of PHANN for finding Pareto-optimal operation assignments are constructed as below:

$$E(x, \alpha) = \sum_{n=1}^{N} \alpha_n F_n(x) + \sum_{l=1}^{L} \lambda_l h_l(x) \tag{6}$$

where

λ_l- a Lagrange multiplier for no satisfaction of the lth constraints,

$h_l(x)$ - a nonnegative penalty function for no satisfaction lth constraint for solution x.

We can calculate the particular weights and particular external inputs for the feasible solutions as shown in [3]. According to (4) synaptic weights and external inputs in network for constraints are multiplied by Lagrange coefficients. Synaptic weights and external inputs in network for minimization of function F_1 are multiplied by systematically increased coefficient α_1. Network for minimization F_2 consists of set of M neurons with external inputs equal to δ_m. There are no synaptic connections. Values of external inputs are normalized to set [0,1] and multiplied by $(1-\alpha_1)$. Finally, all synaptic weights and external inputs are added for each pair of neurons. In this way, the global neural network for finding Pareto-suboptimal solutions (PHANN) is designed.

5. Genetic-neural approach for improving a Pareto-suboptimal solution quality

Neural network PHANN finding Pareto-suboptimal solutions gives the accuracy of obtaining solutions no higher than 15%. It can be improved by an genetic-neural algorithm GNA [3]: In GNA a PHANN population size 2K is chosen. Then, GNA randomly generates the initial states for each PHANN and the neural optimization in the population of PHANN. After the evaluation of the objective function (nonnegative convex combination function) for each PHANN, an crossover operation of K randomly chosen pairs of PHANN from networks population to receive K pairs of offspring by the Goldberg operator. Next, the mutation with the small probability is carried out. The selection, crossovering, and mutation is repeated until GNA can not improve obtained solutions. More details of GNA are given in [3].

5. Two-layer feed-forward networks for controlling of the underwater vehicle motion

In a dynamic control feed-forward neural networks are more suitable then Hopfield's networks. Details of the adaptive control system for underwater vehicle motion in [10] are given. For dynamics equations of the three-dimensional motion of the underwater vehicle an structure and principles of an adaptive control system are submitted. That adaptive control systems gives training patterns for neural networks (fig. 4). This neural control system NCS was trained for several representative conditions to obtain optimal control signal forcing the change of the rudder. Moreover, a comparison of training techniques has been discussed. Finally, the Levenberg-Marquardt optimization technique for training be the designed NCS has been chosen (fig.5).

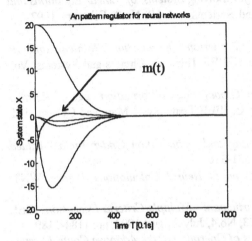

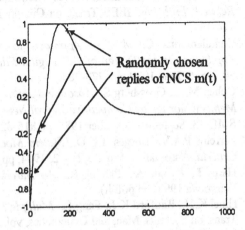

Fig 4. A model trajectory of control forces on the rudder

Fig. 5. Three examples of control by NCS trained by Levenberg-Marquardt method

In the first layer of NCS only five neurons with sigmoid activation function are designed. The second layer of NCS states one linear neuron with the output m(t). This simple network has five inputs representing a change of a trajectory, time of a maneuver, a velocity, a radial velocity, and a rudder state.

6. Concluding remarks

The neural optimal control of resource allocations as a sequence of many static optimization task with different input parameters is a very promising approach for solving many optimization problems. If some robots control problems are transformed to this resource allocation problems, then it is possible to use the above proposed method. Global analog Hopfield's artificial neural networks can combine with genetic algorithm [3] to obtain better accuracy from optimal solutions.

Another neural approach is related with training of multilayer neural networks for an optimal dynamic control. We noted the Levenberg-Marquardt optimization technique is better then the most common used backpropagation method for training of neural control systems.

REFERENCES

1. Aiyer S.V.B., Niranjan M., Fallside F.: *A Theoretical Investigation into the Per- formance of Hopfield Model.* IEEE Trans. on Neural Networks, vol.1, No.2, 1990, pp. 204-215.
2. Ameljańczyk A.: *Multicriterion Optimization.* WAT, Warszawa 1986.(in polish)
3. Balicki J, Kitowski Z.: *Multicriteria Optimization of Computer Resource Allocations with Using Genetic Algorithms and Artificial Neural Networks,* Proceedings of the 12th International Conference on Systems Science, Vol. III, September 1995, Wrocław, Poland, pp. 11-18.
4. Cichocki A., Unbehauen R.: *Neural Networks for Solving Systems of Linear Equations and Related Problems.* IEEE Trans. on Circuits and Systems, vol. 39, No.2, February 1992, pp. 124-137.
5. Charalambous C.: *A New Approach to Multicriterion Optimization Problem and Its Application to the Design of 1-D Digital Filters.* IEEE Trans. on Circuits and Systems, Vol. 36, No. 6, June 1989, pp. 773-784.
6. Cohen M.A., Grossberg S.: *Absolute Stability of Global Pattern Formation and Parallel Memory Storage by Competitive Neural Networks.* IEEE Trans. Syst., Man, and Cybern., vol. SMC-13, September/October 1983, pp.815-825.
7. Ferreira P.A.V., Borges T.C.D.: *System Modeling and Optimization Under Vector- Valued Criteria.* Automatica, Vol. 30, No. 2, 1994, pp. 331-336.
8. Hertz J., Krogh A., Palmer R.: *An Introduction to Neural Calculations Theory.* WNT, Warszawa 1993. (in polish)
9. Hipel K.H., Radford K.J., Fang L.: *Multiple Participant - Multiple Criteria Deci- sion.* IEEE Trans. on Systems, Man, and Cybernetics, vol. 23, No.4, July/August 1993, pp. 1184-1189.
10. Kitowski Z., Garus J.: *A Structure and Principles of Operation of the Adaptive Control System of the Underwater Vehicle Motion.* Proceedings of RoManSy 10: The Tenth CISM-IFToMM Symposium. Springer-Verlag, Wien-New York, 1995, pp. 201-209.
11. Lillo W.E., Hui S., Żak H. : *Neural Networks for Constrained Optimization Problems.* Int. J. of Circuit Theory and Applications, vol.21, 1991, pp.385-399.
12. Seaman C. M., Desrochers A. A.: *A Multiobjective Optimization Approach to Pla- stic Injection Molding.* IEEE Trans. on System, Man, and Cybernetics, Vol. 23, No. 2, March/April 1993, pp.414-425.
13. Sun K.T., Fu H.C.: *A Hybrid Neural Model for Solving Optimization Problems.* IEEE Trans. on Computers, vol.42, No.2, February 1993, pp.219-227.
14. Tarvainen K.: *Generating Pareto - Optimal Alternatives by a Nonfeasible Hierar- chical Method.* JOTA, vol. 80, No. 1, January 1994, pp. 181-185.
15. Tagliarini A., Christ J.F., Page W.: *Optimization Using Neural Networks.* IEEE Trans. on Computers, vol.40, No.12, December 1991, pp.1347-1357.
16. Tank D.W., Hopfield J.J.: *Simple "Neural" Optimization Networks: An A/D Con- verter, Signal Decision Circuit,and Linear Programming Circuit.* IEEE Trans. on Circuits and Systems,vol.CAS-33, May 1986, pp.533-541.
17. Zhang S., Constantindes A.G. : *Lagrange Programming Neural Networks.* IEEE Trans. on Circuits and Systems, vol.39, No.7, July 1992.

Chapter X
Sensing and Machine Intelligence

TYPE SYNTHESIS OF CONTACT SENSING ELEMENTS FOR ROBOTIC FIXTURING

W.W. Nederbragt and B. Ravani

University of California-Davis, Davis, CA, USA

Abstract

This paper uses group theory for type synthesis or enumeration of contacts between geometric elements necessary in the design of tactile sensing mechanical fixtures for robotic applications. Although the scope of the paper is limited to geometric contacts involving points, planes, and spherical surfaces, the techniques developed are general and can be applied to other geometric features and non tactile sensing elements used in robotic referencing and calibration.

1 Introduction

Contact sensing elements with mechanical fixtures are commonly used in robotics and manufacturing (see, for example, [1,2,3]) for part referencing. This is the process of determining the relative location of a part with respect to another part or a world coordinate system. Part referencing is also an essential step in robot calibration (see[4]). In measuring relative locations between two bodies, mechanical fixtures are usually used to simplify the sensing functions and to improve repeatability. Much of the existing mechanical fixtures, for such a purpose, however, are one of a kind ad hoc designs. There exists no work on type synthesis of such fixtures. The only exception to this is the work of McCallion and Pham [2], who used Kutzbach's criteria to aid them in the design of a tactile sensing fixture for robotic assembly. Recently Nederbragt and Ravani [5] have developed a design theory for design of tactile sensing mechanical fixtures. Their design method uses group theory to exploit the symmetry of different measuring arrangements. They have used this theory to design a new and novel tactile sensing fixture for part referencing. In this paper, we extend this work by using group theory for type synthesis or enumeration of all possible sensing-fixture designs. We use the group representations and techniques developed by Hervé [6] for mobility analysis of mechanisms in this study. Much of the background can be found in [5,6].

2 Position Referencing Based on Locations of Geometric Elements

Position referencing involves using coincidence relationships on geometric elements to find the relative location between two bodies. The location of the bodies in question are found using the properties associated with these elements. The location of the geometric elements are found by making contact with other geometric elements. There are in general two categories of contacts. If the position of the contact is known in the local coordinate system of the geometric element we shall refer to the contact as a fixed contact(F). Otherwise, the contact is called a mobile contact(M). For example, if a point comes into contact with a planar surface, the contact is considered fixed if the location of the contact is known in the surface's frame and mobile otherwise. Note, a point is always a fixed contact because the location of a touch to its body must be the point itself. Using points, spherical surfaces, and planar surfaces with both mobile and fixed contacts, a list of all possible contacts is shown in table 1.

Objects in Contact	Type of Contact	Objects in Contact	Type of Contact
Point - Point	(F/M) - (F/M)	Sphere - Sphere	M - M
Point - Sphere	(F/M) - F	Sphere - Plane	F - F
Point - Sphere	(F/M) - M	Sphere - Plane	F - M
Point - Plane	(F/M) - F	Sphere - Plane	M - F
Point - Plane	(F/M) - M	Sphere - Plane	M - M
Sphere - Sphere	F - F	Plane - Plane	(F/M) - (F/M)
Sphere - Sphere	F - M and M - F		

Table 1: All possible contacts between points, spherical surfaces, and planar surfaces.

3 Group Theory Evaluation of Geometric Contacts

In this section, we develop a method for evaluation of contacts between the described geometric elements. The method used involves the use of the Euclidean group and its subgroups. Hence, each of the geometric element contacts is transformed into an equivalent group representation. Using the group representations and techniques described in Hervé [6], a method is constructed for testing combinations of these contacts for their usefulness in measuring the relative position between two bodies (or design of tactile sensing fixtures).

The process of finding the group representation for a contact between two geometric elements is a simple one. First the contact should be described using standard joints(e.g., revolute joints, prismatic joints, and spherical joints). Then each of these joints can be described by their respective subgroups of the Euclidean group [6]. The resulting group representation for a contact is the composition of each of these subgroups. Note, some of the compositions between the subgroups can be joined to form larger subgroups. For example, two linear translational groups(prismatic joints) can sometimes be joined to form one planar translational group, $\{T_P\}$. Moreover, some compositions of groups result in a simplification of the total composition. For example, two linear translational groups that are about the same line (line D) or parallel to line D can be composed into one linear translational group about the same line D.

The relative motion between a point and a mobile planar surface can be broken down into two prismatic joints(their lines of action are perpendicular), a revolute joint(the axis is perpendicular to the plane formed by the translational lines), and a spherical joint(the location of the joint is at a point on the plane formed by the two translational lines). The spherical joint represents the point contact with the planar surface and the other components are due to the planar surface being mobile. Figure 1 shows a picture of the contact and a schematic of the joint motion associated with that contact. The equation showing the composition of the groups and the resulting group representation for this particular contact is

$$(\{T_D\}\bullet\{T_{D'}\})\bullet(\{R_U\}\bullet\{S_O\}) = \{T_P\}\bullet\{S_O\} \tag{1}$$

where D is perpendicular to D', U is perpendicular to both D and D', and O lies on the plane formed by D and D'. As can be seen, there are two simplifications performed on the composition in order to make it "cleaner." The dimension of this group representation is five, two from $\{T_P\}$ and three from $\{S_O\}$. Note, if the planar surface was fixed that the resulting motion would be equivalent to just a spherical joint, $\{S_O\}$.

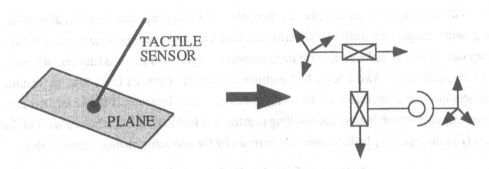

TACTILE
SENSOR

PLANE

Figure 1: point-mobile planar surface contact.

The relative motion between two spherical surfaces can be broken down into two spherical joints, a single spherical joint, or a revolute joint depending on if the surfaces are mobile, fixed, or one is mobile and one is fixed. A two spherical joint motion will occur if both spherical surfaces are mobile because neither surface will know exactly where the contact occurred. A spherical joint motion will occur if one of the spherical surfaces is mobile and one is fixed because one of the surfaces will know where the contact occurred relative to its coordinate system and the other surface will not, therefore, the mobile surface can be rotated about any axis through the center of its spherical surface and no change will be detected by either spherical surface. If both spherical surfaces are fixed then the two surfaces can only rotate about an axis through the center of both spherical surfaces without a change being detected, this results in the equivalent motion of a revolute joint with its axis through the centers of both spherical surfaces. Figure 2 illustrated the three possible contacts between a sphere - sphere contact. Note, the case when both surfaces are mobile results in a spherical joint - spherical joint combination. This combination has only a dimension of five, not six. This is caused by a redundant motion in the combination. Both spherical joints can rotate about an axis through the center of both spherical surfaces, therefore, only one of them is included in the group representation resulting in a decrease in the dimension for the combination.

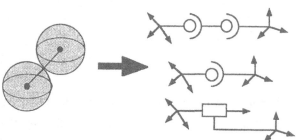

Figure 2: spherical-spherical surface contact, contact where both surfaces are mobile ($\{S_O\} \cdot \{S_O\}$), one is mobile and one is fixed ($\{S_O\}$), and both are fixed ($\{R_U\}$).

With all of the contacts described by their respective group representation and dimension, it is possible to apply the methods Hervé[6] described to combinations of these contacts to find if they can be used to make a complete measurement of the relative position between two bodies. Hervé stated that if two bodies had multiple constraints between them that the resulting constraint is the intersection of the multiple constraints. Moreover, if the dimension of the resulting intersection is zero, the resulting constraint is fixed and this is what we want to find. Let $\{L_1\}$ through $\{L_n\}$ be the constraints imposed by the geometric element contacts, then

$$\text{dimension of } (\{L_1\} \cap \{L_2\} \cap ... \cap \{L_n\}) = 0 \qquad (2)$$

In our case, the constraints are given by the group representations. Hence, if two bodies are separated by two point-point contacts then the overall constraint associated with the bodies is the intersection of the two constraints caused by the point-point contacts. It should be noted that there is a difficulty with using the group representations, they are coordinate dependent.

Objects in Contact	Type of Contact	Group Rep.	Dim.	Description
Point - Point	F - F	$\{S_O\}$	3	Spherical joint (abbr. for this contact is P/F-P/F)
Point - Sphere	F - F	$\{S_O\}$	3	Spherical joint (equivalent to P/F-P/F)
Point - Sphere	F - M	$\{S_O\} \cdot \{S_{O'}\}$	3+3-1=5	Distance from pt. O to pt. O' is the radius of the sphere
Point - Plane	F - F	$\{S_O\}$	3	Spherical joint (equivalent to P/F-P/F)
Point - Plane	F - M	$\{T_P\} \cdot \{S_O\}$	2+3=5	point O lies on the plane P
Sphere - Sphere	F - F	$\{R_U\}$	1	Axis U goes through the center of both spheres
Sphere - Sphere	F - M and M - F	$\{S_O\}$	3	Spherical joint (equivalent to P/F-P/F)
Sphere - Sphere	M - M	$\{S_O\} \cdot \{S_{O'}\}$	3+3-1=5	Distance from pt. O to pt. O' is the sum of the two sphere's radii
Sphere - Plane	F - F	$\{R_U\}$	1	Axis U is perpendicular to the plane
Sphere - Plane	F - M	$\{G_P\}$	3	Plane P is tangent to the sphere
Sphere - Plane	M - F	$\{S_O\}$	3	Spherical joint (equivalent to P/F-P/F)
Sphere - Plane	M - M	$\{T_P\} \cdot \{S_O\}$	2+3	The plane P is displaced from pt. O by the radius of the sphere
Plane - Plane	(F/M) - (F/M)	$\{G_P\}$	3	Planar joint

Table 2: The group representations and dimensions for the geometric contacts.

When finding the intersection between two bodies a coordinate system for each of the bodies must be chosen. Most of the group representations assume that the coordinate frames are located in a specific location. For example, any contact with a group representation $\{S_O\}$ assumes that both bodies have their respective coordinate systems located at the center of the spherical joint, hence, when motion occurs all displacements are part of the spherical joint group. If two contacts are used with spherical group representations then only one can have the coordinate systems placed at its center; the other contact will no longer be represented by a spherical joint but a strange set of displacements, see figure 3. Note, changing the coordinate system does not change the dimension of the motion caused by a geometric element contact.

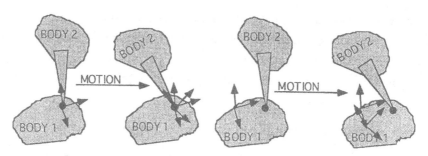

Figure 3: Illustration of group representation dependence on the location of the coordinate systems.

4 Type Synthesis of Various Contact Combinations

Using the group representation obtained in the last section, it is now possible to study geometric element combinations and determine if they can be used for measuring the relative position between two bodies. It should be noted that certain geometric element contact configurations have motions in common with other geometric element contact configurations regardless of the locations at which they are placed. For example, When two point-point contacts are used, the intersection of their group representations results in a subset of both group representations; this subgroup is a revolute group, $\{R_U\}$, with its axis, U, through the center of both point contacts. Therefore, two point - point contacts does not give a complete solution, another piece of information is necessary. A combination of a sphere/fixed - plane/mobile contact with a point - point contact also results in a revolute joint, the axis is through the center of the point and perpendicular to the plane. These cases must be considered when finding complete combinations.

A short coming associated with this method is that if a combination is found with dimension zero it still may have a finite number of possible orientations. For example, the theory suggests that it takes three points to find the position of a sphere of known radius, however, if three points are used, two spheres can fit to that one set of points, see figure 4. In order to remedy this situation another piece of information is necessary. This will be the case with many of the combinations found.

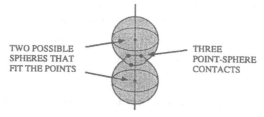

TWO POSSIBLE SPHERES THAT FIT THE POINTS

THREE POINT-SPHERE CONTACTS

Figure 4: Illustration of the two possible solutions when finding a sphere of known radius with three points.

Looking at Table 2 it is apparent that many of the geometric element contact's share the same group representations. This is due to the fact that a point is a special case of a mobile sphere of radius zero. Therefore, all of the possible contacts listed can be reduced from thirteen cases to five. S/F - S/F and S/F - PL/F (we will refer to this set as {R}) both have a group representation of {R_U}. P/F - P/F, P/F - S/F, P/F - PL/F, S/F - S/M, and S/M - PL/F (we will refer to this set as {S}) have a group representation of {S_O}. S/F - PL/M and PL-PL (we will refer to this set as {G}) both have a group representation of {G_P}. P/F - S/M and S/M - S/M (we will refer to this set as {S•S}) both have a group representation of {S_O}•{$S_{O'}$}. Finally, P/F - PL/M and S/M - PL/M (we will refer to this set as {T•S}) both have a group representation of {T_P}•{S_O}. The sets {R}, {S}, {G}, {S•S}, and {T•S} can now be used to make a complete listing of the interactions of each combination. Table 3 is a listing of all possible contacts that will result in a set of dimension zero, as stated earlier, a finite number of possible positions may come out of the combination, these cases will require an additional piece of information for a unique solution.

Combinations of the sets of contacts	Number of geometric element contacts in each comb.	Combinations of the sets of contacts	Number of geometric element contacts in each comb.
{R}, {R}	3	{S}, {T•S}, {T•S}, {T•S}	20
{R}, {S}	10	{G}, {G}, {G}	4
{R}, {G}	4	{G}, {G}, {S•S}	6
{R}, {S•S}	4	{G}, {G}, {T•S}	6
{R}, {T•S}	4	{G}, {S•S}, {S•S}, {S•S}	8
{S}, {S}, {S}	34	{G}, {S•S}, {S•S}, {T•S}	12
{S}, {S}, {G}	18	{G}, {S•S}, {T•S}, {T•S}	12
{S}, {S}, {S•S}	18	{G}, {T•S}, {T•S}, {T•S}	8
{S}, {S}, {T•S}	18	6x {S•S}	7
{S}, {G}, {G}	15	5x {S•S}, {T•S}	12
{S}, {G}, {S•S}	20	4x {S•S}, {T•S}, {T•S}	15
{S}, {G}, {T•S}	20	3x {S•S}, 3x {T•S}	16
{S}, {S•S}, {S•S}, {S•S}	20	{S•S}, {S•S}, 4x {T•S}	15
{S}, {S•S}, {S•S}, {T•S}	30	{S•S}, 5x {T•S}	12
{S}, {S•S}, {T•S}, {T•S}	30	6x {T•S}	7
TOTAL NUMBER OF COMBINATIONS			408

Table 3: Enumeration of possible combinations of points, spheres and planes resulting in a dimension of zero.

It should be noted that only three of the {S•S} elements can be applied to one sphere, after the application of three {S•S} contacts the dimension of the intersection no longer goes down. The same is true with three {T•S} contacts to a plane. These cases occur because the intersection of the first three contact constraints is a subset of the fourth contact constraint. The information provided in table 3 can be used for mechanical design of fixtures using geometric contact elements consisting of points, planes, and spheres.

5 Conclusion

A method for type synthesis of geometric elements composed of contacts between points, spheres, and planes is given. Using this method a list of contact combinations is derived that have the potential to measure the relative position between two bodies. This information can be used for design of mechanical fixtures for part referencing in robotics.

Acknowledgment

This work is supported in part by the California Department of Transportation through the AHMCT research center at the University of California, Davis.

References

[1]   Duffie, N., Bollinger, J., Van Aken, L., Piper, R., Zik, J., Hou, C., and lau, K., "A Sensor Based Technique for Automated Robot Programming," Journal of Manufacturing Systems, Vol. 3, No. 6, pp. 13-26, 1984.

[2]   McCallion, H. and Pham, D. T., "On Machine Perception of the Relative Position of Two Objects   Using Bilateral Tactile Sensing Systems," Proceedings of the Institution of Mechanical Engineers, Vol. 198B, No. 10, pp. 179-186, 1984.

[3]   Slocum, A. H., "Kinematic Couplings For Precision Fixturing -- Part 1 : Formulation of Design Parameters," Precision Engineering, Vol. 10, No. 2, pp. 85-91, April 1988.

[4]   Roth, Z. S., Mooring, B. W., and Ravani, B., "An Overview of Robot Calibration," IEEE Journal of Robotics and Automation, Vol. Ra-3, No. 5, pp. 377-385, Oct. 1987.

[5]   Nederbragt, W. W., and Ravani, "Design of Tactile Fixtures for Robotics and Manufacturing," Proceeding of ASME Mechanisms Conference, Irvine, CA August 1996.

[6]   Hervé, J., "Analyse Structurelle des Mécanismes par Groupe des Déplacements," Mechanism and Machine Theory, Vol. 13, pp. 437-450, 1978.

OBJECT-ORIENTED APPROACH TO PROGRAMMING A ROBOT SYSTEM

C. Zielinski

Warsaw University of Technology, Warsaw, Poland

Abstract: The paper presents an object-oriented approach to the implementation of a software library (MRROC+) which contains building blocks for the construction of multi-robot system controllers tailored to meet specific demands of a task at hand.

1. Introduction

Quite a considerable effort has been concentrated on developing new robot programming languages, both specially defined for robots [3, 10, 11], and computer programming languages enhanced by libraries of robot specific procedures [3, 4, 5, 14]. Specialised languages exhibit a closed structure. If new hardware is to be added to the system, usually some changes to the language itself have to be done. Especially, if new sensors are to be incorporated this problem arises, both because the hardware specific software has to be supplied and because the method of sensor reading utilisation in motion control has to be coded. Those changes have to be reflected in the language and this brings about the necessity of modifying the language compiler or interpreter. Because of this, robot programming languages/libraries submerged in universal computer programming languages are currently favored by robotics research community. Such programming systems have an open structure. Whenever the robot system has to be enhanced new hardware specific procedures are appended to the library/language and the universal language compiler remains unaltered.

The most frequently utilised language platform has been either Pascal, C or C++ very recently. Large libraries of robot specific procedures for the creation of both single- and multi-robot controllers have been designed, e.g. in C: RCCL [5], ARCL [4], RCI [8], KALI [1, 6, 7], RORC [12, 14, 16]; MRROC [17, 16]; in Pascal: PASRO [3], ROPAS [15]: in an object-oriented version of Pascal: ROOPL [13]; and in C++: ZERO++ [9].

While designing a controller for a new experimental manipulator with an arm of serial-parallel structure [2] exhibiting a very rigid structure, the following assumptions were made:

---

[1]The project has been supported by the Program in Control, Information Technology and Automatization at Warsaw University of Technology. The author acknowledges the participation of W. Szynkiewicz in the implementation of MRROC+.

- the controller should have an open structure facilitating any robot control investigations, e.g. external sensor incorporation and utilisation, trajectory planning and generation, research of servo-control algorithms,
- the controller should treat the new manipulator as one of many it will control,
- it should have a user-friendly operator interface,
- it should be easy to create a controller tailored exactly to a research task at hand by building it of ready library blocks (objects) or new procedures and processes that can be easily coded.

To fulfil the above requirements a concurrent version of C++ running on top of a multi-computer real-time operating system QNX-4 was chosen as an implementation platform. A formal approach to designing such controllers presented in [14, 17, 16] has been followed. The paper shows the formal design approach followed, the obtained generic structure of the controller and how this structure is implemented by C++ classes, objects, methods and processes, i.e. object-oriented programming paradigm.

2. Structure of the Multi-Robot Research-Oriented Controller

An open multi-robot system containing cooperating devices and equipped with diverse sensors is considered. No assumption is made as to what tasks will be performed by the system.

Since the data supplied by hardware sensors cannot directly be utilised in motion control it has to be processed to obtain an aggregate that can be used for trajectory modification or generation. This aggregate is named the virtual sensor reading.

$$\mathbf{v} = \mathbf{f}_v(\mathbf{r}, \mathbf{e}, \mathbf{c}) \tag{1}$$

where: $\mathbf{e}$ is the state of effectors, $\mathbf{r}$ is the state of receptors (hardware sensors) and $\mathbf{c}$ is the state of the control subsystem. The control subsystem is responsible for computing motion trajectories for the effectors using its internally stored data, current state of the effectors and the virtual sensor readings.

The system state $\mathbf{s}$ is decomposed by taking into account several distinct effectors and that rather aggregated sensor readings $\mathbf{v}$ than real sensor readings $\mathbf{r}$ are used by the control subsystem to compute a motion trajectory.

$$\mathbf{s} = \ <e_1, \ldots, e_{n_e}, v_1, \ldots, v_{n_v}, \mathbf{c}> \tag{2}$$

where: n_e is the number of distinct effectors in the system ($\mathbf{e} = <e_1, \ldots, e_{n_e}>$) and n_v is the number of virtual sensors ($\mathbf{v} = <v_1, \ldots, v_{n_v}>$). For the purpose of this paper the subsystems and their state are denoted by the same symbols.

To calculate the next effector state some computations have to be done. Those computations are done by the control subsystem. Obviously they can be done by a single centralised control subsystem, but a much better and clearer structure is obtained, if the state of the control subsystem $\mathbf{c}$ is partitioned into $n_e + 1$ parts. As a result the following is obtained:

$$\mathbf{c} = \ <c_0, c_1, \ldots, c_{n_e}> \tag{3}$$

Each subsystem c_l, $l = 1, \ldots, n_e$, is responsible for controlling an effector associated with it, and the subsystem c_0 is responsible for the coordination of all effectors. Hence, with each of the effectors $e_l, l = 1, \ldots, n_e$ an Effector Control Process is associated. Its state is expressed by c_l, $l = 1, \ldots, n_e$. The coordinating process is called the Master Process (MP) and its state is expressed by c_0.

Each control subsystem part c_l in conjunction with the part c_0 is responsible for calculating the next state e_{c_l} (calculated state) of the effector e_l (current state) and causes e_l to become equal to e_{c_l}, i.e. executes a motion step. Treating the system as a discrete time system, the next state of each of the effectors can be computed by a transfer function f_{e_l}:

$$e_{c_l}^{i+1} = f_{e_l}(e_l^i, v_1^i, \ldots, v_{n_v}^i, c_0^i, c_l^i), \quad l = 1, \ldots, n_e \tag{4}$$

The Master Process and each ECP are responsible for computing adequate transfer functions f_{e_l} and executing the related motions. The state of each part of the control subsystem, is the following:

$$
\begin{aligned}
c_0^{i+1} &= f_{c_0}(e_1^i, \ldots, e_{n_e}^i, v_1^i, \ldots, v_{n_v}^i, c_0^i, c_1^i, \ldots c_{n_e}^i) \\
c_l^{i+1} &= f_{c_l}(e_l^i, v_1^i, \ldots, v_{n_v}^i, c_0^i, c_l^i),
\end{aligned} \tag{5}
$$

Each virtual sensor $v_p, p = 1, \ldots, n_v$ is implemented as a process running concurrently to other Virtual Sensor Processes and the Effector Control Processes. In consequence of (1)

$$v_p^i = f_{v_p}(r^i, e_l^i, c_0^i, c_l^i) \tag{6}$$

is obtained, where e_l is the state of the l-th effector (the one associated with v_p). Here it is assumed that only a single effector influences directly a virtual sensor, because only this effector can directly change the state (or rather the configuration) of the real sensors that are mounted on it. It is envisaged here that real sensors fixed to different effectors will not form a single virtual sensor. Usually f_{v_p} depends only on r^i.

Each Effector Control Process creates or kills Virtual Sensor Processes according to the needs of control of motion. The Effector Control Processes in each step i obtain data from the Virtual Sensor Processes. Both kinds of processes can be treated as device dependent drivers. In this way, if only one component of the system is changed the remaining components remain unaltered.

The processes communicate through messages. The communication of each Effector Control Process with the Virtual Sensor Processes it uses can be of two kinds: interactive and non-interactive. In the case of interactive communication the Effector Control Process sends a data request message to an adequate Virtual Sensor Process. The Virtual Sensor Process reads the real sensors, aggregates the obtained data and sends the result to the Effector Control Process. In the case of non-interactive communication the Virtual Sensor Process reads the real sensors, aggregates data and leaves the resulting reading in a buffer without any request from any Effector Control Process. An Effector Control Process can access sensor data immediately by reading the buffer where the aggregated data is stored. A more elegant structure of the software component of the system can be obtained, if each Effector Control Process is partitioned into ECP proper and the Effector Driver Process EDP. The Effector Driver is responsible for:

- transformation of end-effector coordinates into joint coordinates and vice versa,
- transformation of joint coordinates into motor control increments and vice versa,
- transmission of the set-values to the servo-drives,

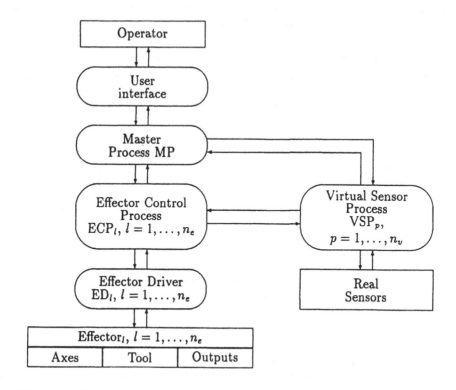

Figure 1: Software structure of `MRROC+`

- transmission of servo status to upper levels of control structure,
- computation of the servo-control algorithm.

The ECP proper, in this case, is responsible for trajectory generation, when the robots loosely interact or are not related to each other. In the case of cooperative action of the robots, the ECP proper simply transmits the commands of the Master Process, which acts as a coordinator. This structure was utilised in the Multi-Robot Research-Oriented Controller `MRROC+` (fig. 1).

3. Object Oriented Programming — OOP

Object-oriented programming (OOP) methodology evolved from structured programming. *Structured programming* is a method of describing a programming task in a hierarchy of modules, each describing the task in increasing detail, until the final stage of coding is reached (programming by step-wise refinement). Strict adherence to modules renders GOTO instructions unnecessary, in effect exhibiting a clear program structure. Nevertheless, initially structured programming treated data and algorithms operating on this data as two separate entities — procedural programming. The object-oriented programming paradigm integrates data and procedures. An *object* is a collection of *data* (variables of appropriate type, which should be treated as *fields* of a *record*) and *procedures* and *functions*, which are called *methods*, operating on these variables. Three main properties characterise an *object-oriented programming* language [18]:

- *encapsulation* — treating *data* and *code* operating on it as one entity – an *object*.
- *inheritance* — defining a hierarchy of *objects* in which each *descendant object* acquires all the properties of the *ancestor objects* (access to *data* and *code* of the *ancestors*) and receives some new properties specific to the newly created *object*.
- *polymorphism* — using the same name for an action that is carried out on different *objects* related by *inheritance*. The action is semantically similar, but it is implemented in a manner appropriate to each of the individual *objects* of the hierarchy.

OOP methodology assumes that certain abstract *objects* will be defined by the programmer. These *objects* have their properties (*data*) and exhibit behaviours (*methods*). The program is written in terms of *objects* behaving in such a way so as to change their properties, i.e. applying *methods* to change *data*.

4. OOP implementation of MRROC+

The overall structure of the system is dictated by theoretical considerations of the previous section which resulted in the division of the system into independent processes running concurrently either on separate computers connected into a network or on a single computer in a time-sharing fashion or both. The choice is made by the programmer implementing a specific task. The OOP paradigm can be used only within the scope of each process separately. Nevertheless the communication buffers treated as *classes* can be shared by two processes communicating with each other. Each process has to do a job of its own and has to communicate with its neighbors obtaining data from them and in return delivering the results of its processing. This will be exemplified by explaining in more detail the functioning of the Effector Driver Process EDP which itself is composed of several sub-processes (fig.2).

The EDP receives commands from its corresponding ECP. The commands are:

- synchronise the robot,
- SET: tool definition, arm pose, outputs,
- GET: tool definition, arm pose, inputs,
- SET and GET simultaneously.

SETting an end-effector pose results in an arm motion and GETting the arm pose results in reading its current position and orientation. The tool definition can be specified in several ways: homogeneous transformation, X, Y, Z coordinates in relation to the wrist flange and several sets of orientation angles (e.g. Euler) or by using an angle and axis notation. The arm pose can be specified in the above manner too and moreover using joint variables and motor increments. The coordinates can be relative or absolute.

This variety of command argument combinations causes the inter-process message to have variable structure. Since at both ends of the communication channel the message buffer has to have the same structure its definition has to be shared by both communicating processes, so a single *class* is defined for both. On both ends of the communication channel different actions are performed on the buffer (one process loads it and the other unloads it and interprets its contents), so the buffer is enclosed in another *class* that is specific to each of the processes. Moreover each process performs its specific actions on the obtained data. This data and the actions (*methods*) are contained in a process specific *class*. Finally a single *class* is derived from the three classes, i.e.: process specific class, input and output buffer classes. Hence, within a process a single *object* exists which inherits its properties from all of the three component *classes*. This method of coding has been replicated for: EDP_MASTER, READING_BUFFER and all of the SERVO processes (fig.3). In the case of EDP_MASTER the *transformer class* is responsible for all coordinate transformations and in the case of SERVO processes the *regulator class* stores control data and computes the servomotor control algorithm. The derived *classes* are responsible for the inter-process communication.

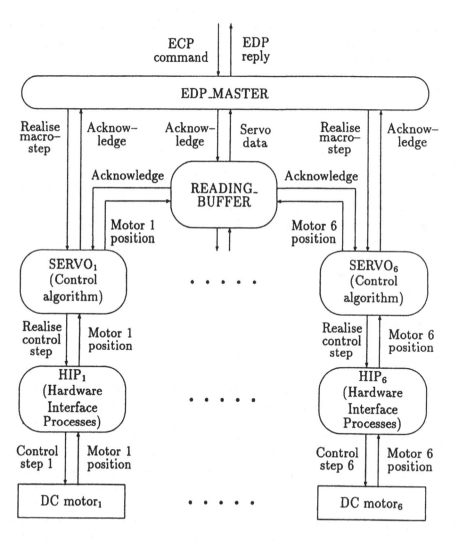

Figure 2: Internal structure of the Effector Driver Process

For a number of reasons errors can occur during computations. Three types of errors can occur: non-fatal errors and two kinds of fatal errors: robot and system fatal errors.

Non-fatal errors are due to invalid parameters of commands external to the driver and result in computation errors. Robot fatal errors are due to improper functioning of robot hardware. From the above mentioned errors the controller must be able to recover and continue functioning. System fatal errors are caused by improper functioning of the computer network or are due to errors in the EDP software itself. In this case the operator has to be informed about the cause of the problem and the system has to be halted. In MRROC+ the first two kinds of errors are treated as *exceptions* and adequate *exception handlers* deal with them separately. Separation of the code dealing with errors from the code handling normal operation again greatly simplifies programming, rendering the code more reliable and easier to debug and modify. Error handling depends not only on the cause of error but also on the place in code that the error has been detected. The same error occuring in different system states might need different actions. The OOP error handling capability is especially well suited to this purpose.

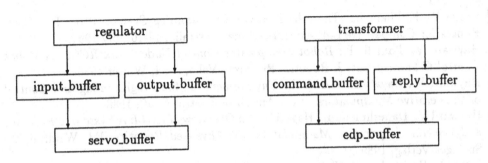

Figure 3: Class inheritance within the SERVO and EDP_MASTER processes

5. Conclusions

The proposed method of coding has several advantages. The change of the robot or of a servomotor brings about only the change in internal functioning of coordinate *transformer* or *regulator class* without changing the inter-process communication. In this way, assessment of the results of changes in the robot arm kinematic structure, servo-control algorithms or a type of an actuator, forces the investigator to change a relatively small portion of the overall controller code, well localised at that, and without unexpected interactions. This feature is due to proper structuring of the code and utilisation of OOP methodology. The structure of the system resulted from formal considerations which took into account the fact that each distinct component of the system should be controlled by its own process, thus limiting inter-process communication and reducing implementation code dependencies. As *classes* contain data and procedures operating on it in one entity the formal parameter lists for *methods* are usually empty, what greatly simplifies coding. Summarising, the experiment with implementing a multi-process concurrent controller for a multi-robot system proved that utilisation of OOP paradigm significantly simplifies both the initial programming effort and reduces the time needed for modifying the code when adding robots of a new type to the system. Currently the upper layer of the controller structure, i.e. ECP and MP, are being recoded using OOP methodology. Although OOP is limited, to a great extent, only to each process, the clarity of code and its simplification due to proper structuring prove beneficial. The system contains two IRb-6 robots (one mounted on a track), conveyor belt, vision system, force/torque sensor, ultrasonic and infra-red proximity sensors. The latest addition to this system is the prototype of the rigid robot [2], which is currently thoroughly investigated by using the described controller.

References

[1] Backes P., Hayati S., Hayward V., Tso K.: *The KALI Multi-Arm Robot Programming and Control Environment.* Proc. NASA Conf. on Space Telerobotics, 1989.

[2] Bidziński J., Mianowski K., Nazarczuk K., Słomkowski T.: *A manipulator with an arm of serial-parallel structure.* Archives of Mechanical Engineering, Vol.34 , 1992.

[3] Blume C., Jakob W.: *Programming Languages for Industrial Robots.* Springer-Verlag, 1986.

[4] Corke P., Kirkham R.: *The ARCL Robot Programming System.* Proc. Int. Conf. Robots for Competitive Industries, Brisbane, Australia, 14-16 July 1993.

[5] Hayward V., Paul R. P.: *Robot Manipulator Control Under Unix RCCL: A Robot Control C Library.* Int. J. Robotics Research, Vol.5, No.4, Winter 1986.

[6] Hayward V., Hayati S.: *KALI: An Environment for the Programming and Control of Cooperative Manipulators.* Proc. American Control Conf., 1988.

[7] Hayward V., Daneshmend L., Hayati S.: *An Overview of KALI: A System to Program and Control Cooperative Manipulators.* In: *Advanced Robotics.* Ed. Waldron K., Springer-Verlag, 1989.

[8] Lloyd J., Parker M., McClain R.: *Extending the RCCL Programming Environment to Multiple Robots & Processors.* Proc. IEEE Int. Conf. Robotics & Automation, 1988.

[9] Pelich C., Wahl F. M.: *ZERO++ – An OOP Environment for Multiprocessor Robot Control.* IASTED Int. J. Robotics and Automation, 1995. (in printing)

[10] Zieliński C.: *TORBOL: An Object Level Robot Programming Language.* Mechatronics, Vol.1, No.4, Pergamon Press, 1991.

[11] Zieliński C.: *Object Level Robot Programming Languages.* In: *Robotics Research and Applications.* Ed.: A. Morecki et.al., Warsaw 1992.

[12] Zieliński C.: *Flexible Controller for Robots Equipped with Sensors.* 9th Symp. Theory and Practice of Robots & Manipulators, Ro.Man.Sy'92, 1-4 Sept. 1992, Udine, Italy, Lect. Notes: Control & Information Sciences 187, Springer-Verlag, 1993.

[13] Zieliński C.: *Robot Object-Oriented Pascal Library: ROOPL.* J. of Theoretical and Applied Mechanics, Vol.31, No.3, 1993.

[14] Zieliński C.: *Controller Structure for Robots with Sensors.* Mechatronics, Vol.3, No.5, Pergamon Press, 1993.

[15] Zieliński C.: *Sensory Robot Motions.* Archives of Control Sciences, Vol.3, no.1, 1994.

[16] Zieliński C.: *Robot Programming Methods.* Publishing House of Warsaw University of Technology, 1995.

[17] Zieliński C.: *Control of a Multi-Robot System*, 2nd Int. Symp. Methods & Models in Automation & Robotics MMAR'95, 30 Aug.–2 Sept. 1995, Międzyzdroje, Poland.

[18] *Turbo C++: Getting Started.* Borland International Incorporated, 1990.

CONTACT TASKS REALIZATION
BY SENSING CONTACT FORCES

B. Borovac, L. Nagy and M. Sabli
University of Novi Sad, Novi Sad, Yugoslavia

1. Introduction

A key issue in robotized realization of any type of the contact task is the control of forces arising during contact between the object and the environment. The most common way of force measurement is with a force sensor, usually placed at the robot wrist, which can measure all six components of force and moment. "The force sensor and the environment are both the systems with natural frequency responses of the same order of magnitude as that of the manipulator. When these are coupled by contact of the manipulator with the environment, then the resulting system is very difficult, if not impossible, to stabilize" [1]. It will be very desirable that the sensor have mechanical compliance properties, i.e. "soft sensors are needed" [2]. The passive compliant behavior of the object itself can be realized by soft grasping of the object, i.e. by eliminating rigid contact surfaces. This can be achieved if soft material [3,4] between the hard object and the gripper "skeleton" is used. Use of a soft material on the fingertips [5,6] can significantly improve behavior of the whole system. Characteristics of various soft materials used on fingertips have been investigated in [7].

In [8], an idea was initially proposed for a sensor capable to measure forces on the contact between the object and the gripper. In [9], we presented an improved design of the sensor, which minimizes the number of necessary sensing elements, and improves the overall information relevant for realization of contact tasks. Further development of this idea [10] leaded to new realization of the sensor where soft and elastic element is whole body of the sensor, and with optocaplers as a non-contact sensing elements.

A number of papers are dealing with tactile sensors. Large portion of research efforts is devoted to investigation of surface characteristics (texture [11-13], crack detection [14], rubbing velocity, deepness of contact [15], friction estimation [16], slip detection [13, 17]).

Measurements of contact forces, extremely important research field, is described in [17-20]. In [18] is described sensor capable of measurement just of normal force, while in [17] is described sensor for measurement of 3-D force acting on the sensor tip. Sensor is very small and convenient for placing on the curved surfaces. Due to its size a number of the sensors may be placed on a single fingertip and activated simultaneously during grasping. Single sensor consists of the hard tetrahedral core covered by the semi-spherical silicon rubber with three knobs pressing three sensing elements. Force is acting on its tip of the tetrahedral core, which transfer force to the knobs. Change in sensing elements resistance correspond to contact force on each knob. This sensor is, actually, not a soft sensor (deflection of the resistive layer below knob may be neglected) and "soft grasping" (i.e. compliant motion of the grasped object) is not possible. Sensor is, also, not capable of moment measurement. In [19] is described very interesting approach to contact forces measurement using "tensor cells" which are distributed inside soft tissue covering gripper finger. "Tensor cell" is capable of measuring complete stress tensor inside flexible body of the soft tissue. In [20] three-axis tactile sensor consisting of CCD camera, light source, acrylic board and silicon rubber sheet as a surface where contact with object occur, is described.

2. Sensors with Soft Contact Surfaces

There are two basic reasons why soft grippers are more appropriate for contact tasks: passive adaptation on positional inaccuracies of objects in contact and variable stiffness of the position of the grasped object. Undesirable contact forces may occur, either, in the initial moment of contact or during task realization. In case the object is grasped with the gripper having soft contact surfaces instantaneous passive adaptation on the positional inaccuracies of the objects in contact is ensured and occurrence of the large contact forces is avoided. Different contact tasks and various objects involved require different grasping stiffness of the object. When object is grasped with gripper having soft contact surfaces desired stiffness can be achieved simply by changing grasping force, even, during task execution.

Using grippers with soft contact surfaces, industrial robots (with rigid links and stiff joints) can be applied in contact tasks where compliant behaviour is a basic requirement.

2.1. Sensor Design and Basic Force Portraits

Sensor consists (Fig. 1.) of two prismatic supports connected by a rigid "bridge" with a layer of soft silicon rubber on top of it, and of the lower part with two rows of appropriate shape. The sensor's basic element is a prism of rigid material [5] of triangular cross-section.

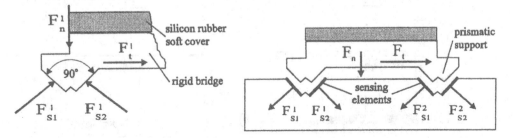

Fig. 1. *Forces on supporting elements and sensor scheme*

Sensing elements (Force Sensing Resistors [21,22]), for measuring normal contact forces, are sandwiched between the prism and the sensor lower part. By measuring forces F^1_{s1}, F^1_{s2}, F^2_{s1} and F^2_{s2} on the sensing elements, the components of normal F_n and tangential F_t, force acting on the top of sensor can be computed. The way of incorporating the sensor into the gripper is shown in Fig. 3. It is clear that the contact between the sensor and object is realized via soft rubber layer, which enables "soft grasping" of the object. When contact force is acting on the object free end, it moves relative to the sensor "bridge", and cause a change in the distribution of pressure over the contact surface. The best way to obtain complete information about forces acting on the grasped object is by forming force portraits.

In Fig. 2 are given three basic force portraits for two sensors which are placed on the opposite fingers of a parallel gripper. The horizontal bars denote intensities of the normal forces on each supporting prism, while vertical bars are proportional to total axial force on each sensor. The calculated position of acting point of total normal force is also denoted.

If no external forces are acting on the tip of the grasped object, intensities of all normal forces on prismatic supports are same and no axial forces exist. In Fig. 2. a. axial force F_a is applied on the object tip, only. Due to deformation of the sensor's soft cover, both upper normal forces increase and both lower normal forces decrease. Axial forces on both sensors increase in same direction. In Fig. 2. b. is shown force portrait corresponding to situation when force F_s, perpendicular to object axis (see Fig.3.) is applied, causing pure inclination of the object. Normal forces on two prismatic supports are increased, while the intensities of the

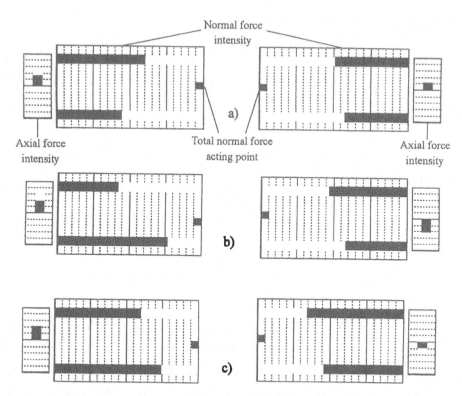

Fig. 2. *Force portraits: a) Axial force F_a applied only, b) Side force F_S applied only, c) Axial and side forces, F_a and F_S, applied simultaneously*

other two decreased. Induced axial forces on each sensor have the same intensity, but opposite directions. In Fig 2. c. both forces, F_a and F_S, are acting simultaneously on the object tip. Thus, it is important to point out that pair of such sensors used simultaneously can, from total force acting on the object tip, extract pure axial component. This is extremely important in applications where axial force, as in "peg-in-hole" task, should be separately monitored.

3.0. The Gripper Prototype and Experiments Realization

A prototype of the gripper equipped with sensors was developed (Fig. 3) by adapting the standard swinging gripper with two fingers. Two plates, each carrying two sensors, were added. Two pairs of sensors on opposite sides of the object lie in two perpendicular planes (Plane 1 and Plane 2), intersecting in the object axis. The sensors LR and RF will simultaneously detect movements of the object in Plane 1, while the sensors RR and LF will detect movements of the object in Plane 2. Any movement of the object which is not strictly in these

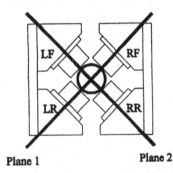

Plane 1   Plane 2

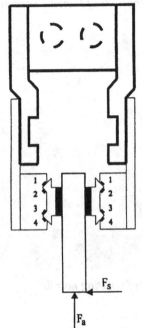

Fig. 3. *Gripper design and sensors disposition*

two planes will be sensed by all four sensors. For each pair of the sensors in one plane we can form patterns of sensed forces analogous to those shown in Fig. 2. A gripper prototype was mounted on the industrial robot ASEA Irb 6. Same gripper was used in all three experiments [23].

For realization of the "peg-in-hole" task the strategy described in [4] were used. It was supposed that in the initial moment tip of the object is at the hole entrance, but certain misalignment of the axes of the object and the hole exists. Two subtasks had to be realized simultaneously. The first one was to keep axial force permanently within predefined range, corresponding to desired insertion force, while the second one was to eliminate relative misalignment of the object and the hole axes. To do this the gripper should move in the plane orthogonal to the hole axis in such a direction to eliminate object inclination due to relative axes misalignment. Due to the fact that axial force is permanently applied, during compensation movement, object will be partially inserted while its axis is passing through the cone where insertion is possible. Procedure should be repeated till object is fully inserted. The axial force was sensed directly, while compensation movement direction was calculated on the basis of the intensities of normal forces on all four sensors. Each force was considered as a vector, and compensation direction was defined by simple vector addition. Experimental setup is shown in Fig. 4.

The second task was handle rotation whose lever length is exactly known, but with unknown exact position of the rotation axis. Handle is grasped (Fig. 5.) in point (x_1, y_1). All normal forces on the sensors are equal. Let rotation center position is known approximatelly and given to be in O_1. When rotation for angle φ is performed, handle arrive in point (x_2, y_2) where equilibrium of normal forces is disturbed. Then, the gripper have to move in such a direction to eliminate disturbance. Direction of compensation motion is determined in similar way as in the

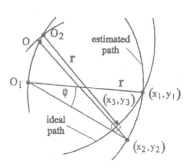

Fig. 5. *Compensation of handle position*

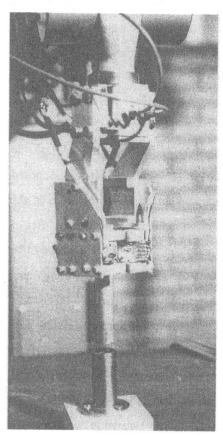

Fig. 4. *"Peg-in-hole" task*

Fig. 6. T*ask of rotating handle*

"peg-in-hole" task. Compensation motion is stopped when balance of forces is established again. Due to the sensor readings inaccuracies gripper will not come exactly to the point on the ideal path, but close to it (point x_3, y_3). New calculated rotation center (which is closer to real center of rotation O) is O_2 which is in intersection of two circles of radii r whose centers are (x_1, y_1) and (x_3, y_3). It is important to point out that, angle of rotation φ is not fixed in advance. Rotation stops when forces overcome a predefined intensity. Experimental setup is shown in Fig. 6.

The third experiment was writing with ball pen on a paper. Relative angle between paper and gripper was fixed in advance. Desired pressure necessary for writing was ensured by pressing pen tip against writing surface what cause pen inclination within gripper. Elastic deformation of the sensor soft cover ensured automatic compensation of small irregularities of writing surface profile. In the horizontal plane robot have to perform such motion to ensure

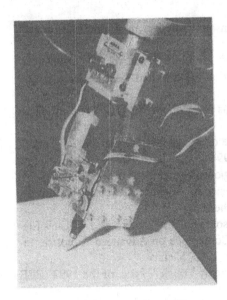

Fig. 7. *Task of writing*

adjusted only in case if pen inclination (what is equivalent to contact force) is out of predefined region. In our experiment, pen path was defined manually by mouse. Experimental setup is shown in Fig. 7.

4.0. Conclusion

A new and improved design of a force sensor with soft contact surfaces for sensing forces in two orthogonal directions relative to the contact surface of the gripper fingers and the grasped object (parallel with the contact surface and orthogonal on it) is described. Use of more sensors simultaneously and forming force portraits, enables deriving a more complex information about forces acting on the object. A prototype of the gripper with four sensors was designed and manufactured. Some basic patterns of normal and tangential forces from sensors which are of crucial importance in such realization of contact tasks are presented. A successful realization of three contact tasks: "peg-in-hole" task, task of rotating handle, and task of writing with the ball pen, using same gripper were described.

5.0. Acknowledgments

Authors are grateful to students V. Stojanović and R. Varga for their excellent work done during manufacturing and testing of sensor prototype and redesigning the gripper..

6.0. References

[1]Paul, R.P.,"Problems and Research Issues Associated with the Hybrid Control of Force and Displacement" Proc. IEEE ICRA, pp. 1966-1971., 1987.

[2]Whitney, D.E. "Historical Perspective and State of the Art in Robot Force Control", The Int. J. of Robotic Research, Vol. 6, No. 1. 1987.

[3]Borovac, B., Šešlija, D, Stankovski, S. "Generalized Approach to the Control of Assembly Process Using Tactile Sensors with Soft Fingers" Proc. of the 6-th International Conference on CAD/CAM, Robotics and Factories of the Future, London, 1991.

[4]Borovac, B., Šešlija, D, Stankovski, S., "Soft Sensored grippers in Assembly Process", Proc. IEEE ICRA, 1992.

[5]Cutkosky, M., Howe, R., "Dynamic Tactile Sensing", Proc. of RO.MAN.SY., 1988.

[6]Howe, R., Cutkosky, M., "Sensing Skin Acceleration for Slip and Texture Perception", Proc. IEEE ICRA, pp. 145-150., 1989.

[7]Shimoga, K. B., Goldenberg, A.A. "Soft Materials for Robotic Fingers. Proc. of IEEE ICRA, 1992.

[8]Borovac, B., Nikolić, M., Nagy, L., "A New Type of Force Sensor for Contact Tasks", Proc. of IEEE ICRA, pp. 1791-1796, San Diego, 1994.

[9]Borovac, B., Nagy, L., " New Design of Two-Component Force Sensor for Contact Sensing", Proc. of The First ECPD International Conference on Advanced Robotics and Intelligent Automation, pp. 265-270, September 1995, Athens

[10]Borovac, B., Nagy, L., Sabli M." New Tactile Sensor with Optocaplers as a Sensing Elements", Submitted to Second ECPD International Conference on Advanced Robotics, Intelligent Automation and Active Systems, September 1996, Vienna

[11]Russell, R. A., Parkinson, S., " Sensing Surface Shape by Touch ", Proc. of the 1993 IEEE ICRA, pp. 423-428, Atlanta.

[12]Shinoda, H, Uehara, M., Ando, S, "A Tactile Sensor using Three-Dimensional Structure", Proc. of the 1993 IEEE ICRA, pp. 435-441, Atlanta.

[13]Shimojo, M., Ishikawa, M., "An Active Touch sensing Method Using a Spatial Filtering Tactile Sensor ", Proc. of the 1993 IEEE ICRA, pp. 948-954, Atlanta.

[14]Patil, R., Basu, A., Zhang, H., " Crack Detection using Contact Sensing", Proc. of IEEE ICRA, pp. 1772-1777, San Diego, 1994.

[15]Shinoda, H, Ando, S, "Ultrasonic Emmision Tactile Sensor for Contact Localization and Characterization", Proc. of the 1994 IEEE ICRA, pp. 2536-2543, San Diego.

[16]Tremblay, M. R., Cutkosky, M. R., "Estimation Friction Using Incipient Slip Sensing During a Manipulation Task ", Proc. of the 1993 ICRA, pp. 429-434, Atlanta.

[17]Yamada, Y., Cutkosky, M. R., "Tactile Sensor with 3-Axis Force and Vibration Sensing Functions and Its Application to Detect Rotational Slip", Proc. of the 1994 ICRA, pp. 3550-3557, San Diego.

[18]Hutchings, B, Grahn, A., Petersen., R, "Multiple-Layer Cross-Field Ultrasonic Tactile", Proc. of the 1994 IEEE ICRA, pp. 2522-2528, San Diego.

[19]Shinoda, H, Uehara, M., Ando, S, "A Tactile Sensor using Three-Dimensional Structure", Proc. of the 1993 IEEE ICRA, pp. 435-441, Atlanta.

[20]Ohka, M., Mitsuya, Y., Tacheuchi, S., Ishihara, H., Kamekawa, O. "A Three-Axis Optical Tactile Sensor (FEM Contact Analyses and Sensing Experiments Using a Large-Sized Tactile Sensor)", Proc. of the 1995 IEEE ICRA, pp. 817-824, Tokyo.

[21]Borovac, B., Šešlija, D, Stankovski, S. "Investigation of Basic Characteristics of Force Sensing Resistor Sensors for Robotic Hands", Proc. Third ISMCR, pp.7-11., Torino, 1993.

[22]Interlink Electronics - FSR^{TM} Integration Guide & Evaluation Parts Catalog, 1993.

[23]Nagy, L., Borovac, B., Sabli M. " Contact Tasks Experiments Based on an Improved Two-Component Force Sensor ", Video Proc. of the 1996 IEEE ICRA, Atlanta.

Chapter XI
Application
and Performance Evaluation

APPLICATION OF FORCE CONTROL IN TELEROBOTICS

R. Bicker, D. Glennie and Ow Sin Ming
University of Newcastle upon Tyne, Newcastle upon Tyne, UK

Abstract

A force reflecting telerobotic system has been developed at the University of Newcastle upon Tyne which incorporates a man-machine interface that permits the user to configure the 'slave robot' to be operated manually, as a conventional teleoperator, or autonomously as a sensor driven robot [1]. The system uses a Puma 260 robot as a force-reflecting hand controller, and a Puma 762 industrial robot has been configured to operate as the 'slave'. A network of parallel processors is used for both high and low level control. The implementation permits the operator to manually control the position (and/or velocity) of selected axes, whilst the remaining axes are controlled autonomously, using force control, in either world or tool coordinates.

Introduction

Mechanical master-slave teleoperator systems are widely used in the nuclear industry in applications involving relatively low-activity radioactive waste handling. The intrinsic force-reflection yielded by the mechanism provides the operator with a sense of *feel*, and permits dextrous manipulations, albeit in relatively close proximity to the hazard. Electrically-linked and hydraulic master-slave teleoperator systems enable the operator to be situated further away from the danger, and are thus to be preferred in applications which are extremely hazardous. However, it is considered essential that some form of force-reflection be introduced, which must be augmented via the servo-system. Such bilateral systems can exhibit a high mechanical stiffness and when coupled with low servo-update rates and the operator's relatively poor neuro-muscular response, can result in sustained oscillations when the slave arm comes into contact with a non-compliant or rigid surface - particularly when high gain force reflection is used.

Most teleoperator systems have directly coupled joint-based servoes, using either position-position feedback, or position-force feedback based on motor current or hydraulic pressure sensing. Bilateral control schemes based on these techniques have a relatively low bandwidth of operation, with a consequential reduction in fidelity. Nevertheless, the ability of operators to routinely perform hazardous tasks relies on intelligence and well-integrated sensory and manipulative characteristics to recognise problems and to plan the necessary corrective action.

This present paper describes some of the main features of a tele-robotic system which is being developed at Newcastle University that makes use of, and takes advantage of, the technologies developed in the field of robotics, such as, pre-programmed operation and teach-and-repeat, with fully integrated force-reflection [2]. The system provides the operator with both bilateral force-feedback when operating in teleoperator mode, and can also make use of an autonomous force control when the slave is used robotically. The implementation also incorporates *shared control* [3], which permits the operator to manually control the position (and/or velocity) of selected axes on the slave robot, whilst the remaining axes are controlled autonomously, using an active compliance force control algorithm, to maintain the slave end-effector in contact with the workpiece at a prescribed force (or torque) along selected axes.

Force-reflecting *Master* arm

A six-axis Puma 260 robot has been configured as an *active joystick*, with a miniature six-axis force-torque transducer mounted between the wrist of the robot arm and the hand-grip (see Fig. 1). The force-torque sensor provides feedback of both magnitude and direction of the applied forces/torques exerted by the operator's hand, and this information is used to *backdrive* the 260 in the direction of the applied forces/torques. The VAL controller has been adapted to servo the robot via a commercial multi-axis motion controller, which has a servo-update period of 1 ms.

The 260 has been used to teleoperate both a large 5-axis computer controlled gantry serving manipulator, and a Puma 762 'slave' robot which was suspended from the gantry mast [1]. In rate mode the 260 behaves as an active 'spring', and the cartesian velocity of the slave axes are proportional to the displacement of the master arm from its null position. In view of the relatively small size of the master, this mode of control is considered particularly useful when large spatial movements are necessary.

In unilateral position mode, selected axes of the slave robot are driven with a one-to-one correspondence, i.e. without force reflection. When in bilateral mode, the forces experienced at the end-effector of the slave robot are reflected back to the operator, using one of three pre-set gain ratios. The operating envelope of the hand controller is limited by software to a cube of size 200mm for translation, and ±45° for wrist rotations. Whilst these limits prevent excessive movement of the master arm, they are also used to facilitate soft switching from position to rate control if any axis encounters its limit.

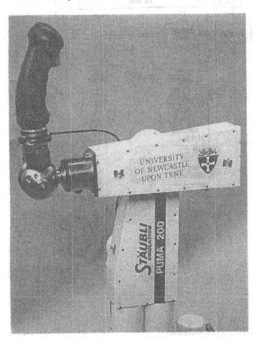

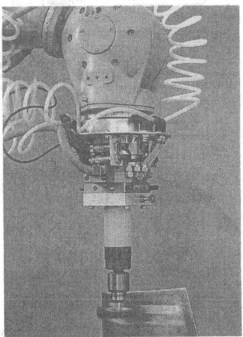

Fig.1 *Master* Controller   Fig.2 *Slave* Manipulator

Puma 762 Slave Robot

A six-axis Staübli-Unimation Puma 762 industrial robot has been configured as the 'slave' robot. Fig. 2 illustrates a force-controlled polishing operation on a contoured surface. The 760 VAL controller has been fitted with a dedicated Ethernet interface, called E-Slave, which facilitates real-time servoing of the joint axes every 3.6ms. A six-axis force/torque sensor is mounted on the wrist of the slave robot, to which is attached a tool-change adaptor. Several tools have been developed for use on the facility, including an electric drill, reciprocating saw, impact wrench, angle grinder and a pneumatic parallel-action gripper. When not in use, the tools are stationed in a tool change magazine mounted on the base of the robot, and can be automatically selected as and when required. The range of tools has been designed to investigate a number of tasks which by their nature involve hard/hard contact with the environment. Passive compliance has been built into the design of the impact wrench, whereas all other tools are rigidly mounted onto the tool adaptor.

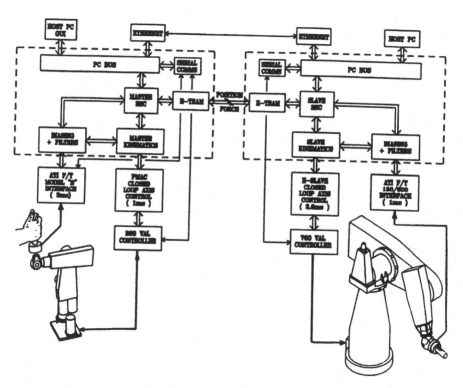

Fig. 3. System Hardware Architecture

Control System Architecture

The system has been designed to facilitate true remote operation, and is configured using two sub-system controllers, as shown in Fig. 3. The *Master* sub-system controller accommodates supervisory communication, based around a central or host computer, and also incorporates the real-time control processes for the high-level servoing of the master arm, in conjunction with the multi-axis motion controller. The *Slave* sub-system controller provides autonomous task-based control of the slave arm when operating in *robotic* mode, and also serves to provide coordinate and axis control of the slave servo system. The supervisory level monitors and schedules the different processes occurring during task execution, and enables different modules to share computing resources. The networked architecture provides multiple independent communication paths between processors, each capable of operating at data rates up to 20 Mbits/sec. The sub-system controllers are linked to each other via supervisory and real-time Ethernet connections.

Both forward and inverse kinematic solutions and singularity protection algorithms are employed for both the 260 master and 760 slave robots. These are computed in real-time to provide joint servo demands to their respective joint controllers. The 260 is capable of driving the slave manipulator using a variety of operating modes, in either joint, world or tool coordinates. Co-ordinate transformations, including forward and inverse kinematics, and singularity protection processes are embedded on a single T805 floating-point transputer processor in each sub-system controller. Joint axis demands are subsequently translated to the master sub-system using dual-ported ram, and an Ethernet tram (*transputer module*) is dedicated to servicing the E-slave communications on the slave sub-system. Running average filters and force-sensor biasing algorithms are also implemented on both sub-systems. Status checking is executed in real-time to facilitate a safe and systematic close-down in the event of a fault being detected on either robot controller or force sensing system.

In unilateral teleoperator mode, path modification data for the 'slave' robot is computed using position data derived from the 260 input device. In bilateral control mode, path modification data for both the master-arm and the slave-arm are computed, based on the difference between the forces (and torques) applied by the operator at the master arm, and at the tool/task interface on the slave.

Man-Machine Interface

A graphic user interface has been developed to provide the operator with a relatively simple, yet powerful, means of controlling the telerobotic system via a single console. The operator is able to select one of four different terminal emulator routines, one for each of the two VAL controllers, which allows off-line communication via the terminal inputs. Terminal emulators for the force/torque sensors are used primarily for monitoring the functioning of the sensors and their status. The operator is able to view both raw sensor data, and resolved force/torque data. Other functions are available to facilitate automatic tool changing operations and a control interface for initialising and executing autonomous task routines.

In tele-operator mode the operator can select from a comprehensive range of different operating modes. *Soft-switching* is used to activate (or deactivate) axis control in either rate, position (unilateral or bilateral) and autonomous modes. Radio buttons allow different gains to be selected for each axis. To assist the operator during task execution a real-time force/torque bar-graph display gives the operator an indication of the magnitude and direction of each component of force/torque, experienced by either the master-arm or slave-arm force sensors. The joint angles of the slave robot are also displayed, in relation to their allowable range of travel, along with a 'singularity-index' which provides the operator with an indication of the proximity of the slave robot to a singular position. Radio buttons provide functional control of individual tools, opening/closing of the gripper, a system pause function, switching of the slave controllers discrete I/O lines, and indexing/resetting of the 260 hand controller.

Shared Compliant Control

An effective task-based shared control scheme must be flexible and facilitate direct operator intervention at any time. The merging and extraction of any direct human input command must be carried out in a smooth fashion, such that the influence of any autonomous command is transparent to the operator. In tasks, such as drilling and cutting, the operator manually positions the tool (using teleoperation) at the desired location, and then an automatic alignment procedure can be executed to ensure the tool is accurately aligned, relative to the workpiece. Once completed, the operator can then initiate the machining operation. At any time during this procedure it is possible to pause, or stop the sequence. The operator acts in a supervisory capacity, whilst the system operates in a semi-autonomous mode.

Most assembly operations are rather similar, e.g. inserting an electrical plug into a socket is quite similar to inserting a peg into a hole, although the pins in the plug must be accurately aligned with the socket. Whilst a specific search algorithm can be used to autonomously carry out the assembly operation, a shared control strategy can be adopted by the operator to easily accommodate the task requirements, and also allow intervention during real-time execution. The task-based scheme uses the real-time graphic force/torque display to allow the operator to monitor the progress of task execution.

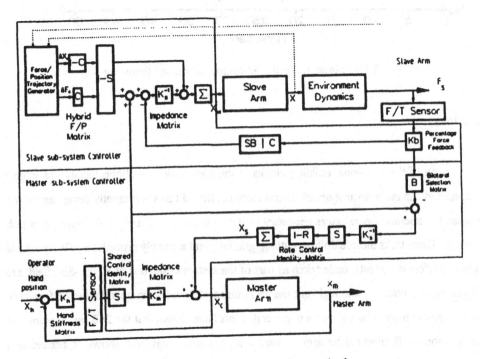

Fig. 4 Block diagram of shared control scheme

Shared control has recently been applied to cutting, grinding and polishing tasks. A task sequence is initiated which first instructs the operator to manually position the tool adjacent to the workpiece. Next a guarded move is executed which brings the tool into contact with the workpiece, after which point the contact force is then increased to a pre-set magnitude. The tool is switched on, and the operator is then able to programme the tool to move at a nominally constant tangential speed along a contoured surface, whilst the control system regulates the normal force, and simultaneously follows the surface contour using a hybrid force/position strategy . Fig. 4 illustrates the architecture of the shared control scheme, and Fig. 5 shows the results of a polishing test on a contoured surface, illustrating the regulation of contact force and contour tracking.

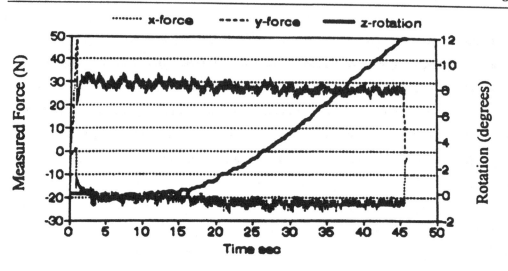

Fig. 5 Test results using shared control in contour following

Conclusions

The integration of the functional building blocks in the Newcastle telerobotic facility has largely been completed. An in-depth analysis of several force controlled tasks is presently being carried out to establish the operating parameters necessary to define both manual and semi-autonomous task descriptions. These tasks include drilling, cutting, grinding, and assembly procedures. The results of preliminary performance tests, undertaken as part of the system evaluation at BNFL-Sellafield, are considered most promising, and highlight the importance of establishing quantitative performance criteria for remote manipulator systems in general. Tests have shown that the 260 can operate as a generalised, force-reflecting master arm. In particular, the device has been shown to behave in a smooth, stable manner, which helps to minimise operator fatigue, and the ability to drive it relative to world or tool-based coordinate frames ensures that cross-coupling effects are kept to a minimum.

References

[1]   TELEMAN Programme on remote handling in hazardous or disordered nuclear
        environments, Commission of the European Communities, 'TELEMAN TM48
        Intelligent Nuclear Gantry Robot Integrated Demonstrator (INGRID)', 1995
[2]   Ow SM, 'Task Based Shared Compliant Control of a Telerobotic System',
        University of Newcastle upon Tyne, PhD Thesis, to be submitted in March 1996.
[3]   Hayati S & Venkataraman S T, 'Design and implementation of a robot control system with
        traded and shared control', Proc. of 1989 IEEE Int. Conf. on Robotics & Automation.

DEVELOPMENT AND APPLICATION OF THE TELBOT SYSTEM
A NEW TELE ROBOT SYSTEM

W. Wälischmiller and Hong-You Lee
Hans Wälischmiller GmbH, Meersburg, Germany

Abstract

This paper presents a new tele robot system, trade named TELBOT. The robot is developed by Hans Wälischmiller GmbH and has been used for steam generator maintenance in Canada and for decommissioning glove boxes in Japan. This is a robot with six revolution joints and a unique drive system that allows unrestricted motion of each joint. All of the electronics, cables, motors, reduction gears and encoders are contained within the unique base of the manipulator. The transmission of motions of the arm are translated from the base drive system by a set of concentric tubes that make up the links and sets of bevel gears that make up the joints. There is no electrical parts in the arm itself, therefore, all the revolution joints can rotate over 360 degrees continuously and the arm itself is relatively compact and lightweight. There is no closed-form solution of the inverse kinematics, because the wrist joint axes do not intersect at one point. The method to solve the inverse kinematics problem of series-chain manipulators with general geometry proposed in the earlier papers of the second author is used to find all the 16 possible configurations. It greatly facilitates selecting the optimal configuration for a given workspace and avoiding obstacles.

Fig. 1 TELBOT

1 Introduction

In recent years, the acceptable levels of worker exposure to environmental hazardous such as radiation, toxic fumes and asbestos has steadily decreased. As a result, the use of both tele-operated and computer-controlled robots has increased dramatically.

A general purpose tele robot system, trade named TELBOT, has been developed by Hans Wälischmiller GmbH and used for nuclear steam generator maintenance in CANADA and for decommissioning glove boxes in Japan. The Telbot robot has six degrees of freedom and a unique drive system that allows unrestricted motion of each joint (Fig. 1). All the motors, measuring system and the transmission gear box of the manipulator are located in the robot base. There is not any electrical cables from the base to the end-effector. The transmission of the motion from the motors to the end-effector is realized via concentric tubes and bevel gears, so that all the revolution joints can be rotated over 360 degrees continuously. The end-effector or the tool can reach its work position in any orientation. Furthermore, due to there is no electrical cable in the whole arm, the manipulator can be easily protected from the dirty environment, and also some of the parts can be easily changed if necessary.

The TELBOT robot has been successfully used for cleaning steam generator tubes at the primary side in CANADA (Bains, N., Majarais, B. and Scott, D.A., 1995). This was the first time in the world that such an advanced mechanical cleaning process had been used on the primary side tube cleaning of steam generators in the nuclear industry.

The present paper will focus on the novel mechanical design, the complete kinematic solutions and the application in hazardous environments.

2 Mechanical Design

The Telbot manipulator has six degrees of freedom and a unique drive system that allows unrestricted motion of each joint (Fig. 1). The first three joint axes of the TELBOT robot are in a standard configuration with one vertical and two parallel horizontal axes. The three wrist joint axes have an offset between the fourth and sixth joint axis, so that all the three wrist joints can also be rotated over 360 degrees continuously. All of the electronics and cables are contained within the base (Fig.2) of the manipulator with no electrical parts in the arm itself. The modular design of the gear block allows for adaptation to various transmission ratios and moments. An overload coupling with status indicator is housed in an intermediate block.

Fig.2-5 shows the inside of the joints of TELBOT robot. The rotation of all the revolution joints and the opening and closing movement of the gripper are transmitted with the concentric tubes in the arm and bevel gears in the joints. Tubes and joints with different sizes are designed for meeting the requirements of different payloads.

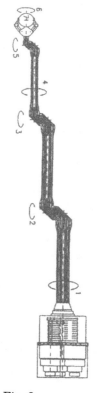

Fig. 2:
Diagram of Tubes

Fig. 3: Robot Base

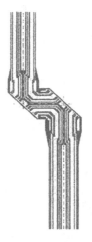

Fig. 4: Joint Design

Fig. 5: End-Effector

3   Kinematics

It was shown by Lee(Li) and Liang (1988), and by Lee (Li), Woernle and Hiller (1991) that there are up to 16 solutions of the inverse kinematics for robots with 6R joints. The TELBOT robot with non-zero offset of the 5th joint has no closed-form solution of the inverse kinematics either, although it still has special geometry. It will be shown that the inverse kinematics problem of the TELBOT manipulator can be reduced to a 16th polynomial in one unknown and five linear equations in the other five unknown joint angles. The univariate polynomial is derived from only three constraint equations. Once the Polynomial is solved for the unknown joint angle numerically, all the other five unknown joint angles are determined by solving linear equations. The real solutions correspond to the different robot configurations. The optimal configuration for a given task and limited workspace can be chosen once all the possible solutions are found. Because all the joint rotations are transmitted by concentric tubes in the arm and concentric bevel gears in the joints, the joint angles must be then converted in the rotation angles of the tubes and finally rotation of the motors.

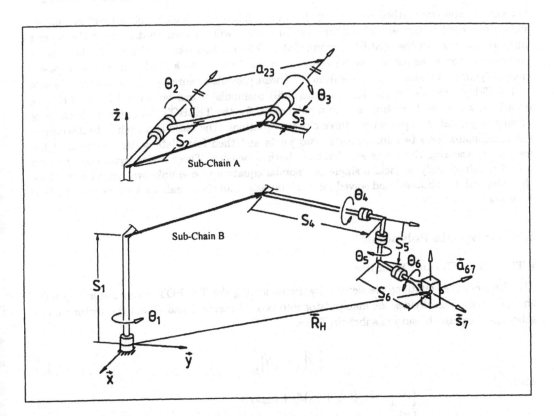

Fig. 6: Kinematic Chain and Denavit-Hartenberg Parameters

3.1   Geometrical Modeling

A kinematic diagram of the TELBOT manipulator showing all link parameters is given in Fig. 6. The directions of the joint axes are sequentially labeled with the unit vectors $\vec{s}_i$, $i = 1, 2, \ldots, 6$.

The directions of the common normals between two successive joint axes $\vec{s}_i$ and $\vec{s}_j$ are labeled with the unit vectors $\vec{a}_{ij}$, $ij = 12, 23, \ldots, 56$, and the lengths of the common normals (link lengths) with a_{ij}. The mutual perpendicular distances between pairs of successive links $\vec{a}_{ij}$ and $\vec{a}_{jk}$ along the joint axis $\vec{s}_j$ are denoted by S_j, $j = 1, 2, \ldots, 6$, and are called joint offsets. The joint angles θ_i and the twist angles α_{ij} are measured by right-hand rotations about $\vec{s}_i$ and $\vec{a}_{ij}$, respectively.

A robot base frame is located with $\vec{z}$ axis coaxial with $\vec{s}_1$, the axis of the first joint. The position of the end-effector is specified in the robot base frame by $\vec{R}_H$ while the orientation is defined by the pair of orthogonal unit vectors $\vec{a}_{67}$ and $\vec{s}_7$.

It is important to note that only the joint angles (θ_j) are unknown quantities; all other Denavit-Hartenberg parameters, i.e., twist angles (α_{ij}), offsets (S_j), and link lengths (a_{ij}) have constant values that are listed in Table 1.

Table 1: Link Parameters

i	S_i	$a_{i(i+1)}$	θ_i	$\alpha_{i(i+1)}$
1	S_1	0.0	θ_1	$\pi/2$
2	S_2	a_{23}	θ_2	0.0
3	S_3	0.0	θ_3	$\pi/2$
4	S_4	0.0	θ_4	$\pi/2$
5	S_5	0.0	θ_5	$\pi/2$
6	S_6	0.0	θ_6	0.0

The general analysis method for solving the inverse kinematics problem of serial-chain manipulators proposed in the second author's earlier papers will be used to determine the inverse kinematic solution of the TELBOT manipulator. The method reduces the inverse kinematics problem for any 6 degree-of-freedom serial-chain manipulator to a single univariate displacement polynomial of minimum degree from the fewest possible closure equations. It is well known that a 16th degree polynomial for the general 6R manipulator can be derived from 14 closure equations, while the derivation of such a polynomial for the TELBOT manipulator, because of its special geometry, requires only three closure equations. The basic approach is to disconnect the manipulator into two subchains at two joints and then formulate scalar products of the vectors representing the joints and links for both subchains. These scalar equations are then solved simultaneously to yield a single polynomial equation in one unknown joint angle. Once the polynomial is obtained and solved for the joint angle, all the remaining joint variables follow from linear equations.

3.2   Univariate Polynomial

● **The Three Closure Equations:**

The three closure equations are derived by disconnecting the TELBOT manipulator at joint 1 and 4 (Fig. 6). This divides the manipulator into two subchains A and B and the vector sum of subchain A equals the sum of subchain B, i.e.

$$\left[\vec{R}\right]_A = \left[\vec{R}\right]_B \tag{1}$$

where

$$\left[\vec{R}\right]_A = ((S_2 + S_3)\vec{s}_2 + a_{23}\vec{a}_{23}) \tag{2}$$

$$\left[\vec{R}\right]_B = -(S_4\vec{s}_4 + S_5\vec{s}_5 + S_6\vec{s}_6 - (\vec{R}_H - S_1\vec{s}_1)) \tag{3}$$

Note that the subscripts A and B on a bracketed quantity indicates that all vectors within the brackets must be expressed using the vectors and rotations associated with that subchain. Three scalar constraint equations are now obtained by formulating two scalar and one triple scalar products for both subchains A and B:

$$[\vec{s}_4 \cdot \vec{s}_1]_A = [\vec{s}_4 \cdot \vec{s}_1]_B \tag{4}$$

$$\left[\vec{R} \cdot (\vec{s}_1 \times \vec{s}_4)\right]_A = \left[\vec{R} \cdot (\vec{s}_1 \times \vec{s}_4)\right]_B \tag{5}$$

$$\left[\vec{R} \cdot \vec{R}\right]_A = \left[\vec{R} \cdot \vec{R}\right]_B \tag{6}$$

Although it may not be immediately apparent, these three scalar equations contain only θ_5, θ_6 and $(\theta_2 + \theta_3)$, as unknowns. It should, therefore, be possible to solve this system. A key to efficient formulation is to express vectors in the most convenient coordinate frame before the scalar and triple scalar product operations are performed. A change in reference frame is easily accomplished by using standard joint and twist rotation matrices. A rotation about the local joint axis $\vec{s}_i$ can be expressed as

$$M_i = \begin{bmatrix} c_i & -s_i & 0 \\ s_i & c_i & 0 \\ 0 & 0 & 1 \end{bmatrix} \tag{7}$$

where $c_i = \cos\theta_i$, $s_i = \sin\theta_i$.

Similarly, a twist between joint i and $(i+1)$ can be expressed as a rotation about the local axis $\vec{a}_{i(i+1)}$, or

$$M_{i(i+1)} = \begin{bmatrix} 1 & 0 & 0 \\ 0 & c_{i(i+1)} & -s_{i(i+1)} \\ 0 & s_{i(i+1)} & c_{i(i+1)} \end{bmatrix} \tag{8}$$

where $c_{i(i+1)} = \cos\alpha_{i(i+1)}$, and $s_{i(i+1)} = \sin\alpha_{i(i+1)}$.

Because all the twist angles $\alpha_{i(i+1)}$ are equal to zero or ninety degrees, all the entries of $M_{i(i+1)}$ are 1 or 0.

Note also that the unit vector $\vec{s}_i$ and $\vec{a}_{i(i+1)}$ can always be expressed in their local frame as

$$\vec{s}_i = \hat{z} = [0 \ 0 \ 1]^T \tag{9}$$

$$\vec{a}_{i(i+1)} = \hat{x} = [1 \ 0 \ 0]^T \tag{10}$$

The given end-effector position vector and the orientation matrix are defined in the robot base frame as follows:

$$\vec{R}_{H0} = \vec{R}_H - S_1\vec{s}_1 = [r_x \ r_y \ (r_z - S_1)]^T \tag{11}$$

$$M_H = [\vec{a}_{67}, \ (\vec{s}_7 \times \vec{a}_{67}), \ \vec{s}_7] = \begin{bmatrix} h_{00} & h_{01} & h_{02} \\ h_{10} & h_{11} & h_{12} \\ h_{20} & h_{21} & h_{22} \end{bmatrix} \tag{12}$$

The First Closure Equation:

Evaluating the each side of Eq. (4) by using the rotation matrices with the given twist angles. This results in the following:

$$[\vec{s}_4 \cdot \vec{s}_1]_A = \hat{z}^T M_{34}^{-1} M_3^{-1} M_{23}^{-1} M_2^{-1} M_{12}^{-1} \hat{z}$$

$$= -\cos(\theta_2 + \theta_3) \tag{13}$$

$$[\vec{s}_4 \cdot \vec{s}_1]_B = \hat{z}^T M_{45} M_5 M_{56} M_6 M_H^{-1} \hat{z}$$

$$= h_{20}s_5c_6 - h_{21}s_5s_6 - h_{22}c_5 \tag{14}$$

Equation (4) may now be written as

$$c_5 h_{22} + s_5(-h_{20}c_6 + h_{21}s_6) = \cos(\theta_2 + \theta_3) \tag{15}$$

The Second Closure Equation:

Developing Eq. (5) requires the evaluation of triple scalar products. The left-hand side is calculated as following:

$$\begin{aligned}
\left[\vec{R} \cdot \vec{s}_1 \times \vec{s}_4\right]_A &= ((S_2 + S_3)\vec{s}_2 + a_{23}\vec{a}_{23}) \cdot \vec{s}_1 \times \vec{s}_4 = (S_2 + S_3)(\vec{s}_2 \times \vec{s}_1 \cdot \vec{s}_4) \\
&= (S_2 + S_3)(-\vec{a}_{12} \cdot \vec{s}_4) = -(S_2 + S_3)\hat{x}^T M_2 M_{23} M_3 M_{34}\hat{z} \\
&= -(S_2 + S_3)\sin(\theta_2 + \theta_3)
\end{aligned} \tag{16}$$

Where, it was obvious that the triple product $(\vec{a}_{23} \cdot \vec{s}_1 \times \vec{s}_4)$ equals to zero, because the three vectors lie on the parallel planes.

The right-hand side of Eq. (5) is written as

$$\begin{aligned}
\left[\vec{R} \cdot \vec{s}_1 \times \vec{s}_4\right]_B &= -(S_4\vec{s}_4 + S_5\vec{s}_5 + S_6\vec{s}_6 - \vec{R}_{HO}) \cdot \vec{s}_1 \times \vec{s}_4 \\
&= -(S_5\vec{s}_5 + S_6\vec{s}_6 - \vec{R}_{HO}) \cdot \vec{s}_1 \times \vec{s}_4
\end{aligned} \tag{17}$$

The first triple scalar product is calculated as follows:

$$\begin{aligned}
\vec{s}_5 \cdot \vec{s}_1 \times \vec{s}_4 &= \vec{a}_{45} \cdot \vec{s}_1 = \hat{x}^T M_5 M_{56} M_6 M_H^{-1}\hat{z} \\
&= (h_{20}c_6 - h_{21}s_6)c_5 + h_{22}s_5
\end{aligned} \tag{18}$$

In order to calculate the second triple scalar product $(\vec{s}_6 \cdot \vec{s}_1 \times \vec{s}_4)$ in this equation, a coordinate frame for calculations must be chosen. If the coordinate system shares the axis with the unit vector $\vec{s}_6$, which is in the center of the subchain containing the three vectors $\vec{s}_6$, $\vec{s}_1$ and $\vec{s}_4$, the result will be free of any product terms $c_i s_i$, c_i^2 and s_i^2 (Lee and Reinholtz, 1996). Evaluating the triple product $(\vec{s}_6 \cdot \vec{s}_1 \times \vec{s}_4)$ in the reference frame attached at joint 6 results in the following:

$$\begin{aligned}
[\vec{s}_6 \cdot \vec{s}_1 \times \vec{s}_4]_B &= \det\left[\hat{z} \quad M_6 M_H^{-1}\hat{z} \quad M_{56}^{-1}M_5^{-1}M_{45}^{-1}\hat{z}\right] \\
&= -s_5(h_{21}c_6 + h_{20}s_6)
\end{aligned} \tag{19}$$

The third triple scalar product of Eq. (17) is calculated in the fixed frame, which results in the following:

$$\begin{aligned}
&\left[\vec{R}_{HO} \cdot \vec{s}_1 \times \vec{s}_4\right]_B \\
&= \det\left[\vec{R}_{HO} \quad \hat{z} \quad M_H M_6^{-1}M_{56}^{-1}M_5^{-1}M_{45}^{-1}\hat{z}\right] \\
&= r_x[s_5(h_{10}c_6 - h_{11}s_6) - h_{12}c_5] + \\
&\quad r_y[s_5(-h_{00}c_6 + h_{01}s_6) + h_{02}c_5]
\end{aligned} \tag{20}$$

Therefore, the Eq. (5) may now be rewritten as the following:

$$\begin{aligned}
&c_5\left[c_6(h_{20}S_5) - s_6(h_{21}S_5) + (-h_{12}r_x + h_{02}r_y)\right] + \\
&s_5\left[c_6(h_{10}r_x - h_{00}r_y - h_{21}S_6) + s_6(-h_{11}r_x + h_{01}r_y - h_{20}S_6) + (h_{22}S_5)\right] \\
&= (S_2 + S_3)\sin(\theta_2 + \theta_3)
\end{aligned} \tag{21}$$

The Third Closure Equation:

Equation (6) is evaluated in the same manner. For convenience, we first divide the equation (6) by 2 and then expand the left- and right-hand sides. This results in the following:

$$
\frac{1}{2}\left[\vec{R}\cdot\vec{R}\right]_A = \frac{1}{2}((S_2+S_3)\vec{s}_2 + a_{23}\vec{a}_{23})\cdot((S_2+S_3)\vec{s}_2 + a_{23}\vec{a}_{23})
$$
$$
= \frac{1}{2}((S_2+S_3)^2 + a_{23}^2) \tag{22}
$$

$$
\frac{1}{2}\left[\vec{R}\cdot\vec{R}\right]_B = \frac{1}{2}(S_4\vec{s}_4 + S_5\vec{s}_5 + S_6\vec{s}_6 - \vec{R}_{HO})\cdot(S_4\vec{s}_4 + S_5\vec{s}_5 + S_6\vec{s}_6 - \vec{R}_{HO})
$$
$$
= \frac{1}{2}(S_4^2 + S_5^2 + S_6^2 + R^2) + S_4S_6(\vec{s}_4\cdot\vec{s}_6) - S_4(\vec{s}_4\cdot\vec{R}_{HO})
$$
$$
-S_5(\vec{s}_5\cdot\vec{R}_{HO}) - S_6(\vec{s}_6\cdot\vec{R}_{HO})
$$
$$
= \frac{1}{2}(S_4^2 + S_5^2 + S_6^2 + R^2) + S_4S_6\hat{z}^T M_{45}M_5M_{56}\hat{z} -
$$
$$
S_4\hat{z}^T M_{45}M_5M_{56}M_6M_H^{-1}\vec{R}_{HO} - S_5\hat{z}^T M_{56}M_6M_H^{-1}\vec{R}_{HO} - S_6\hat{z}^T M_H^{-1}\vec{R}_{HO}
$$
$$
= \frac{1}{2}(S_4^2 + S_5^2 + S_6^2 + R^2) - S_4S_6c_5 - S_4(H_x s_5 c_6 - H_y s_5 s_6 - H_z c_5) -
$$
$$
S_5(H_x s_6 + H_y c_6) - S_6 H_z \tag{23}
$$

where

$$
[H_x,\ H_y,\ H_z]^T = M_H^{-1}\vec{R}_{HO} \tag{24}
$$
$$
R^2 = r_x^2 + r_y^2 + (r_z - S_1)^2 \tag{25}
$$

After moving $\frac{1}{2}((S_2+S_3)^2 + a_{23}^2)$ to the right-hand side, and rearranging, the closure equation (6) is now written as

$$
c_5 S_4(H_z - S_6) + s_5(-c_6 S_4 H_x + s_6 S_4 H_y) - c_6 S_5 H_y + s_6 S_5 H_x +
$$
$$
\left[(1/2)(S_4^2 + S_5^2 + S_6^2 + R^2 - (S_2+S_3)^2 - a_{23}^2) - S_6 H_z\right]
$$
$$
= 0 \tag{26}
$$

• The 16th Degree Polynomial:

The closure equations (15), (21) and (26), which contain all the information needed to derive the 16th degree polynomial, now can be written in the following form:

$$
F_1(\theta_5, \theta_6) = \cos(\theta_2 + \theta_3) \tag{27}
$$
$$
F_2(\theta_5, \theta_6) = (S_2 + S_3)\sin(\theta_2 + \theta_3) \tag{28}
$$
$$
F_3(\theta_5, \theta_6) = 0 \tag{29}
$$

Where the functions $F_1(\theta_5, \theta_6)$, $F_2(\theta_5, \theta_6)$ and $F_3(\theta_5, \theta_6)$ are linear combinations of the terms $(c_5 c_6,\ c_5 s_6,\ c_5,\ s_5 c_6,\ s_5 s_6,\ s_5,\ c_6,\ s_6)$.

Eliminating $(\theta_2 + \theta_3)$ from Eqs. (27) and (28) gives

$$
F_1^2 + [F_2/(S_2 + S_3)]^2 = 1 \tag{30}
$$

The next step is to eliminate θ_5 (or θ_6) from Eqs. (29) and (30). For convenience, the two equations can be expressed as polynomials in $x_5 = \tan(\theta_5/2)$ and $x_6 = \tan(\theta_6/2)$ as follows:

$$A_0 + A_1 x_5 + A_2 x_5^2 = 0 \qquad (31)$$
$$B_0 + B_1 x_5 + B_2 x_5^2 + B_3 x_5^3 + B_4 x_5^4 = 0 \qquad (32)$$

where A_0, A_1 and A_2 are quadratic polynomials in x_6, and B_0, B_1, B_2, B_3 and B_4 are quartic polynomials in x_6.

Multiplying Eq. (31) by x_5, x_5^2 and x_5^3, and Eq. (32) by x_5, we obtain four additional equations. Together with the two original equations (31) and (32), a total of six equations are obtained. These can be expressed in the matrix form:

$$
\begin{bmatrix}
A_2 & A_1 & A_0 & 0 & 0 & 0 \\
0 & A_2 & A_1 & A_0 & 0 & 0 \\
0 & 0 & A_2 & A_1 & A_0 & 0 \\
0 & 0 & 0 & A_2 & A_1 & A_0 \\
B_4 & B_3 & B_2 & B_1 & B_0 & 0 \\
0 & B_4 & B_3 & B_2 & B_1 & B_0
\end{bmatrix}
\begin{bmatrix}
x_5^5 \\
x_5^4 \\
x_5^3 \\
x_5^2 \\
x_5 \\
1
\end{bmatrix}
=
\begin{bmatrix}
0 \\
0 \\
0 \\
0 \\
0 \\
0
\end{bmatrix}
\qquad (33)
$$

In order for this system to have a nontrivial solution, the coefficient matrix must be singular, i.e., the determinant of the coefficient matrix must be equal to zero. The expansion of the determinant yields a 16th degree polynomial in x_6:

$$\sum_{i=0}^{16} e_i x_6^i = 0 \qquad (34)$$

This is the desired 16th degree univariate polynomial in x_6 with constant coefficients e_i.

3.3   Determination of All the Joint Variables

Once the desired hand position and orientation is specified, all e_i coefficients of Eq. (34) can be calculated. The roots of the polynomial give the values of θ_6 which correspond 16 possible configurations of the manipulator. Each value of θ_6 will give a unique value of x_5 by solving the two equations (31) and (32). For each pair of θ_5 and θ_6, a value of $(\theta_2 + \theta_3)$ can be easily obtained from equations (27) and (28). Then for each set of $(\theta_2 + \theta_3)$, θ_5 and θ_6, the joint angles θ_1 and θ_4 can be determined from the following two equations:

$$M_1 M_{12} M_2 M_{23} M_3 M_{34} \hat{z} = M_H M_6^{-1} M_{56}^{-1} M_5^{-1} M_{45}^{-1} \hat{z} \qquad (35)$$
$$M_4^{-1} M_{34}^{-1} M_3^{-1} M_{23}^{-1} M_2^{-1} M_{12}^{-1} \hat{z} = M_{45} M_5 M_{56} M_6 M_H^{-1} \hat{z} \qquad (36)$$

The equations (35) and (36) can be formulated so that their components are linear in terms of sines and cosines of θ_1 and θ_4 respectively, thus can be solved for directly.

The joint angles θ_3 can be easily determined from the following equations, respectively:

$$\left[\vec{R} \cdot \vec{a}_{34} \right]_A = \left[\vec{R} \cdot \vec{a}_{34} \right]_B \qquad (37)$$

$$\left[\vec{R} \cdot (\vec{s}_3 \times \vec{a}_{34}) \right]_A = \left[\vec{R} \cdot (\vec{s}_3 \times \vec{a}_{34}) \right]_B \qquad (38)$$

The two equations (37) and (38) are linear in terms of sines and cosines of θ_3. If θ_3 is found, the joint angle θ_2 can be determined using the value of $(\theta_2 + \theta_3)$.

It should be noted that if $(\theta_2 + \theta_3)$ near to zero or 180 degrees, the proposed equations (35) and (36) cannot be used. The joint angles θ_1 and θ_4 must be calculated by using some other closure equations.

4   Application

An important application of the TELBOT system is for nuclear steam generator maintenance. Fig. 8 shows the steam generator primary head tube cleaning system block diagram. The steam generator contains a shell-and-tube type heat exchanger with thousands of tubes. The plurality of the tubes is encased in a vertical cylindrical pressure vessel with a hemispherical lower end (bowl). The tube ends are inserted through a plate (tube-sheet) and welded and forms a planar cap over the bowl. The tubes are bent into an upside down "U" shape. These tubes must be inspected and repaired from within the bowl quarter spheres.

The steam generator bowl is a high radiation area. Some robotic systems were developed and used for cleaning the tubes, for example, COBRA robotic system developed by B&WNT (Tidwell and Glass, et al, 1991). An advantage by using TELBOT system is that the motors, gear block and all the electrical cables are OUTSIDE of the bowl (Fig. 7). The robot arm parts which are inside the bowl can be very easily protected from dirty radioactive materials. Some arm parts can be quickly changed if necessary, because of the modular design. This can greatly reduce the cost of nuclear steam generator maintenance.

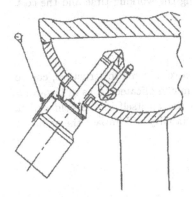

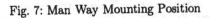

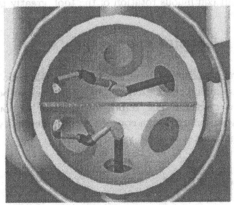

Fig. 7: Man Way Mounting Position   Fig. 8: Steam Generator Maintenance

Once the manipulators were installed at the man-ways, the operator could initiate the calibration program. An initializing routine was used to calibrate the manipulator base coordinates with respect to the overall tube-sheet reference coordinates. The manipulator automatically moved to a pre-programmed location at the tube-sheet underneath a tie-rod location which was convenient to use for calibration as its position was known and was visually different from the tubes. They were also well spaced. For this initial calibration the stand-off distance of the tool to the tube-sheet was greater than the working distance. This allowed for any misplacement of the manipulators during the manual setting up operation. In order to ensure the manipulators moved in a plane parallel to the tube-sheet, the tool had a metallic "finger" that was used to touch the tube-sheet in three places each time the "finger" touched the tube-sheet a circuit was closed and the operator was alerted that the manipulator was in contact with the tube-sheet. From the three locations the manipulator base frame was calculated through the software and referenced to the tube-sheet plane. The manipulator could now be operated in straight line interpolation mode with respect to the tube-sheet. This made the tele-operation of the manipulator much easier than using the individual joint control. Once the manipulator was in this mode it was referenced to the actual tubes. Again the manipulators were initiated to move to the tie-rod locations. The manipulator was then tele-operated so that the center of the tube adjacent to the tie rod location was in the center of the field of view of the cameras. This was simply carried out using a set of cross hairs over-layed on the monitors. Once this was achieved for two or more tie rod locations the manipulator was referenced with respect to the tubes. During the cleaning process this type of calibration was repeated several times.

This means that the robot base frame is determined with respect to the world frame (attached to the bowl) in two steps. First, only the plane of tube-sheet is calibrated to the robot base frame, and then the position of the tubes with respect to the robot base frame. This greatly simplified the process of the calibration of the robot base frame with respect to the world frame, because there are not three or more sharp points in the bowl whose position is known exactly and could be touched by tool tip of the robot. The problem to plug the finger into the tubes is that the depth of the tool finger in the tube is very difficult to be exactly measured by using normal camera. Therefore, the plane of the tube-sheet is first determined , and then the position of tubes.

Another important application of the TELBOT robot is for decommissioning glove boxes in Japan. The robot is used for dismantling boxes in radioactive environment and for loosing screws on the boxes and then to cut the boxes. The novel properties of the robot, such as no limit continuous rotation of the tool, greatly facilitate reducing the working time and the cost.

5   Conclusions

The TELBOT system is a universal tele robotic system. Its novel kinematic structure, control system and simulation system allow itself to be able to perform complicated tasks in hazardous environments. The Telbot manipulator without a doubt has proven itself in the field and its revolutionary design coupled with its flexibility and proven reliability will impact the advanced manipulator field for years to come.

References

[1] Bains, N., Majarais, B. and Scott, D.A., 1995, "Cleaning of Steam Generator Tubes at the Primary Side Using a General Purpose Manipulator", ANS conference, San Francisco, October.

[2] Lee (Li), H.-Y. and Liang, C.-G., 1988, "Displacement Analysis of the General Spatial 7-Link 7R Mechanism", International Journal of Mechanism and Machine Theory, Vol. 23, pp.219-226.

[3] Tidwell, P. H., Glass, S. W. and et al, 1991, "COBRA - Design and Development of a Manipulator for Nuclear Steam Generator Maintenance", Proceedings of the 2nd National Applied Mechanisms and Robotics Conference, USA, Vol. 1, pp. IVB1-1 to IVB1-10.

[4] Wälischmiller, W. and Frager, O., 1995, "TELBOT - A Modular Tele-Robot System for Hostile Environments with Unlimited Revolutionary Joints and Motors, Drive Sensors and Cables in the Base", Proceedings of the ANS 6th Topical Meeting on Robotics and Remote Systems, Monterey, California, USA, Vol. 1, pp.274-280.

A DECISION-MAKING SYSTEM FOR THE INITIAL STAGE IMPLEMENTATION OF ASSEMBLY STRATEGIES

C.J. Tsaprounis, N.A. Aspragathos and A. Denstoras
University of Patras, Patras, Greece

Abstract

In the present paper, a decision making system is presented for a successful implementation of the initial stage of assembly strategies. The system is part of an integrated knowledge based system for automatic generation of assembly strategies currently under development.

The derivation of robot motions sequence in the initial stage of an assembly strategy is a very critical problem in the case of uncertainty existence. The corresponding developed decision making process consists of two main subjects: First, a method is introduced to infer on the assemblability of the task based on fuzzy logic tolerance analysis. The second subject refers to the identification of the contact points when the moving and the stationary parts come in contact for the first time.

The above methods are presented in detail and the operation of the decision making system within the framework of the integrated knowledge based system for automatic generation of assembly strategies is explained.

Introduction

The systematic development of strategies for robot fine motion in assembly tasks has attracted great attention among the researchers in the robotics community. The most important contribution in this systematic approach has been based on the concept of force/position hybrid control [1] and the identification of natural and artificial constraints in the case of contact between two mating parts [2]. They have introduced the Compliant Frame (C-Frame) where the constraints are defined along or about its Cartesian axes. In [3], the authors introduce a method to classify the mating conditions for the extraction of the degrees of freedom of the assembled parts, during several stages of assembly cases. Their system can be used as a basis for automating the fine motion control phase of assembly operations in the ideal domain. So the authors conclude that there are many other factors such as friction, wedging and jamming to be considered in the development of assembly strategies in the real domain.

To cope with robot inaccuracies, Aspragathos[4] introduced a systematic way to generate assembly strategies, which is consistent with the idea of the C-Frames. In this paper, the author describes an integrated system that defines the C-Frame in each step of the assembly process and produces the natural and artificial constraints.

Kosei, Kitagaki et all [5] analyze the identification of contact points from the force sensor signals. Initially they formulate the static equilibrium equations. The equations are linear and the coefficient matrix of this linear system is singular. To overcome this problem, the least squares' method is applied. In the same paper, a mathematical formulation for the identification of the transitional motion of the grasped object is introduced. This method is based on the force sensor signals and is assumed that the grasped object is always in contact with the environment.

A very important subject for the generation of appropriate fine motion of a robot in the initial stage of an assembly strategy is the relation among the tolerance of the assembled parts - with respect to the repeatability error of the robot - and the assembly feasibility. Whitney and Gilbert [6] introduced a tolerance representation for assemblies compatible with tolerance analysis, accompanied by a method to describe standard tolerances using homogeneous matrix transforms. So their method calculates a statistical estimate of the location of the last assembled part starting from the fixture. Their work does not goes on with the inference of the feasibility of an assembly task but can be used as a basis for the development of techniques to judge on the degree of difficulty of an assembly task.

The main subject of the present paper is the development of an inference mechanism able to search for the appropriate sequence of fine motions of a robot at the initial stage of a robot assembly task. The initial stage is defined as the motions from the non-contact to first contact situation of an assembly task. The first problem that an intelligent control system has to solve is to judge on the feasibility of a given assembly task. This can be carried out by comparing the clearance between the mating parts and the repeatability of the robot. In the present paper an analysis based on fuzzy logic has been introduced, which can be used to guide the next movements of the robot.

The next task of an intelligent robot controller, in the case of uncertainty, is to identify the exact position and orientation of the moving part when the first contact between the mating parts occurs, in order to establish the appropriate C-frame to define the artificial constraints of the next movement towards to the insertion stage. A novel analysis is presented for the identification of the contact points and the orientation of the moving part.

In the following section, a brief overview of the knowledge based system is given. The part of decision making for the initial stage of the assembly strategies is analyzed in the rest sections in details.

DESCRIPTION OF THE KNOWLEDGE- BASE SYSTEM

In Fig.1, an overview of the knowledge-based system for the extraction of the assembly strategies is shown in the form of a block diagram. It must be noticed that while some modules of the system apply Artificial Intelligence techniques for deducting results that are mostly qualitative in nature, there are other parts where conventional algorithms are used, especially when the available knowledge can be explicitly expressed in terms of mathematical formulas and equations.

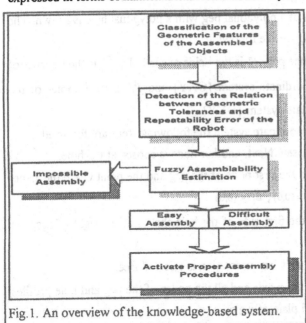

Fig.1. An overview of the knowledge-based system.

At the top, there is a part where a classification of the geometric features of the objects to be assembled takes place. Here a set of "IF-THEN" rules are used in order to represent the available knowledge about the object features and to deduct conclusions concerning the subsequent assembly strategy to be followed.

In some cases the assembly process can be easily implemented and, therefore, there is no need to continue with the next steps. However, for the rest of the cases, there must be an identification of the relation between the geometric tolerances of the objects and the repeatability error of the robot. This relation is vital for both the feasibility of the assembly as well as for the subsequent robot motions. Through a fuzzy assemblability estimation process, an assemblability index is determined which can alternatively guide to an "impossible", "easy", or "difficult" assembly path. While for the first case the assembly is no more feasible, for the rest two cases additional procedures must be activated. Especially for the "difficult" assembly case, the existence of chamfer must be further investigated, since under certain circumstances, can guide to an easier assembly process. If no chamfer exists, then an initial misalignment procedure must be implemented which, eventually, will guide to both part insertion and satisfaction of the final constraints.

Tolerance Analysis

In the general peg and hole problem, the tolerance analysis with respect to the repeatability error of the robot is vital for the feasibility of the assembly task and for the selection of the strategies. For encountering this problem and deciding what has be done further, a fuzzy decision system has been developed. The mathematical formulation of the decision making system covers the general case of peg and holes with polygonal profiles. Due to this generalization, the round peg and hole profiles can be examined as two regular polygons with infinite number of sides.

In Fig.2 the basic configuration of a polygonal peg with a polygonal hole is shown. The symbols assigned at the drawing are the following:

l_i^p is the length of the i-side of the peg, l_i^h is the i-side of the hole, $\bar{S}_i^p$ is the i-corner of the peg with respect to the coordinate system {OXY} and $\bar{S}_i^h$ is the i-corner of the hole with respect to the coordinate system {OXY}

In order to define the {OXY} coordinate system the following steps are followed:

- First, a node of the peg is chosen. Next, another node is chosen at the hole.
- The first node of the hole and the first node of the peg and the point O are the same. So it can be written that: $\bar{S}_1^p = (0,0)$, $\bar{S}_1^h = (0,0)$
- The x axis is defined along the first side of the hole that is at the right of the first node of the hole's polygon.
- The y axis is vertical to the x axis. Its orientation is selected freely.
- The first side of peg is along the x axis and all the nodes of the peg and hole profiles must belong to the same semi plain of y axis.

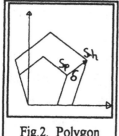

The dimensions of the parts vary around the nominal ones due to machines uncertainties. For this reason, the lengths of the polygonal profiles of both the peg and hole follow the relations $l_i^p \in [l_i^{p,min}, l_i^{p,max}]$ and $l_i^h \in [l_i^{h,min}, l_i^{h,max}]$. For each pair of nodes of the peg and the hole, the clearance is defined as the vector:

$$\bar{\delta} = \bar{S}_i^h - \bar{S}_i^p \qquad (1)$$

Fig.2. Polygon positioning

(At the following equations when min or max is the top index of a vector then this refers to the situation of $l^{p,max}$ or $l^{p,min}$ respectively).

The values of the clearance that refers at the minimum and maximum lengths of the polygon sides are the following:

$$\vec{\delta}_i^{max} = \vec{S}_i^{h,max} - \vec{S}_i^{p,min}, \qquad \vec{\delta}_i^{min} = \vec{S}_i^{h,min} - \vec{S}_i^{p,max} \qquad (2)$$

If $\qquad \vec{a}_i = \dfrac{\vec{S}_i^{p} - \vec{S}_{i-1}^{p}}{\left|\vec{S}_i^{p} - \vec{S}_{i-1}^{p}\right|}, \qquad \vec{b}_i = \dfrac{\vec{S}_i^{p} - \vec{S}_{i+1}^{p}}{\left|\vec{S}_i^{p} - \vec{S}_{i+1}^{p}\right|} \qquad (3)$

then overlapping between the two polygons appears when:

- Case1: Convex polygon $\vec{\delta}_i \cdot \vec{a}_i < 0$ $\quad$ and $\quad$ $\vec{\delta}_i \cdot \vec{b}_i < 0$

- Case 2: Non Convex polygon $\vec{\delta}_i \cdot \vec{a}_i > 0$ $\quad$ or $\quad$ $\vec{\delta}_i \cdot \vec{b}_i > 0$

In order to define the assemblability of each task due to the tolerances, the following fuzzy mechanism is applied.

If $\qquad t_i^{min} = \dfrac{\pm\left|\vec{\delta}_i^{min}\right|}{r_e}, \qquad t_i^{max} = \dfrac{\pm\left|\vec{\delta}_i^{max}\right|}{r_e} \qquad (4)$

where r_e is the repeatability error of the robot then the sign at the expressions above depends upon the appearance of overlapping among the polygonal profiles. The final formulations of the expressions are the following:

$$t_i^{min} = sgn[max\{(\vec{\delta}_i^{min} \cdot \vec{a}_i^{max}), (\vec{\delta}_i^{min} \cdot \vec{b}_i^{max})\}]\dfrac{\left|\vec{\delta}_i^{min}\right|}{r_e}$$

$$t_i^{max} = sgn[min\{(\vec{\delta}_i^{max} \cdot \vec{a}_i^{min}), (\vec{\delta}_i^{max} \cdot \vec{b}_i^{min})\}]\dfrac{\left|\vec{\delta}_i^{max}\right|}{r_e} \qquad (5)$$

where:

$$\vec{a}_i^{max} = \dfrac{\vec{S}_i^{p,max} - \vec{S}_{i-1}^{p,max}}{\left|\vec{S}_i^{p,max} - \vec{S}_{i-1}^{p,max}\right|} \qquad \vec{b}_i^{max} = \dfrac{\vec{S}_i^{p,max} - \vec{S}_{i+1}^{p,max}}{\left|\vec{S}_i^{p,max} - \vec{S}_{i+1}^{p,max}\right|}$$

$$\vec{a}_i^{min} = \dfrac{\vec{S}_i^{p,min} - \vec{S}_{i-1}^{p,min}}{\left|\vec{S}_i^{p,min} - \vec{S}_{i-1}^{p,min}\right|}, \qquad \vec{b}_i^{min} = \dfrac{\vec{S}_i^{p,min} - \vec{S}_{i-1}^{p,min}}{\left|\vec{S}_i^{p,min} - \vec{S}_{i+1}^{p,min}\right|} \qquad (6)$$

Finally the variables t_{max}, t_{min} are defined by the following expressions:

$$t_{min} = \min_{i=2,n}\{t_i^{min}\}, \qquad t_{max} = \max_{i=2,n}\{t_i^{max}\} \qquad (7)$$

These variables are the inputs to the fuzzy scheme. The possible values of the mentioned variables belong to the fuzzy subsets TS = too small clearance $x \in (-\infty, 1]$, M = medium clearance $x \in [0.5, 1.5]$, TB = too large clearance $x \in [1.0, \infty)$. In Fig.5 the graphs of the membership functions of the subsets are shown.

A set of 'IF-THEN' rules is used in order to produce values for the assemblability index (A.I). This index may get one of the following possible values:

- N = No assembly, E = Easy assembly, D = Difficult assembly

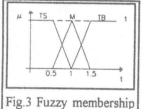

Fig.3 Fuzzy membership functions

A rule producing an 'Easy Assembly' has the form:

IF [$t_{min} = TS$ AND $t_{max} = TB$] then [AI=E]

The value of the membership function of each rule is calculated using the minimum operation for the membership functions of t_{min}, t_{max} it is shown below:

$$\mu(Rule_i) = \min\{\mu(t_{min}), \mu(t_{max})\} \tag{8}$$

The result of the inference mechanism is the one that refers to the rule with the maximum value of membership function. The relative FAM -fuzzy associative memory- is shown in the following table:

t_{min} \ t_{max}	TS	M	TB
TS	N	N	D
M		D	E
TB			E

In the special case of round peg and hole the values of t_{min}, t_{max} are given as:

$$t_{min} = \frac{D_{min} - d_{max}}{2 * r_e}, \quad t_{max} = \frac{D_{max} - d_{min}}{2 * r_e} \tag{9}$$

where: D = the diameter of the hole, $D \in [D_{min}, D_{max}]$, d = the diameter of the peg, $d \in [d_{min}, d_{max}]$. The rest of the parts of the fuzzy system remains the same.

Initial Misalignment And Chamfer Existence

During manual assembly operations, the human operator exploits information produced by sources as the reacting forces applied on his wrist and the visual perceptions of the results produced by his own actions.

At any artificial system, the feedback signal comes from the force sensor at the wrist of the robot arm. This means that for the first assembly motion there are not any data available. Additionally, the repeatability error of the robot adds uncertainty in the subtask of first positioning. To overcome these difficulties the idea of initial misalignment has been used.

The existence of a chamfer, available for assembly operations, reduces the uncertainty due to the repeatability error. For this reason the system under consideration identifies the existence of a chamfer. A chamfer is available for assembly process when its length is greater than the repeatability error.

When no chamfer exists, the whole process becomes more complicated. The analysis for the misalignment in the case of round peg and hole when chamfer does not appear follows.

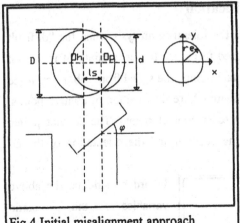

Fig.4 Initial misalignment approach

For having high possibility of success, the initial position of the peg is designed to be in the area of the hole entrance. The best case for initial misalignment where high accuracy from robot is not needed is shown in Fig.4. The center of the bottom of the peg lies on the plain of hole's entrance presenting two contact points with it. This case is shown in Fig.4.

The symbols used are the following: Op : center of the bottom surface of peg, Oh : center of the entrance of hole, ls : distance between Op and Oh, d : diameter of the peg and D : diameter of the hole

As can be extracted from the above figure, the following expression is true:

$$l_s = \frac{1}{2}\sqrt{D^2 - d^2} \qquad (10)$$

If the repeatability error of the robot is r_e then the following possible cases of bad misalignment may arise

- along x^+, y axis

Here only one contact point exists among peg and hole. This is the lower point of the peg where is positioned at the limits of the hole's entrance. For this case, the repeatability error of the robot is:

$$r_{ec} = \frac{D}{2} - \frac{1}{2}\sqrt{D^2 - d^2} + \frac{d}{2}\cos(\varphi) \qquad (11)$$

In order to avoid the inappropriate misalignment along x^+, the repeatability error of the robot must be smaller than r_{ec}. Since the repeatability error is standard for each robot type, the correct selection of the approach angle φ may produce a critical repeatability error bigger than the robot's one, so the possibility of inappropriate misalignment tends to zero.

- along x^-

For this case, the critical value of the repeatability error is: $r_{ec} = \frac{D}{2} + l_s$ (12)

The worst case for inappropriate misalignment is the first one. For this reason the calculations for the approach angle are based on this case. According to these calculations the lower point of the peg will be always under the entrance of the hole.

Initial Misalignment Contact Identification

The initial misalignment of the peg follows the tolerance analysis when the "difficult assembly case" appears. This approach is used to reduce the uncertainty of the robot positioning. For the identification of the contact points, the signal from robot arm force sensor is used. The variables that must be identified are the position of contact points, the orientation of peg and the reaction forces at contact points. The contact points must be defined since they are related to the position and the orientation of the C-Frames.

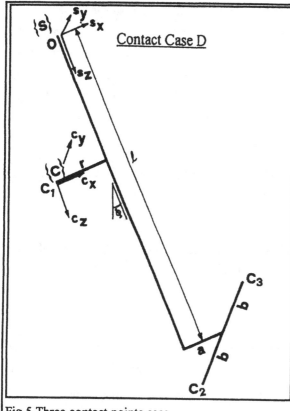

Contact Case D

Fig.5 Three contact points case

In order to define the above variables the static equilibrium equations are formulated. In some contact cases the unknowns are more than the equations. In such situations a redundancy arises and the only way to be reduced is the extraction of additional equations. Geometrical compatibility - due to peg shape- equations and energy theorems from mechanics are employed.

At the initial misalignment four are the possible cases of contact among the peg and hole. These are, no contact points (A), one contact point (B), two contact points (C) and three contact points (D).

In the present paper, the case with three contact points will be analyzed. It should be mentioned that the following analysis is applied both in round and in prismatic pegs. Although friction does not appears in this case, the problem is still redundant.

In Fig.5 a simple drawing of the symmetry axis of the round peg and hole and the contact points on the external surface of the cylinder are shown. The coordinate system $\{O\}$ is positioned at the force sensor and the contact points are the $c1$, $c2$, $c3$. From the application of the static equilibrium conditions the following equations are produced:

$$f_{c1} - mg\sin(\varphi) + F_x = 0$$

$$F_z - f_{c3} - f_{c2} + mg\cos(\varphi) = 0$$

$$f_{c3}b - f_{c2}b - M_x = 0 \tag{13}$$

$$f_{c1}h + f_{c3}a + f_{c2}a - mg\sin(\varphi)l_w + M_y = 0$$

where $[F_x, F_y, F_z, M_x, M_y, M_z]$ the force sensor components. The number of equations are four and the unknowns are seven ($\varphi, a, b, h, f_{c1}, f_{c2}, f_{c3}$). So a redundancy arises and three more equations are needed.

From the geometrical compatibility due to peg geometry the following equations are extracted:

$$d = 2\sqrt{a^2 + b^2}, \quad \tan(\varphi) = \frac{l - h}{\frac{d}{2} + a} \tag{14}$$

From the unit displacement theorem [7] can be derived that:

$$\vec{F} \cdot \vec{\delta s} = \iiint \sigma \, \delta \varepsilon \, dV \tag{15}$$

where the integrals are referred to the volume of the sensor. If the components of the force sensor are $\vec{F}_s$ and $\vec{M}_s$ then the above equation can be written in the form:

$$\vec{F} \cdot \vec{\delta s} = \left(\vec{F}_s \cdot \delta \vec{y} + \vec{M}_s \cdot \delta \vec{\theta} \right) \tag{16}$$

$$\vec{F} // \vec{\delta s} \quad \text{and} \quad \|\vec{\delta s}\| = 1$$

where $\delta \vec{\theta}$ is the rotation caused by the unit displacement of the structure at the point that force $\vec{F}$ is applied. Applying this equation to the structure of the peg the following equation is derived:

$$f_{c1} = \frac{1}{1 + \frac{g_t}{g_T}h} F_x + \frac{\frac{g_t}{g_T}h}{1 + \frac{g_t}{g_T}h} M_y \tag{17}$$

where g_t is the gain of the sensor in tension along x and g_T is the gain of the sensor in torsion along y.

This is the last equation that is needed for the reduction of redundancy. Now the unknowns can be defined by the solution of the above non linear system.

In order to formulate the above equation, it was assumed that the force sensor has finite stiffness and the external work done by the contact forces has been saved as strain energy on it.

The other three contact cases are easier - lower order of redundancy - than the previous. After the above calculations, all data needed for the positioning and the orientation of the C-Frames are available.

Conclusions

Within the framework of a knowledge based system for generation of assembly strategies, a decision making system is presented that is able to plan the appropriate sequence of the robot fine motions at the initial stage of an assembly strategy. This module is very critical for the generation of a successful assembly strategy.

In the development of this module, qualitative and quantitative knowledge has been combined in such a manner that mechanical, manufacturing and geometrical data are numerically processed to help qualitative knowledge processing.

The other modules of the knowledge based system for automatic generation of assembly strategies is under development. By its integration, the system will be able to manipulate the features of the mated parts for each assembly task and their secondary characteristics such as dimensions, tolerances, and will generate the appropriate assembly strategy for each task. Nominally the knowledge based system will generate the sequence of robot fine motions, the C-frame and the artificial constraints for each motion. The artificial constraints will feed to the force/position hybrid control system of the robot as reference values. Thus an integrated intelligent robot control system will be finally developed.

REFERENCES

1. M. Raibert and J. Craig, "Hybrid Position Force Control of Manipulators", Trans. ASME, J. Dynamic Syst., Measurement, and Control, pp 126-133, June 1981.
2. Tomas Lozano-Perez, M. T. Mason R. H. Taylor, "Automatic Synthesis of Fine-Motion Strategies for Robots", The International Journal of Robotics Research, Vol. 3, No. 1, Spring 1984.
3. Chi-Haur Wu and Myong Gi Kim, "Modeling of Part Mating Strategies for Formating Assembly Operations for Robots", Transactions On Systems Man And Cybernetics, Vol 24, No 7, July 1994
4. N. A. Aspragathos "Assembly Strategies for parts with a plane of symmetry", Robotica (1991), Vol. 9, pp. 189-195.
5. K. Kitagaki, T. Ogasawara and T. Suehiro, "Methods to Detect Contact State by Force Sensing in an Edge Mating Task", Proceedings of IEEE Conference on Robotics and Automation, 1993.
6. D. E. Whitney, O. L. Gilbert, "Representation of Geometric Variations Using Matrix Transforms for Statistical Tolerance Analysis in Assemblies", Proceedings of IEEE Conference on Robotics and Automation, 1993.
7. J.S. Przemieniecki, "Theory of Matrix Structural Analysis", Dover Publications, Inc. New York, Dover 0-486-64948

Acknowledgment: This work has been developed in the EE research project ROBAS funded under COPERNICUS frame. 3

Appendix A
Programme
and
Organising Committee

PROGRAMME AND ORGANISING COMMITTEE

Chairmen

Prof. Giovanni BIANCHI
  CISM, Palazzo del Torso
  Piazza Garibaldi 18
  33100 Udine
  ITALY
  Tel.: +39 432 294989 or 508251
  Fax: +39 432 501523
  E-mail: cism@cc.uniud.it

Prof. Adam MORECKI
  Warsaw University of Technology
  Institute of Aeronautics and Appl. Mechanics
  Nowowiejska 24
  00-665 Warsaw
  POLAND
  Tel.: +48 22 660 7513
  Fax: +48 22 622 38 76
  E-mail: amorecki@meil.pw.edu.pl

Members

Prof. A.P. BESSONOV
  Russian Academy of Sciences
  Griboedova 4
  Moscow-Centre, 101000,
  RUSSIA
  Fax: +7 095 200 42 39

Prof. Jean-Claude GUINOT
  Laboratoire de Robotique de Paris
  Centre Universitaire Technologique de Vélizy
  10-12 avenue de l'Europe
  78140 Vélizy
  FRANCE
  Tel.: +33 1 39254965
  Fax: +33 1 39254967
  E-mail: guinot@robot.uvsq.fr

Prof. Bodo HEIMANN
  Institut für Mechanik
  Universität Hannover
  Appelstrasse 11
  30167 Hannover 1
  GERMANY
  Tel.: +49 511 7624161
  Fax: +49 511 7624164
  E-mail: heimann@ifm.uni-hannover.de

Prof. J.R. HEWIT
    Dept. of Applied Physic
    and Electronic & Manufacturing Engineering
    University of Dundee
    Dundee
    DD14HN Scotland
Prof. Oussama KHATIB
    Stanford University
    Computer Science Department
    Stanford, CA 94305 2085
    U.S.A.
    Tel.: +1 415 7239753
    Fax: +1 415 7251449
    E-mail: khatib@cs.stanford.edu
Prof. A.I. KORENDYASEV
    Russian Academy of Sciences
    Griboedova 4
    Moscow-Centre, 101000,
    RUSSIA
    Fax: +7 095 200 42 39
Prof. W.O. SCHIEHLEN
    Universität Stuttgart
    Inst. B für Mechanik
    Pfaffenwaldring 9
    70550 Stuttgart 80
    GERMANY
    Tel.: +49 711 6856388
    Fax: +49 711 6856400
    E-mail: wos@mechb.uni-stuttgart.de
Dr. Eng. K. TANIE
    Biorobotics Division Robotics Dept.
    Mechanical Eng. Laboratory
    1-2 Namiki, Tsukuba
    Ibaraki, 305
    JAPAN
    Tel.: +81 298 587088
    Fax: +81 298 587201
    E-mail: m1750@mel.go.jp
Prof. M. VUKOBRATOVIĆ
    Institute "Mihajlo Pupin"
    Volgina 15 - P.O. Box 906
    11000 Beograd
    YUGOSLAVIA
    Fax: +381 11 77870 or 773074

Prof. K.J. WALDRON
    Dept. of Mechanical Engineering
    The Ohio State University
    206, W.18th.Ave.
    Columbus, OH 43210
    U.S.A.
    Fax: +1 614 2923163
    E-mail: kwaldron@magnus.acs.ohio-state.edu

Scientific Secretary

Dr. Cezary RZYMKOWSKI
    Warsaw University of Technology
    Institute of Aeronautics and Appl. Mechanics
    Nowowiejska 24
    00-665 Warszawa
    POLAND
    Tel.: +48 22 660 7992
    Fax: +48 22 628 2587
    E-mail: czarek@meil.pw.edu.pl

Secretary

Dr. Paola AGNOLA
    (*Secretary*)
    CISM, Palazzo del Torso
    Piazza Garibaldi 18
    33100 UDINE
    ITALY
    Tel.: +39 432 294989 or 508251
    Fax: +39 432 501523
    E-mail: cism@cc.uniud.it

Appendix B
List of Participants

LIST OF PARTICIPANTS

BELGIUM

Ir.   Hans DE MAN
Vrije Universiteit Brussel
Faculty of Applied Sciences
Dept. of Mechanical Engineering
Pleinlaan 2
1050 Brussels
BELGIUM
Tel.: +32 2 6292808
Fax: +32 2 6292865
hdman@vnet3.vub.ac.be

BULGARIA

Petko KIRIAZOV
Institute of Mechanics
Bulgarian Academy of Sciences
Acad. G. Bonchev Str., bl. 4
1113 Sofia
BULGARIA
Tel.: +359 2 7135213
Fax: +359 2 702056
dors@BGCICT.ACAD.BG

CANADA

Prof. Jorge ANGELES
Mc Gill University
Mechancal Engineering Dept.
817 Sherbrooke St. W.
Montreal Quebec H3A 2K6
CANADA
Fax. +1 (514) 398 7348
Tel. +1 (514) 398 6313
angeles@cim.mcgill.ca

Prof. Marek KUJATH
Mech. Eng., TUNS
Halifax, NS, B3J 2X4
CANADA
Tel.: +1 902 420 7811
marek@poisson.me.tuns.ca

FRANCE

Dr.   Lotfi BEJI
CEMIF Centre Etude en Mecanique
d'Ile de France
40 rue de Pelvoux
91020 Evry Cedex
FRANCE
Tel.: +33 1 69477531
Fax: +33 1 69477599
beji@cemif.uni-evry.fr

Dr.   Philippe BIDAUD
L.R.P. 10 Av. Europe
78140 Velizy
FRANCE
Tel.: +39 1 39254351
Fax: +39 1 39254967
bidaud@robot.ursq.fr

Sylvie BOUDET
EDF-DER
6 quai Watier
78401 Chatou Cedex
FRANCE
Tel.: +33 1 30877609
Fax: +33 1 30878434
sylvie.boudet@der.edfgdf.fr

Ph.D. Student Olivier BRUNEAU
Laboratoire de Robotique de Paris
10-12 Avenue de L'Europe
78-140 Velizy
FRANCE
Tel.: +33 1 39254956
Fax: +33 1 39254967
bruneau@robot.uvsq.fr

Prof. Jean-Claude GUINOT
Laboratoire de Robotique de Paris
Centre Univ. Technologique de Velizy
10-12 avenue de l'Europe
78140 Velizy
FRANCE
Tel.: +33 (1) 39254965
Fax: +33 (1) 39254967
guinot@robot.uvsq.fr

Dr.   Patrick HENAFF
       Laboratoire de Robotique de Paris
       10-12 Avenue de l'Europe
       78-140 Velizy
       FRANCE
       Tel.: 33 1 39254975
       Fax: 33 1 39254967
       henaff@robot.uvsq.fr

Dr.   Jean Pierre MERLET
       INRIA Sophia-Antipolis
       BP 93
       06902 Sophia-Antipolis Cedex
       FRANCE
       Tel.: +33 93 657761
       Fax: +33 93 657643
       Jean-Pierre.Merlet@sophia.inria.fr

Dr.   Guillaume MOREL
       EDF DER EP SDM P 29
       6, QUAI Watier
       78401 Chatou Cedex
       FRANCE

Ph.D. Pascal RUAUX
       Laboratoire de Robotique de Paris
       10-12 Avenue de L'Europe
       78-140 Velizy
       FRANCE
       +33 1 39254972
       +33 1 39254967
       ruaux@robot.uvsq.fr

Nelly TROISFONTAINE
       Laboratoire de Robotique de Paris
       10-12 Avenue de l'Europe
       78-140 Velizy
       FRANCE
       Tel.: +33 1 39254958
       Fax: +33 1 39254967
       nelly@robot.uvsq.fr

GERMANY

Dr.   Manutschehr DAEMI
       Institut für Mechanik
       Universität Hannover
       Appelstrasse 11
       30167 HANNOVER 1
       GERMANY
       Fax: +49 511 7624164
       mdaemi@ifm.uni-hannover.de

Prof. Bodo HEIMANN
       Institut für Mechanik
       Universität Hannover
       Appelstrasse 11
       30167 Hannover 1
       GERMANY
       Tel.: +49 511 7624161
       Fax: +49 511 7624164
       heimann@ifm.uni-hannover.de

Max L. KELER
       Parkstr. G.
       D 82065 Baierbrunn
       GERMANY
       Tel.: +49 89 7930772
       Fax: +49 89 7930772

Dr.   Hong You LI
       Tele Robot Engineering
       Klingleweg 8
       88709 Meersburg
       GERMANY
       Tel.: +49 7532 432031
       Fax: +49 7532 432099
       hongyou@mee.hwm.com

Dipl. Ing. Wolfgang RISSE
       Institut für Mechatronik IMECH GmbH
       Eurotec Ring 15
       47445 Moers
       GERMANY
       Tel.: +49 2841 101276
       Fax: +49 2841 101251
       risse@mail.imech.uni-duisburg.de
       http://www.imech.uni-duisburg.de

Prof. W.O. SCHIEHLEN
       Universität Stuttgart
       Inst. B für Mechanik
       Pfaffenwaldring 9
       70550 Stuttgart 80
       GERMANY
       Tel.: +49 (711) 6856388
       Fax: +49 (711) 6856400
       wos@mechb.uni-stuttgart.de

Dipl. Ing. Martin SCHNEIDER
    Gerhard Mercator Universität Duisburg
    Fachgebiet Mechatronik
    47048 Duisburg
    GERMANY
    Tel.: +49 203 3792476
    Fax: +49 203 3789143
    schneider@mechatronik.uni-duisburg.de
    http://
    www.mechatronik.uni- duisburg.de

GREECE

Prof. Nicos ASPRAGATHOS
    University of Patras
    Mechanical and Aeron. Eng. Dept.
    PATRAS T.K. 261100
    GREECE
    Tel.: +30 61 997268
    Fax: +30 61 991626

HUNGARY

Prof. Elisabeth FILEMON
    Technical University of Budapest
    Dept. of Technical Mechanics
    Muegyetem rkp 3
    1111 Budapest
    HUNGARY
    Tel.: +36 1 4631370
    Fax: +36 1 4633471

Prof. Gabor STEPAN
    Dept. of Applied Mechanics
    Technical University of Budapest
    1521 Budapest
    HUNGARY
    Tel.: +36 1 463 3740
    Fax: +36 1 4633741

ITALY

Prof. Giovanni BIANCHI
    CISM, Palazzo del Torso
    Piazza Garibaldi 18
    33100 Udine
    ITALY
    Tel.:+39 (432) 294989 or 508251
    Fax:+39 (432) 501523
    cism@cc.uniud.it

Dr.   Francesca COSMI
    Politecnico di Milano
    Dip. di Meccanica
    via Bonardi 9
    20133 Milano
    ITALY
    cosmi@picolit.diegm.uniud.it

Giorgio FIGLIOLINI
    Dept. of Industrial Engineering
    University of Cassino
    via di Biasio 43
    03043 Cassino (Fr)
    ITALY
    +39 776 299662
    +39 776 310812
    figliolini@ing.unicas.it

Prof. Bruno SICILIANO
    Dip. di Informatica e Sistemistica
    Universitá di Napoli Federico II
    via Claudio 21
    80125 NAPOLI
    ITALY
    Tel.: +39 81 7683179
    Fax: +39 81 768 3186

JAPAN

Prof. Kozo MATSUSHIMA
    Toin University of Yokohama
    Dept. of Control and System Eng.
    1614 Kurogane - Cho, Aoba-Ku
    Yokohama 225
    JAPAN
    Tel.: +81 45 974 5065
    Fax: +81 45 9725972
    matusima@aoba.toin.ac.jp

Kiyoko NAKAMOTO
    1614 Kurogane Aoba
    Toin University
    Yokohama
    JAPAN
    Tel.: +81 467 274 5065

Prof. Yoshiyuki SANKAI
University of Tsukuba
Institute of Engineering Mechanics
Tsukuba 305
JAPAN
Tel.: +81 298535151
Fax: +81 298 535207
sankai@kz.tsukuba.ac.jp
http://sanlab.kz.tsukuba.ac.jp

Koji SHIBUYA
Sugano Lab., Dept. of Mech.
Engineering
School of Science and Eng.
Waseda University
3-4-1 OKUBO Shinjuku-ku
169 TOKYO
JAPAN
Tel.: +81 3 52863264
Fax: +81 3 52720948
koji@paradise.mech.wased.ac.jp

Prof. Atsuo TAKANISHI
Waseda University 59-308
Shinjuku-ku, Okubo 3-4-1
Tokyo 169
JAPAN
Tel.: +81 3 52863257
Fax: +81 3 52732209
takanisi@cfi.waseda.ac.jp

Hideaki TAKANOBU
59-308 Dept. of Mechanical
Engineering
Waseda University
Okubo 3 4 1 Shinjuku - ku
Tokyo 169
JAPAN
Tel.: +81 3 5286 3257
Fax: +81 3 5273 2209

Dr. Eng. K. TANIE
Biorobotics Division Robotics Dept.
Mechanical Eng. Laboratory
1-2 Namiki, Tsukuba
Ibaraki, 305
JAPAN
Tel.:+81 298 587088
Fax: +81 298 587201
m1750@mel.go.jp

Dr.   Jinichi YAMAGUCHI
Humanoid Research Laboratory
room no. 55 5 0 707
Adv. Research Center for Science
and Engineering
Waseda University
3-4 -1 Okubo, Shinjunku-ku
Tokyo 169
JAPAN
Tel.: +81 3 3208 8714
Fax: +81 3 3208 8714
yamajin@cfi.waseda.ac.jp

POLAND

Dr.   Jerzy BALICKI
The Naval Academy
Śmidowicza St.
81-919 Gdynia 19
POLAND
Tel.:+48 58 262532
Fax: +48 58 253881
amw@beta.nask.gda.pl

Radosław CIEŚLAK
Technical University of Wrocław
1 - 16 Wyb. Wyspiańskiego 27
50-370 Wrocław
POLAND
Tel.: +48 71 212546
Fax: +48 71 227645
rado@sun1000.ci.pwr.wroc.pl

Dr.   Piotr GAWŁOWICZ
T.U. of Zielona Góra
Dept. of Organisation Prod.Systems
ul. Podgórna 50
65-246 Zielona Góra
POLAND
Tel.: +48 68 254831 ext. 426
Fax: +48 68 253944
p.gawlowicz@oiz.wsi.zgora.pl

Prof. Zyqmunt KITOWSKI
AMW
ul. Śmidowicza
81-019 Gdynia 19
POLAND
Tel.: 0048 58 262625

Dr.   Krzysztof ŁASIŃSKI
       T.U. of Zielona Góra
       ul. Podgórna 50
       65-246 Zielona Góra
       POLAND
       Tel.: +48 68 254831
       Fax: +48 68 253944
       p.gawlowicz@oiz.wsi.zgora.pl

Prof. Adam MORECKI
       Warsaw University of Technology
       Inst. of Aeronautics and Appl. Mech.
       Nowowiejska 24
       00-665 Warszawa
       POLAND
       Tel.: +48 22 21007513
       Fax:+48 22 622 38 76
       amorecki@meil.pw.edu.pl

Dr.   Cezary RZYMKOWSKI
       Warsaw University of Technology
       Inst. of Aeronautics and Appl. Mech.
       Nowowiejska 24
       00-665 Warszawa
       POLAND
       Fax:+48 22 628 2587
       czarek@meil.pw.edu.pl

Prof. Tadeusz UHL
       University of Mining and Metallurgy
       Dept. of Mechanical Eng. & Robotics
       ul. Mickiewicza 30
       30-059 Kraków
       POLAND
       Tel.: +48 12 173128
       Fax: +48 12 343505
       tuhl@rob.wibro.agh.edu.pl

Prof. Teresa ZIELIŃSKA
       Warsaw University of Technology
       Inst. of Aeronautics and Appl. Mech.
       ul. Nowowiejska 24
       00-665 Warszawa
       POLAND
       Fax: +48 22 6282587
       teresaz@meil.pw.edu.pl

Prof. Cezary ZIELIŃSKI
       Warsaw University of Technology
       Inst. of Control and Computational Eng.
       ul. Nowowiejska 15/19
       00-665 Warszawa
       POLAND
       Tel.: +48 22 6607123
       Fax: +48 22 253719
       c.zielinski@ia.pw.edu.pl

UNITED KINGDOM

Dr.   Robert BICKER
       Dept. of Mechanical Engineering
       Stephenson Building
       University of Newcastle upon Tyne
       Claremont Road
       Newcastle upon Tyne
       England NE1 7RU
       U.K.
       Tel.: +44 191 2226219
       Fax: +44 191 2228600
       robert.bicker@ncl.ac.uk

Henning BUSCH
       Dept. of Applied Physics
       Electronic and Mechanical Eng.
       University of Dundee
       Dundee DD1 4HN
       U.K.
       Fax: +44 1382 202830

U.S.A.

Prof. Steven A. VELINSKY
       Dept. of Mech., Aeronautical Eng.
       University of California, Davis
       U.S.A.
       +1 916 752 4166
       +1 916 752 6714
       savelinsky@ucdavis.edu

Prof. Gregory S. CHIRIKJIAN
       124 Latrobe Hall
       Dept. Mechanical Engineering
       Johns Hopkins University
       3400 N. Charles st.
       Baltimore MD 21218
       U.S.A.
       Tel.: +1 410 5167127
       Fax: +1 410 5167254
       greg@geronimo.me.jhu.edu
       http://caesar.me.jhu.edu

Prof. K. Joseph DAVIDSON
    Dept. of Mechanical & Aerospace
    Engineering
    Arizona State University
    Tempe Arizona 85287 6106
    U.S.A.
    Fax: +1 602 9652412
    j.davidson@asu.edu

Prof. Steven DUBOWSKY
    Room 3-469A
    MIT Dept. of Mechanical Engineering
    Cambridge MA 02139
    U.S.A.
    Tel.: +1 617 253 2144
    Fax: +1 617 258 7881
    dubowsky@mit.edu

Prof. Joseph DUFFY
    Dept. Mechanical Engineering
    University of Florida Center F.
    Intelligence
    Machines and Robotics
    Gainesville 32611 FL
    U.S.A.
    Tel.: +352 352 392 0814
    Fax: +352 352 392 1071

Prof. Oussama KHATIB
    Stanford University
    Computer Science Department
    Stanford, CA 94305 2085
    U.S.A.
    Tel.: +1 415 7239753
    Fax:: +1 415 7251449
    khatib@cs.stanford.edu

Prof. Bahram RAVANI
    Dept. of Mechanical
    & Aeronautical Engineering
    University of California
    Davis CA 95616
    U.S.A.
    Tel.: +1 916 752 4986
    Fax: +1 916 752 6714
    bravani@ucdavis.edu

Prof. Bernard ROTH
    Dept. of Mech. Engineering
    Stanford University
    Stanford CA 94305
    U.S.A.
    Tel.: +1 415 723 3657
    Fax: +1 415 7233521
    roth@flamingo.stanford.edu

Prof. K.J. WALDRON
    Dept. of Mechanical Engineering
    The Ohio State University
    206, W.18th.Ave.
    Columbus, OH 43210
    U.S.A.
    Fax: +1 (614) 2923163
    kwaldron@magnus.acs.ohio-state.edu

YUGOSLAVIA

Prof. Branislav BOROVAČ
    Faculty of Technical Sciences
    TRG D. Obradovica 6
    21000 Novi Sad
    YUGOSLAVIA
    Tel.: +381 21 55011
    Fax: +381 21 59536
    borovac@uns.ns.ac.yu

Prof. M. VUKOBRATOVIČ
    Institute "Mihajlo Pupin"
    Volgina 15 - P.O. Box 906
    11000 Beograd
    YUGOSLAVIA
    Fax: +381 11 77870 or 773074

Printed in the United States
By Bookmasters